I0897045

Causes and Consequences of Forest Growth Trends in Europe

The Scientific Advisory Board

Prof. Dr. E.P. Farrell, Ireland, Chairman

Prof. Americo Carvalho Mendes, Portugal
Dr. Emil Cienciala, Czech Republic
Prof. Dr. Hubert Hasenauer, Austria
Dr. Eeva Hellström, Finland
Dr. Antoine Kremer, France
Prof. Michael Köhl, Germany
Prof. Leena Paavilainen, Finland
Prof. Jean-Luc Peyron, France
Dr. Viktor Teplyakov, Russian Federation

VOLUME 21

Causes and Consequences of Forest Growth Trends in Europe

European Forest Institute Research Report 21
Results of the RECOGNITION Project

Edited by

Hans-Peter Kahle
Timo Karjalainen
Annette Schuck
Göran I. Ågren
Seppo Kellomäki
Karl Mellert
Jörg Prietzel
Karl-Eugen Rehfuess
Heinrich Spiecker

BRILL

LEIDEN • BOSTON
2008

This book is printed on acid-free paper.

Library of Congress Cataloging-in-Publication Data

A C.I.P. record for this book is available from the Library of Congress.

ISSN: 1238-8785

ISBN-13: 978 90 04 16705 6

© Copyright 2008 by Koninklijke Brill NV, Leiden, The Netherlands.
Koninklijke Brill NV incorporates the imprints BRILL, Hotei Publishing, IDC Publishers, Martinus Nijhoff Publishers and VSP.

All rights reserved. No part of this publication may be reproduced, translated, stored in a retrieval system, or transmitted in any form or by any means, electronic, mechanical, photocopying, recording or otherwise, without prior written permission by the publisher.

Authorization to photocopy items for internal or personal use is granted by Koninklijke Brill NV provided that the appropriate fees are paid directly to Copyright Clearance Center, 222 Rosewood Drive, Suite 910, Danvers, MA 01923, USA. Fees are subject to change.

PRINTED IN THE NETHERLANDS

Table of Contents

Foreword

The current report, "Relationships Between Recent Changes of Growth and Nutrition of Norway Spruce, Scots Pine, and European Beech Forests in Europe – Recognition", follows in the footsteps of the study "Growth Trends in European Forests – Studies From 12 Countries" edited by Spiecker et al. (1996). Whereas Spiecker et al. identified growth changes, the Recognition project aimed at elucidating the underlying causes of the observed changes in the growth of European forests and to analyse the implications for their sustainable management.

The forest research network of EFI made it possible to bring together experts in the fields of forest ecology and management, forest growth, forest nutrition and soil and forest ecosystem modelling from all over Europe to join the ambiguous research project. The project was unique in combining empirical studies based on experimental and observational data with process-based modelling approaches. The Recognition project (1999–2002) was carried out with the financial support from the Commission of the European Communities, Agriculture and Fisheries (FAIR CT98-4124) DG Research – Quality of Life Programme – Unit B 1.2 and RTD Cooperation with third countries and international organizations – INCO (IC20-CT98-0309).

For further details on the project, please visit www.efi.int/portal/project/recognition/

I would like to express my sincere thanks to the editors and authors of this report, to the various project partners that contributed valuable data and expertise. In addition, I'd like to thank EFI Communications team for preparing the report for publishing.

12 December 2007
Risto Päivinen
Director
European Forest Institute

Acknowledgements

The authors like to thank all partners, associated partners and collaborators that have contributed to the implementation of the project. The support of the European Commission (DG-AGRI) and ICP Forests (International Co-operative Programme on Assessment and Monitoring of Air Pollution Effects on Forests) in giving access to information of the monitoring of forests in Europe in general and of FIMCI (Forest Intensive Monitoring Coordinating Institute) namely of Evert Vel, Gert-Jan Reinds and Wim de Vries in providing data of the forest intensive monitoring programme as well as the N-deposition scenarios are gratefully acknowledged. We thank numerous local experts associated with the National Focal Centres for assistance in finding suitable sites and stands and the forest owners for the allowance to cut sample trees for stem analysis.

David Viner (Climatic Research Unit, University of East Anglia, CRU) provided HadCM3 data. The 'CRU 0.50 lat/lon gridded monthly climate data set' has been supplied by the Climate Impacts LINK Project (UK Department of the Environment Contract EPG 1/1/16) on behalf of CRU. Maria Noguer (Working-Group I of the IPCC, Hadley Centre, UK) provided the CO_2 scenario, and Jean-Michel Terres and Alfred de Jager (EU Joint Research Centre) provided daily weather data from the MARS Meteorological Database.

Hubert Hasenauer and Wim de Vries contributed valuable and useful comments in the mid-term review of the project. We are indebted to sixteen independent reviewers of this book for providing constructive comments which helped to considerably improve the manuscripts.

The project "Relationships Between Recent Changes of Growth and Nutrition of Norway Spruce, Scots Pine, and European Beech Forests in Europe – RECOGNITION" was carried out with the financial support from the European Community under FAIR6-CT98-4124 and INCO20-CT98-0309.

We are indebted to Ian Hunter, former project coordinator and former director of the European Forest Institute (EFI) for creating a productive and congenial work environment during the proposal and initialization phase of the project. The support by Uli Riemer and the student helpers who did a great job in managing the numerous field campaigns is gratefully acknowledged.

Finally, many thanks are accorded to Tim Green (EFI) for the careful reading and language revision and to Minna Korhonen (EFI) for her technical assistance and to both for their special diligence in finalizing this volume.

December 2007
The Editors

Executive Summary

Forest growth in Europe has been increasing during the last decades. The possible causes suggested to explain this phenomenon have been elevated atmospheric carbon dioxide concentration, temperature and precipitation climate beneficial to tree growth, increasing nitrogen deposition and better management. In this book complementary approaches are adopted to discriminate between the levels of importance of these factors. Investigations over large geographical areas are used to separate current spatial variability while studies of the growth of individual trees allow time trends to be evaluated. Four different mechanistic forest ecosystem models supplement the empirical investigations.

This study gives empirical evidence for the prominent role of increasing nitrogen supply as the key driver of the observed forest site productivity increases. However, to which part increased nitrogen availability is driven by recovery of soil from former often devastative land use, and to which extent it is attributed to anthropogenically increased nitrogen deposition has not been unequivocally clarified. The process-based modelling gives clear evidence that nitrogen played the most important role for forest site productivity changes in the past, but in the future, climate change and especially CO_2 fertilization will be of increasing importance.

The Recognition project:
Together with 24 partners from across Europe, the European Forest Institute launched in April 1999 with financial support from the European Union the EC-FAIR project "Relationships Between Recent Changes of Growth and Nutrition of Norway Spruce, Scots Pine, and European Beech Forests in Europe – Recognition".

The objectives of the research project were:

- to identify potential causes of recent growth trends in European forests;
- to investigate the relative importance of nutrients, climate, and land-use changes; and
- to analyse the long-term consequences and risks of observed changes for sustainability and to assess the implications for future forest management.

The methods applied in the research project were:

- correlative statistical approaches consisting of (i) the analysis of time-series from long-term forest research plots (the Historical Development Investigation – HDI), and (ii) the analysis of temporal trends and spatial patterns of the variables of interest (the Present State Analysis – PSA);
- a mechanistic (process-based) modelling approach based on the adaptation and use of four different ecophysiological and ecosystem models; and
- a combined correlative and modelling approach.

The following data from sites across Europe were used for the different research approaches:

- The Historical Development Investigation was applied to 12 sites with Scots pine and 7 sites with Norway spruce. For an additional 16 Scots pine sites and 16 Norway spruce sites only time-series of foliar nutrition have been analysed.
- For the Present State Analysis data from 19 Scots pine, 17 Norway spruce and 9 European beech sites have been employed. These sample sites were selected among the Level II plots of the EC-UN/ECE Forest Intensive Monitoring Programme in Europe.
- The process-based models simulated forest growth at 22 sites: 13 covered with Scots pine and 9 with Norway spruce. These sites largely overlap with the sites used for the Historical Development Investigation.

For the mechanistic modelling approach four process-based models were selected:

- EFIMOD – developed by the European Forest Institute;
- EFM – developed by the Natural Environment Research Council, Centre for Ecology and Hydrology, Edinburgh, United Kingdom;
- FinnFor – developed by the University of Joensuu, Faculty of Forest Sciences, Joensuu, Finland;
- Q – developed by the Swedish University of Agricultural Science, Department of Ecology and Environmental Research, Uppsala, Sweden.

Main results achieved by the Historical Development Investigation:

- During recent decades most Scots pine stands in Central Europe showed significantly faster height growth as compared to previous decades.
- This increase was predominantly caused by improved nitrogen nutrition which is the combined result of a recovery of sites from former devastative land-use practices and anthropogenically increased atmospheric nitrogen deposition.
- Nitrogen nutrition played a prominent role on sites where nitrogen deposition was high. On the sites in Northern Europe and Great Britain, increased nitrogen supply was less important for long-term growth changes of Scots pine stands than in Central Europe.
- Precipitation and thus water availability are key factors for the short-term fluctuations of growth, but not for the long-term growth trends.
- No significant direct effect of changes in air temperature on growth changes of Scots pine and Norway spruce stands could be detected.
- It is likely that increased atmospheric CO_2 concentration has only been of secondary importance for the tree growth changes in the past.

Main results achieved by the Present State Analysis:

- During the period 1960–2000 average annual height growth increased by +23% for spruce, and by +25% both for pine and beech.
- During the first 50 years of their lifetime the average annual diameter growth of trees with germination dates in the 1940s and 1950s was on average increased by +16 % for spruce, +13 % for pine, and +27 % for beech as compared to

trees which started to grow between 1900 and the 1920s. This indicates that the stand management adapted to the growth increases through earlier and more frequent and/or intensive thinnings.
- Based on species specific functional relations between stand height and total wood volume production potential validated for German conditions the observed changes in height growth gave rise to increases in total wood volume production over the period 1960–2000 of +56 % for spruce, +45 % for pine, and +74 % for beech, which is equivalent to average annual productivity increases of +1.4 %, +1.1 %, and + 1.8 % for spruce, pine and beech respectively.
- Pine and beech sample trees with high foliar nitrogen nutrition at the end of the observation period showed a significantly lager increase in height growth than those with low nitrogen nutrition.
- Pine and beech sample trees from sites with low initial site fertility showed a significantly larger increase in height growth than those from sites with high initial site fertility.
- Confounding of the analysed causative factors (i) change in seasonal air temperature, (ii) change in seasonal precipitation, and (iii) level of foliar nitrogen nutrition complicated identification of the independent effects these variables have on growth change.

Main results achieved by the mechanistic modelling approach:

- All four process-based models identified increasing nitrogen supply as the major cause of observed changes in European forest growth during the 20th century. This finding is mainly the result of the large increase in nitrogen deposition rates during that period.
- The process-based model simulations showed that the relative importance of the different changing environmental factors might change over time: Further changes in forest growth (i.e. in the simulation time-periods 1960–2040 and 2000–2080) are more likely to be caused by increasing atmospheric CO_2 concentration and, especially in northern latitudes, by increasing temperature.
- In the model simulations, the effects of the environmental factors air temperature, moisture availability, availability of nitrogen and CO_2 were largely additive.
- Uncertainties regarding the response of nitrogen uptake and nitrogen use efficiency to environmental change evident in diverging model results need to be resolved in experiments and models.

Implications and consequences of the growth changes for sustainability of European forests:

- There are no indications from the project, at least in the short term, that the growth changes themselves represent any major threat to the sustainability of the forest ecosystems, and their provision of services. In a longer time perspective, the situation might be different and increased water stress and deficiencies of nutrients other than nitrogen might occur.
- Increased growth and accumulating wood volume in the forests indicate that if the structure of the forests should be maintained, changes in the intensity of

management practices, e.g. more frequent and/or intensive thinnings, are needed. This also refers to ensuring the stability of the stands, e.g. with respect to storm.

- The model simulations indicate that the time for timber production can be shortened; a finding which especially refers to Central Europe where increases in forest site productivity have been largest during recent decades, whereas in the future tree growth and consequent profitability of forestry may increase most in Northern Europe.
- Sustainable forest management in the future may require more attention to the nutritional budget of forest ecosystems. The Pan-European EC-UN/ECE Programme for Intensive and Continuous Monitoring of Forest Ecosystems is one of the tools that can provide early-warning signals if large-scale changes in European forests are occurring.

1. Introduction

1. Introduction

Timo Karjalainen[1] and Annette Schuck[2]

[1] *European Forest Institute and Finnish Forest Research Institute, Joensuu, Finland*
[2] *European Forest Institute*

Background

Growth of European forests has been of particular interest: are forests growing slower or faster than earlier, and if so, what could be the reasons? Many publications have reported increased growth rates of forests (e.g. Pretzsch, 1985; Andersson et al., 1988; Kenk and Fischer, 1988; Becker et al., 1990; Makkonen-Spiecker, 1991; Kenk, 1993; Tomter, 1994; Elfving and Tegnhammar, 1996). In a few cases a reduction of growth has been reported as a result of local high levels of pollution (Kenk et al., 1983; Röhle, 1987; Pretzsch, 1996). The book "Growth Trends in European Forests" (Spiecker et al., 1996) with contributions of researchers from several European countries presents an analysis of a large amount of growth data. An increasing growth trend was observed in most cases, while for specific sites a decreasing trend was observed. Despite the fact that growth trends have been identified and described, the causes of growth increases have not been identified. Several variables influencing growth are known to affect forest growth and could be responsible for the changes. These include increased anthropogenic N deposition, increased atmospheric CO_2, changes in land use, and changes in physical climate (e.g. increases in winter temperature and precipitation). It is important to understand which factors or factor combinations are responsible for the observed spatial and temporal pattern of growth changes. The long-term consequences and the possible ameliorative measures required will vary with the differing causes.

Tree growth in general is regulated by the availability of light, water, nutrients, and by the length of the growing season. The presence or absence of potentially damaging abiotic factors (storm, frost, snow or noxious chemical agents), and/or biotic factors (insect pests, fungal diseases or other animal pests) also influences growth. Thus, the increment changes reported may be due to various and variable growth-limiting factors.

Among the various factors that have been proposed as being responsible for the observed increases in growth are, for example:

1. *Increased site fertility/productivity*. The fertility/productivity of a site may increase as a consequence of changes in forest management and land-use practice (Dyck et al., 1994). During earlier centuries, much of the European forests were nitrogen depleted due to various forms of land use such as litter-raking and forest pasture (Glatzel, 1999; Hunter and Schuck, 2002).

Nowadays, site fertility might have increased as a result of: (i) enhanced anthropogenic nitrogen inputs; (ii) higher rates of humus mineralization; and (iii) forest fertilization with limiting elements and through liming. On the other hand, decreasing site fertility has been reported regionally, and this should result in decreased nutrient availability and reduced forest growth.

2. *Climate conditions.* Effects of climate fluctuations are site- and species-specific (Spiecker, 1987; Kenk, 1993). The long-term trend of climate in Central Europe shows a significant increase of spring and autumn temperature and precipitation (IPCC, 2001). This trend could result in a growth increase of forests due to prolonged growth periods (Menzel and Fabian, 1999; Myneni et al., 1997), and reduced periods of water shortage. Therefore, a possible cause for the recent growth increases is – at least for some sites – favourable weather conditions during the last decades.
3. *CO_2 fertilization.* In addition to the climate change, a direct fertilization effect of increased CO_2 concentration in the atmosphere has been suggested to be responsible for at least part of the recent growth increases of forests, particularly if the N nutrition is also improved.
4. *Improved silvicultural management and tree breeding.* This should also result in more vital trees with increased biomass production.

It can be summarized that the interrelation between forest growth and soil fertility, tree nutrition, the changing physical and chemical climate (including CO_2 concentration) are manifold and complex. It matters very much what is causing these growth increases. Some causal processes could decrease long-term sustainability of forests and forestry in Europe.

The international seminar "Causes and consequences of accelerating tree growth in Europe" was organised in Nancy, France on 17–19 May 1998, with an objective to learn to what extent the findings of increased tree growth and its consequences in Europe are identified, quantified and understood (Karjalainen et al., 1999). Through the communications made by participants, the seminar concluded that there is evidence of accelerating forest growth in Europe, but that this is only based on case studies. Therefore, there was a need for studies combining different approaches, to utilise national forest inventory data, data from trials on permanent plots and tree analysis, as well as modelling. It was also concluded that since accelerating forest growth seems to be a pan-European phenomenon, it requires contributions from scientists all over Europe to analyse possible causes.

A pilot project was initiated in 1997–1998 to study the possible causes (Rehfuess et al., 1999). The study used one ecological model to clarify whether the magnitude of the increase could be confidently assigned to one or more causes. The conclusion was that hypothesised causes such as nitrogen deposition, CO_2 fertilisation, increases in temperature and precipitation and changes in land use could be responsible for the observed changes. However, the study could not reveal the primary factor or factor combination or account for the full magnitude of the observed increases.

Collaborators from across Europe approached the European Union with a research proposal to study the matter in depth. The EC-FAIR Programme agreed to fund the research activity. The project "Relationships between Recent Changes of Growth and Nutrition of Norway Spruce, Scots Pine, and European Beech Forests in Europe" – Recognition – started in April 1999 and this report presents the outcomes of the project.

Objectives and hypothesis

The following specific objectives were formulated for the Recognition project:

- to identify potential causes of recent growth trends in European forests and to investigate their interactions with focus on the relative importance of nutrients (primarily nitrogen and CO_2), climate (primarily temperature and precipitation), and land-use changes;
- to focus on growth analyses on selected sites where conditions and availability of historical data allow testing of specific hypotheses about causes of changes in site productivity;
- to analyse the long-term consequences and risks of observed changes for sustainability and to assess the implications for future management strategies.

When the site-specific interrelations between the growth of three main European forest tree species – Norway spruce (*Picea abies* (L.) H. Karst.), Scots pine (*Pinus sylvestris* L.), and beech (*Fagus sylvatica* L.) – and changes in soil fertility, climate, and management are analysed, and once the governing key variables are detected and quantified, a regionalization of the observed trends in forest growth, soil properties, and in forest nutrition should be possible. It should also be possible to predict future growth trends with appropriate models. This knowledge would provide a basis for a sustainable forest management including adoption of site-specific harvesting and forest regeneration techniques as well as fertilization in order to prevent nutrient imbalances and to maintain site fertility and ecosystem stability. Definitions of key terms used in Recognition are provided in Table 1.

As noted earlier, the observed changes in site productivity, and thus on growth trends, are the result of complex interactions between a number of factors. These changes take place with different intensities in different regions. Nutritional factors have been selected as a focus by the project partners; yet other factors had to be considered, particularly in relation to nutritional aspects. The problem can be tackled by analysing observations and measurements, and also by modelling.

Changes in site productivity are related to changes in nutrient and/or water availability for tree growth. The changes in nutrient availability lead to changes in nutrient uptake and are manifested in new patterns of carbon allocation and changes in nutrient efficiencies. Nutrient availability is a result of decomposition and mineralization of organic matter, deposition, and weathering of primary minerals.

Changes in the temperature and precipitation regime, and as a consequence water availability, will affect the availability of nutrients for uptake by trees and other vegetation as well as tree growth directly. Finally, the long-term increase in atmospheric CO_2 should have a direct effect on trees and growth, which will itself be mediated by the availability of nutrients in the ecosystem, but long-term effects remain unclear. The availability of mineral nutrients is also a function of previous land-use history and forest management.

The overall objectives have been approached by testing a number of specific hypotheses:

Table 1. Key terms and their definitions as used in the Recognition project.

Biomass	Amount of living plant material (dry matter in kg or ton per unit area). It can be given per tree or per stand, and as a total or by fraction (i.e. stem, bark, branches, foliage, stump, roots – including fine roots).
Growth	Change (usually increase) of diameter (1.3 m above ground), basal area (1.3 m above ground), height, wood volume or total biomass of a tree or stand.
Growth change	Temporal variation in the rate of growth. Patterns of growth changes are growth fluctuations or growth trends. Growth changes can be random variations and might be disrupted by irregular and discontinuous growth anomalies. Growth fluctuations are NOT the focus of this report.
Growth trends	Deviation from average/recorded growth in the past induced by changes in site productivity. Growth trends are the focus of this report.
Site	Sum of the environmental conditions (biotic, edaphic, topographic and climate conditions, including atmospheric composition) at a particular location.
Site productivity	Growth or biomass production potential per unit time of a site for a particular tree species or forest type. Site productivity is determined by climate and inherent soil properties (site quality). Site productivity usually refers to small spatial scales (fractions of a stand or individual stands) whereas forest productivity refers to larger units on the landscape level (entire ecosystems or management units). Forest productivity is determined by site quality but it can be readily manipulated by silvicultural management (Dyck et al., 1994).
Tree nutritional status	A property of a tree usually described by the concentration of mineral nutrients in the current year leaves or needles, in a reference period.

Hypotheses related to potential ***causes***

- changes in site productivity can be accounted for by trends in temporal and spatial patterns of anthropogenic nitrogen deposition interacting with different site conditions, for example:
  - inherent soil nitrogen availability;
  - availability of other nutrients as determined by soil fertility;
  - availability of water as determined by soil water storing capacity, precipitation and evaporation rates; and
  - nutritional status of stands.

However, changes in nitrogen availability may not be sufficient to explain changes in site productivity:

- CO_2 concentration may be important; and
- silvicultural management and land-use history may be important at many sites.

Hypotheses related to potential ***consequences***

- increases in growth will continue until about 2100 as CO_2 concentration and temperature continue to increase;

- precipitation changes in the future will significantly modify growth trends;
- the effects will be species-specific;
- some sites in Europe may be able to maintain increased tree growth with no adverse effects, but on other sites tree growth may encounter severe site limitations, e.g. deficiency of nutrients other than nitrogen.

A large pool of long-term growth data from permanent research plots was identified by the team. We selected a subset from these plots that spans gradients in N nutritional status, temperature and precipitation. Some of the data were still in unprocessed paper form before the project, but have now been put into suitable electronic form for processing.

Most of these plots are control plots of long-term fertilizer experiments (i.e. unfertilized controls) for which long-term series of foliage analysis are available. For these plots we have identified temporal changes of nutrient concentrations in foliage. For each of the plots to be included we identified: (1) possible temporal changes in local climate conditions; and (2) the sensitivity of increment variations to temporal climate variations.

Three approaches were used to test the above-mentioned hypotheses.

A. A multivariate correlative approach consisting of:
   - chronosequence analysis (Historical Development Investigation);
   - identification of causative elements in the current spatial distribution of the variables of interest (Present State Analysis);

B. The adaptation and use of ecophysiological and ecosystem models (mechanistic modelling approach); and

C. A combined correlative and modelling approach

The **Correlative Approach** used the available data regarding forest growth, stand nutrition, soil chemistry, climate, and atmospheric deposition determined for a great variety of European forest sites. With appropriate statistical techniques, the temporal changes during the recent decades (*Historical Development Investigation – HDI*), as well as the present spatial pattern of these variables and their interrelationships (*Present State Analysis – PSA*) were analysed. For the retrospective study of historical changes, control plots (but in addition also the manipulated plots) of long-term fertilization and amelioration experiments seemed most suitable. For the investigation of present spatial patterns, additional data from the European Union Level II network was used for gradient analyses.

The **Modelling Approach** consisted of the adaptation and utilization of existing sophisticated models integrating various ecosystem levels (physiological level, tissue level, tree level, ecosystem level, region level). The adequate functioning and the validity of the models to simulate and predict real changes were tested by a comparison of model results with observations from control plots of long-term fertilization experiments. The future development of forest growth, stand nutrition, and soil chemistry for various deposition, climate, and atmospheric CO_2 concentration scenarios was forecast for some sites.

The correlative and the modelling approaches were combined **(Combined Approach)** in order to validate and to further improve the models. In addition, prognoses of the future development of the variables of interest (e.g. stand growth, stand nutrition, soil chemistry) were made and the risk of a future disharmonic forest nutrition was assessed for different sites. Furthermore, the results were extended to describe the growth changes at larger regional units (e.g. site types, biogeoclimatic zones).

These approaches interacted with one another throughout the project. The Modelling Approach required additional empirical data to improve its predictive power. The gradients for field analysis needed to be selected with the modellers' requirements in mind. All three approaches are necessary and complementary. The Historical Data Investigation is the most robust for detecting change on one site, but is limited by data availability and consistency of measurement methodology over time. The Present State Analysis allowed for the selection of those sites that maximized the gradient in any one candidate causal factor (e.g. N deposition), but was limited because correlation does not prove causation. Data collected for this activity provided a broader basis for generalisation. Although ecophysiological models simplify physiological processes, they offer the best means of investigating hypotheses regarding multiple causative factors.

After the analytic phase we compared the predicted relationships between the various putative causal factors and growth with a view to:

- establishing the generality of the change in growth rates;
- identifying geographical patterns that may exist;
- explaining those patterns by causal factors.

We attempted to generalize the findings regarding specific site types or geographic areas and in particular to identify sites where forest sustainability may be at risk. In the context of risk assessment, special emphasis has been given to possible nutrient imbalances, possible changes in drought, storm and frost incidence, and insect resistance of forest stands.

There were 24 collaborators – 6 partners and 18 associated partners – involved in this project (Table 2). The partners were involved in both research and coordination tasks, whereas the associated partners were mainly involved in research tasks.

The Correlative Approach was coordinated together by the Chair of Soil Science of the Munich University of Technology (focusing on nutrient related matters) and the Institute for Forest Growth of the Albert-Ludwigs-University Freiburg (investigating growth related matters). The Modelling Approach was coordinated by the Department of Ecology and Environmental Research of the Swedish University of Agricultural Sciences. The Combined Approach was coordinated by the European Forest Institute.

The following chapters describe in detail the outcome of the Recognition project. Chapter 2 outlines the material and methods for the Correlative Approach, describing collection, selection, processing of data, sampling, measurement and analysis methods. Chapter 3 provides descriptions of the models applied in the Modelling Approach, and describes their application to explain and predict changes in European forest growth. Chapter 4 presents the results and has been divided into the following sections (the first three are empirical):

1. Historical development of nutrition and climate, and their relationships with tree growth for three European tree species;
2. Temporal trends and spatial patterns of height growth changes in relation to changes in air temperature and precipitation, and in relation to levels of foliar nitrogen nutrition and nitrogen deposition;

Table 2. Project partners and associated partners.

Partners

- European Forest Institute (EFI), Joensuu, Finland
- Munich University of Technology (former Ludwig Maximilians University) (CSS), Chair of Soil Science, Munich, Germany
- Albert-Ludwigs-University Freiburg (IWW), Institute for Forest Growth, Freiburg, Germany
- Swedish University of Agricultural Sciences (SLU), Department of Ecology and Environmental Research, Uppsala, Sweden
- Natural Environment Research Council (NERC), Center of Ecology and Hydrology (CEH, former Institute of Terrestrial Ecology, ITE), Edinburgh, United Kingdom
- University of Joensuu (JOY), Faculty of Forest Sciences, Joensuu, Finland

Associated partners

- Austrian Forest Research Centre (FBVA), Vienna, Austria
- University of Ghent (UG1), Laboratory of Forestry, Ghent, Belgium
- Forest Establishment Eberswalde (FFE), Eberswalde, Germany
- Forest Ecosystems Research Center (GAU), Göttingen, Germany
- Forest Research Institute of Lower Saxony (NFV), Göttingen, Germany
- Technical University of Cottbus (BTUC), Chair of Soil Protection and Recultivation, Cottbus, Germany
- Saxon State Institute for Forestry (LAF), Graupa, Germany
- National Institute of Agronomical Research (INRA), Champenoux, France
- Norwegian Forest Research Institute (NLH), Ås, Norway
- Finnish Forest Research Institute (Metla), Vantaa, Finland
- Danish Forest and Landscape Research Institute (DFLRI), Hørsholm, Denmark
- University of Ghent (UG2), Laboratory of Plant Ecology, Ghent, Belgium
- Forestry Commission Research Agency (FCRA), Northern Research Station, Midlothian, United Kingdom
- National Institute of Agriculture Research (INIA), Madrid, Spain
- University of Padova (PADUA), Department of Agricultural Biotechnology, Padova, Italy
- Swiss Federal Institute for Forest, Snow and Landscape Research (WSL), Birmensdorf, Switzerland
- University of Agriculture (IFE), Institute of Forest Ecology, Brno, Czech Republic
- Slovenian Forestry Institute (SFI), Ljubljana, Slovenia

3. Short-, medium-, and long-term variation in radial growth, and the role of changes in the climatic water balance for the growth of three tree species in Europe; and
4. Evaluation of past and future changes in European forest growth by means of four models (including underlying processes and uncertainties).

Chapter 5 includes results of the Combined Approach and has been divided into the following sections:

1. Comparison of empirical and process-based modelling methods for analysing changes in European forest growth;

2. Analysis of the relative importance of nutrients, climatic factors and CO_2 as causes of observed changes in growth;
3. Analysis of the long-term consequences for sustainability of observed growth changes of the European forests; and
4. Implications and consequences for future management strategies.

Chapter 6 contains conclusions and recommendations.

References

Andersson, F., Aronsson, A., Popovic, B. and Tamm, C.O. 1988. Swedish forests in a changing environment, effects on soils, tree nutrition, and forest growth. Symp. Forest health and productivity, StoraSkog, Sept. 1988. Falun, Sweden. 35 p.

Becker, M., Bräker, O.U., Kenk, G., Schneider, O. and Schweingruber, F.H. 1990. Kronenzustand und Wachstum von Waldbäumen im Dreiländereck Deutschland - Frankreich - Schweiz in den letzten Jahrzehnten. Allg. Forst Zeitschr. 45: 263–274.

Dyck, W.J., Cole, D.W. and Comerford, N.B. (eds.) 1994. Impacts of forest harvesting on long-term site productivity. Chapman & Hall, London. 371 p.

Elfving, B. and Tegnhammar, L. 1996. Trends of tree growth in Swedish forests 1953–1992: an analysis based on sample trees from the National Forest Inventory. Scan. J. For. Res. 11: 38–49.

Glatzel, G. 1999. Historic forest use and its possible implications to recently accelerated tree growth in Central Europe. In: Karjalainen, T., Spiecker, H. and Laroussinie, O. (eds.). Causes and consequences of accelerating tree growth in Europe. EFI Proceedings 27. European Forest Institute. Pp. 65–74.

Hunter, I. and Schuck, A. 2002. Increasing forest growth in Europe – possible causes and implications for sustainable forest management. Plant Biosystems 136: 133–141.

IPCC. 2001. Climate change 2001: Synthesis report. A contribution of working groups I, II, and III to the third assessment report of the Intergovernmental Panel on Climate Change. [Watson, R.T. and the core writing team (eds.)] Cambridge University Press, Cambridge, United Kingdom, and New York, NY, USA. 398 p.

Karjalainen, T., Spiecker, H. and Laroussinie, O. (eds.). 1999. Causes and Consequences of Accelerating Tree Growth in Europe. EFI Proceedings 27. European Forest Institute. 286 p.

Kenk, G., Evers, F.H., Unfried, P. and Schröter, H. 1983. Düngung als Therapie gegen Imissionswirkungen in Tannen-Fichten-Beständen. Allg. Forst- und Jagdztg. 154: 153–170.

Kenk, G. 1993. Growth in 'declining' forests of Baden-Württemberg (Southwestern Germany). In: R.F. Huettl and G. Müller-Dombois (eds.): Forest decline in the Atlantic and Pacific Region. Springer-Verlag, Berlin. Pp. 202–215.

Kenk, G. and Fischer, H. 1988. Evidence from nitrogen fertilisation in the forests of Germany. Environ. Poll. 54: 199–218.

Makkonen-Spiecker, K. 1991. Wachsen in Finnland und Schweden die Bäume in den Himmel. Allg. Forst Zeitschr. 46: 585.

Menzel, A. and Fabian, P. 1999. Growing season extended in Europe. Nature 397: 659.

Myneni, R.B., Keeling, C.D., Tucker, C.J., Asrar, G. and Nemani, R.R. 1997. Increased plant growth in the northern high latitudes from 1981 to 1991. Nature 386: 698–702.

Pretzsch, H. 1985. Wachstumsmerkmale oberpfälzischer Kiefernbestände in den letzten 30 Jahren. Vitalitättszustand - Strukturverhältnisse - Zuwachsgang. Allg. Forst Zeitschr. 42: 1122–1126.

Pretzsch, H. 1996. Growth trends of forests in Southern Germany. In: Spiecker, H., Mielikäinen, K., Köhl, M. and Skovsgaard, J.-P. (eds.). Growth trends in European forests. Studies from 12 countries. EFI Research Report 5. Springer Verlag, Berlin.

Rehfuess, K.-E., Ågren, G.I., Andersson, F., Cannell, M.G.R., Friend, A., Hunter, I., Kahle, H.-P., Prietzel, J. and Spiecker, H. 1999. Relationships between recent changes of growth and nutrition of Norway spruce, Scots pine, and European beech forests in Europe. EFI Working Paper 19. European Forest Institute. 94 p.

Röhle, H. 1987. Entwicklung von Vitalität, Zuwachs und Biomassestruktur der Fichte in verschiedenen bayerischen Untersuchungsgebieten unter dem Einfluß der neuartigen Walderkrankungen. Forst. Forschungsber. München 83. 122 p.

Spiecker, H. 1987. Düngung, Niederschlag und der jährliche Volumenzuwachs einiger Fichtenbestände Südwestdeutschlands. Allg. Forst- und Jagdztg 159: 70–76.

Spiecker, H., Mielikäinen, K., Köhl, M. and Skovsgaard, J. (eds.) 1996. Growth trends in European forests. Studies from 12 countries. EFI Research Report 5. Springer Verlag, Berlin. 372 p.

Tomter, S.M. 1994. Skog 94. Statistikk over skogforhold og ressurser i Norge. NIJOS Report. 103 p.

2. Material and methods for the Correlative Approach

2.1 Collection, selection, and processing of data

Jörg Prietzel[1], Karl-Heinz Mellert[1], Ralf Straussberger[1], Hans-Peter Kahle[2], Heinrich Spiecker[2] and Karl-Eugen Rehfuess[1]

[1]*Chair of Soil Science, Technical University Munich, Germany*
[2]*Institute for Forest Growth, University Freiburg, Germany*

2.1.1 Introduction

The analysis of potential driving forces hypothesised as the causes for widespread recent deviations of measured growth from expected growth of European forests as described in Chapter 1 is limited by the temporal and spatial coverage of data describing the different driving forces. A comprehensive analysis of the causes of these deviations is possible only for those forest stands and those time periods, where all important growth-influencing factors can be quantified reliably, preferably by direct measurement on site. For variables that have not been measured directly, available data of nearby sites can be adopted by appropriate regionalisation techniques in some cases.

Testing the general hypotheses of the Recognition project requires reliable data concerning:

1. stand growth (the target or dependent variable);
2. climate (basically temperature and precipitation);
3. stand nutrition (primarily CO_2 and N; additionally P and nutrient cations).

Reliable data concerning the following variables that affect tree and stand growth would also be beneficial:

4. atmospheric deposition;
5. soil chemistry;
6. land-use changes;
7. forest management; and
8. frequency and intensity of damage caused e.g. by insect infestation or severe storms.

However, since factors 4–8 generally affect tree growth indirectly through modification of factors 2 and 3, availability of data regarding factors 4–8 is not obligatory.

Fortunately data on forest growth, climate, stand nutrition, soil chemistry, and atmospheric deposition have been monitored intensively and over a long time period in Europe. However, even in Europe the evaluation of the causes of recent growth changes is considerably restricted by the limited knowledge about important growth factors. The potential historical availability of important variables for European forests, based on published information, is summarised in Figure 1.

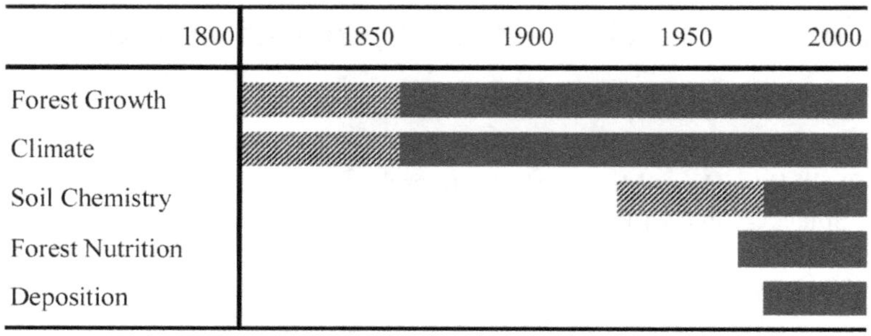

Figure 1. Maximum historical availability of data with relevance for *Recognition.* Hatched bars represent time intervals with sparse data of limited reliability.

Long-term (i.e. covering a period of >10–20 years) data series regarding soil chemistry, forest nutrition, and deposition exist for only a small number of forest stands (see below).

The implementation of the intensive monitoring of forest ecosystems in Europe in 1994 (the so-called Level II network) by the European Commission and the UN/ECE (EC-UN/ECE, 1998; de Vries et al., 2000) resulted in a significant increase in the number of forests, where all important environmental factors (climate, atmospheric deposition, soil chemistry, foliar nutrient levels), indicators of stand vitality (foliar nutrient levels, crown condition), and also stand growth are intensively investigated using standardised methods. During recent years the number of forest stands included in the Level II network has steadily increased in number. At present, the Level II network consists of more than 800 forest ecosystem monitoring plots distributed over 30 countries in Europe. The plots ideally represent typical forest ecosystems of the surrounding region.

2.1.2 The variables

2.1.2.1 Tree and stand growth

Forest tree or stand growth can be quantified using various methods (Spiecker, 1999). Each method requires specific preconditions concerning site, stand, and forest management. One of the most reliable and best-suited methods for the retrospective assessment of the growth of forest trees is stem analysis. This method provides height, diameter, and volume increment data on an annual basis, and most important for the Recognition project, it can be performed retrospectively. Thus the availability of historical growth data is not a critical precondition for the suitability of any European forest stand for evaluation, as long as old enough sample trees are available and costs for retrospective tree analysis can be covered.

2.1.2.2 Climate

Daily temperature measurements were already being made in 1659 in central England. An uninterrupted record of daily temperature for this location is available (Jones and Bradley, 1992). In the 18th century, regular measurements of temperature and precipitation were begun in many European cities (Manly, 1974), and have been continued until today by the national weather services. For Europe as a whole, there is a reliable and dense network of stations measuring air temperature and precipitation. Using spatial interpolation techniques, the data gathered at meteorological stations can be regionalised to provide information also for sites where direct measurements are not available. Historical meteorological data are available from the Joint Research Centre (JRC) for the whole of Europe on a daily basis starting from 1975 in a 50 km × 50 km grid net (MARS project, Monitoring Agriculture with Remote Sensing; http://mars.jrc.it/), and on a monthly basis starting in 1901 in a 0.5° latitude × 0.5° longitude grid net from the Climate Research Unit (CRU) (http://www.cru.uea.ac.uk; New et al., 1999 and 2000). These consistent and homogeneous data sets can be used for sites where no appropriate climate data had been measured either on site or at nearby meteorological stations.

Most conventional weather stations are not located in forested areas. However, it can be accepted that the long-term changes and trends of climate variables recorded at these stations resemble the respective changes in nearby forest stands. Direct measurements of meteorological data in forests began at a few places in the late 1960s. The number increased slowly during the 1970s and 1980s. The initiation of the Level II programme in 1994 resulted in a rapid increase in the number of meteorological stations in forests. At present meteorological data are collected at 188 Level II plots distributed all over Europe (EC-UN/ECE, 1998; de Vries et al., 2000). In contrast to conventional stations, the latter provide data regarding typical climate conditions in forest areas, resembling the climate experienced by the trees.

2.1.2.3 Stand nutrition

The increase in atmospheric CO_2 concentration is considered by many scientists as a major cause of recent forest growth changes. Because of the fast atmospheric mixing the concentrations that have been measured at the Mauna Loa since 1959 (Houghton et al., 1996) can be used directly in empirical models and multivariate statistical analyses for European forests.

The status of other nutrients in forest stands can be assessed most properly and reliably by chemical analysis of the concentrations of the nutrients of interest in the needles or leaves of representative trees (Diagnostic Foliar Analysis; Lundegårdh, 1951; Wehrmann, 1959 and 1963; Tamm, 1964; Fiedler et al., 1973; Evers, 1986; Brække, 1996). However, for a reliable data assessment providing comparable results of different studies, specific rules concerning sampling and analysis of foliage have to be considered (Wehrmann, 1959 and 1963; Evers, 1972 and 1988; Hüttl, 1985 and 1991).

Unfortunately, the availability of data characterising the nutritional status of forests is restricted to the past 40 years at a maximum, even though Saussure and Liebig had already stated in the 19th century that the nutritional status of plants is reflected by the concentrations of nutrients in plant shoots, and particularly in foliage. Yet, it was not until the 1950s that Diagnostic Foliar Analysis had been sufficiently developed and adopted for forest tree species. Since then long-term data sequences have been collected, often continuing uninterrupted until today. The earliest foliar nutritional data are available from the early 1960s for control plots of forest fertilisation experiments (e.g. Hausser et al., 1971; Sauter, 1991; Hippeli and Branse, 1992; Prietzel et al.; 1997; Spiecker et al., 1992). Data from forest sites other than fertilisation experiments cover at most the last 30 years (e.g. Solling), and often periods of less than 20 years. A broad assessment of forest nutrition in Europe was started in 1994 with the Level II network. Within that programme foliar nutrient concentrations are now measured biannually at 864 plots in 30 European countries, with first and second year needles/leaves collected for evergreen trees (EC-UN/ECE, 1998; de Vries et al., 2000).

At present there is no reliable method for retrospective assessment of the nutritional status of forest trees or stands. Some efforts have been made to use element concentrations in tree rings to address the historical availability of nutrients or noxious elements (e.g. Zech et al., 1990; Trüby, 1994; Hagemeyer, 1997; Nebe et al., 1999) or as proxy variables to reconstruct soil chemical properties (Guyette et al., 1992; Guyette and Cutter, 1994; Levy et al., 1996). However, the applicability of these methods is compromised by radial translocation of some elements of interest within the stem wood (Hagemeyer, 1997).

The nutritional status of forest stands is related to the chemical properties and nutrient contents of the rooted soil zone (e.g. Miller et al., 1977; Ke and Skelly, 1990/91; Liu and Trüby, 1989; Nohrstedt and Jakobson, 1994; Brække, 1996). Thus, changes in soil chemical properties – besides other exogenous factors such as weather, water supply, fungal diseases, insect pests, or endogenous factors such as ageing and fruit production – can affect the nutritional status of trees and forest stands. Therefore, soil chemistry should be considered when time-trends of stand nutrition and growth have to be explained, discussed or predicted. Changes in soil chemistry may also – paying attention to a number of caveats – be used as a surrogate variable describing changes in the nutritional status of forests for sites where direct information about stand nutrition is lacking. However, the availability of soil chemical data or archived soil samples is subject to similar limitations as the availability of forest nutrition data.

2.1.2.4 Atmospheric deposition

There is not much data about the historical atmospheric deposition into forests. For the period before 1970, there are almost no deposition records for European forests. Systematic and continuous measurements of atmospheric deposition into forests started at some places in the late 1960s. Thus, the longest continuous data sequences cover about 30 years, and are only available for a few sites. One of them is the Solling site in Germany. In the early 1980s the 'forest dieback' phenomenon in Central Europe was the subject of a great deal of public attention. This resulted in the establishment of the Level-II programme of EC-UN/ECE, comprising a network of plots, where deposition into forest ecosystems was measured. In 1994 the first deposition measurements were conducted within that programme. Since then the number of deposition monitoring stations has increased steadily. Within the framework of the Level-II programme, atmospheric deposition is currently measured inside and outside 494 forest stands (EC-UN/ECE, 1998; de Vries et al., 2000).

Some old deposition data exist for agricultural areas. At the Rothamsted experimental station in the United Kingdom, N and S deposition has been recorded regularly since 1853 (Blake et al., 1999). N deposition has also been assessed at other agricultural research stations in the 19th century (Skeffington and Wilson, 1988). Historic trends in the deposition of various air pollutants in Europe have also been reconstructed; for example, from the analysis of ice cores from glaciers in the high Alps (Asman et al., 1988).

Based on measurements from stations monitoring deposition all over Europe, the spatial pattern of the atmospheric deposition of N, S and acid (H^+) equivalents has been modelled by the European Co-operative Programme for Monitoring and Evaluation of the Long-Range Transmission of Air Pollutants in Europe (EMEP) (http://www.emep.int/; EMEP, 2001). Historical annual rates of N deposition are available since 1985.

2.1.2.5 Land-use changes/past forest utilisation

In many European regions, humans for centuries have significantly influenced the morphology and chemistry of soils by logging, agriculture, pasture, litter raking, manuring, fertilisation, and amelioration (Kreutzer, 1972; Glatzel, 1991 and 1999; Kirby and Watkins, 1998). Driven by an increasing human population and by technological advance, the anthropogenic impact on the chemical status of European forest ecosystems including their soils has increased greatly since the 15th century. It was at this time that intensive forest exploitation began (including activities such as litter raking, intensive use of forest for a large number of reasons, including pasture, exploitation for glass production, firewood and construction timber, shipmaking), and the primary forest was replaced by secondary forests of different species composition and structure in some areas (Johann et al., 2004). Depending on the intended use different management regimes including decisions on species and rotation, were carried out. Intensive forest management (including prescribed burning, liming, amelioration, and fertilisation) was common particularly in the 20th century. Moreover, exotic species (e.g. Douglas fir, Sitka spruce) have been introduced with enormous effects on the ecosystem and the environment. Finally, large areas have been cleared in the past – either on purpose for agriculture or by abuse, fire, etc. – and later were reforested again. Pre-industrial anthropogenic changes of soil properties can only be assessed and estimated very roughly, and for most forests there are no records that describe the intensity and duration of historical forest utilisation and land-use changes.

2.1.2.6 Forest management

Forest management is also usually well documented for forest stands where data about growth, nutrition, or soil chemistry are available. Furthermore, height growth of dominant trees, which is regarded as an indicator for forest site productivity in the Recognition project, is considered to be largely unaffected by forest management (Spiecker, 1999).

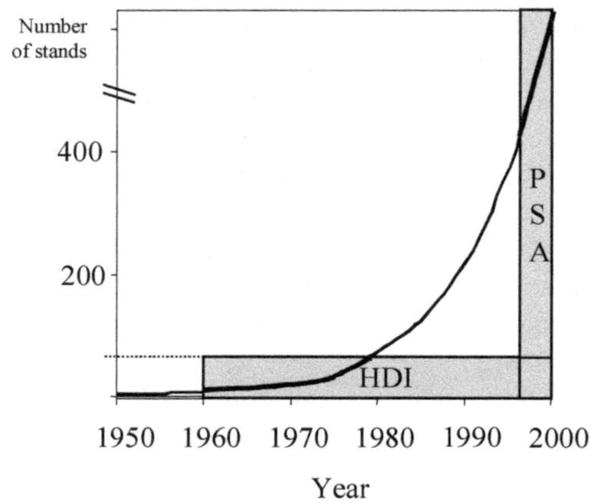

Figure 2. Number of forest stands in Europe with available data on stand nutrition and design of the Correlative Approach, consisting of HDI (Historical Development Investigation) and PSA (Present State Analysis).

2.1.2.7 Summary

A comprehensive evaluation of the contribution of different growth factors and their changes to recent growth changes of European forests is limited by the number of sites with available long-term data on stand nutrition, rather than the number of candidate study sites with available data on forest growth or climate. There are very few sites where stand nutrition data have been recorded before the 1980s (Figure 2). During recent years the number of sites with available data has increased by more than two orders of magnitude, mainly due to the implementation of the Level II network of the EC-UN/ECE (EC-UN/ECE, 1998; de Vries et al., 2000).

2.1.3 Research concepts

Because of the limited availability of sites with long-term data for stand nutrition, the Correlative Approach of the Recognition project has been divided into two research concepts (Figure 2): (1) Historical Development Investigation (HDI); and (2) Present State Analysis (PSA).

2.1.3.1 Historical Development Investigation

In the Historical Development Investigation (HDI) concept, data from long-term forest nutrition and yield studies in various European countries are analysed and compared with each other. The studies are mostly control plots of fertilisation and amelioration experiments. Basic requirements for the suitability of a plot to be included in the HDI are: (i) frequently repeated assessments of the stand's nutritional status by foliar analyses of important elements for a period of at least 15–20 years; and (ii) a controlled and well-documented silvicultural management history of the stands in the study. For each plot, relationships between the temporal variation and trend of growth and of the nutritional status of the stand can be quantified statistically. If in addition time series of soil and climate characteristics and deposition data exist, effects of these factors as well as their interactions on forest growth can also be analysed. Based on the results of these calculations, hypotheses on site-specific cause–effect relations can be derived. According to an inquiry conducted among 25 research institutions, 56 forest stands in eight European countries meet the requirements mentioned for the tree species of interest – Scots pine (*Pinus sylvestris*; n=28), Norway spruce (*Picea abies*; n=23), and European beech (*Fagus sylvatica*; n=5). This data collective was used to obtain an overview on the influence of different site-specific factors on forest growth and stand nutrition in different regions of Europe.

The HDI allows a direct formulation of site-specific cause–effect relations for each plot, as well as an integrated comprehensive evaluation of all sites.

2.1.3.2 Present State Analysis

In the Present State Analysis (PSA) concept, current growth and recent growth changes of forest stands throughout Europe are assessed together with a determination of important variables describing their present nutritional status as well as soil chemical variables, atmospheric deposition, and climatic conditions. The first step is an adequate stratification of the available data (e.g. by species, stand age, climatic conditions of the sites together with the water-storing capacity of the soil, and levels of atmospheric deposition – particularly N deposition). Then, patterns in the spatial distribution of the present growth or growth trends and the nutritional status of different stands are identified and compared. Finally, various methods of multivariate statistical data analysis (e.g. multiple regression analysis, and geostatistical methods) are used to isolate and quantify relationships between present forest growth, recent deviations of measured growth from expected growth, and site fertility.

The Level II forest intensive monitoring plots of the EC-UN/ECE (UN/ECE, 1998, de Vries et al., 2000) provide ideal material for the PSA. The Level II network consists of more than 800 forest ecosystem monitoring plots distributed over 30 countries in Europe. The plots represent typical forest ecosystems of the surrounding region. The Level II sites are supplemented by sites that have been monitored intensively in forest ecosystem research projects by various European universities and research stations.

The combination of the HDI with the PSA in the Correlative Approach is a powerful tool to assess past and recent relationships between site fertility, forest nutrition, climate, and forest growth in Europe. The combination of both research concepts is particularly promising, since some sites included in the PSA (e.g. the Solling Level II sites) have been monitored for >20 years and up to 30 years and are also studied in the HDI. These sites constitute an interface between the two research concepts, allowing a direct combination of the results achieved with different methods.

2.1.4 Collection of data

The aim of the Recognition project was to utilise and integrate existing data rather than to conduct new studies and make new measurements. For the HDI concept, this philosophy was compelling, since many relevant data cannot be measured retrospectively. The exceptions are: (i) the supplementation of existing long-term data sequences with additional data covering the period 1998–2000; and (ii) the replacement of earlier growth measurements with data from retrospective stem analysis.

A questionnaire was produced and distributed to relevant research teams all over Europe in the project preparation phase in 1998. Research teams were asked to identify forest stands that might be suitable for inclusion in the Recognition project. A preliminary list was created of potentially available data for the Recognition project. This list was updated at the Recognition Start-up Meeting in Freiburg in April 1999. Subcontracts for data submission were formulated, and additional data that should be gathered during the Recognition project were identified.

A framework for the co-operation between the ICP Forests/FIMCI and Recognition was established in winter 1999/2000, which allowed the data of the Level-II network to be used.

Data was transferred from the Recognition partners to the Correlative Approach database in digital form via standardised Data Submission Files (DSFs) in Microsoft Excel format, consisting of several spreadsheet templates, in which the different types of data (e.g. general plot information, foliage data, soil data) had to be filled. The DSFs included an example data set as a blueprint in order to facilitate correct completion of the spreadsheets.

For some forest stands included in the Recognition project, no time series of measured data of important climate variables or N deposition are available. Missing climate data were supplemented with data modelled by the Climatic Research Unit of the University of East Anglia (CRU; http://www.cru.uea.ac.uk). Missing data on N deposition were supplemented with data modelled by the European Co-operative Programme for Monitoring and Evaluation of the Long-Range Transmission of Air Pollutants in Europe (EMEP; http://www.emep.int/).

2.1.5 Evaluation and selection of data

Both the HDI and the PSA approaches consist of two steps of evaluation, which are described in detail in Chapter 2.2. The first step of each approach involves the formulation and calibration of empirical models to describe temporal and/or spatial relationships between levels/changes of growth-regulating factors (independent variables) and levels/changes of tree/stand growth (dependent variable) for the various study sites. This step was carried out for 89 sites (25 sites of Recognition partners, and 64 Level II plots) that had been selected from the available sites according to certain criteria (see below and Chapter 2.2). For these sites, stand growth was generally reassessed by stem analysis, irrespective of whether data from former growth measurements already were available or not. The second step consists of the validation of the empirical models. This step is carried out for the remaining 187 study sites using either existing growth data from measurements of the Recognition partners or growth data from retrospective tree analysis.

The basic requirements for inclusion in the HDI were met by 56 stands of Scots pine, Norway spruce or European beech, for which data were available for the Recognition project (Figure 3). A list of the stands selected for HDI analysis, including information about data availability and important site and stand variables, is presented in Table 1. The HDI sites were divided into two categories: category 1 (filled symbols in Figure 3) includes those sites (n=25) used for the formulation and calibration of empirical models to describe the temporal relationships between levels/changes of growth-regulating factors and levels/changes of forest growth; and category 2 (open symbols in Figure 3) includes the sites (n=31) used for validation of the empirical models.

For inclusion in the PSA approach there are no particular data requirements, except the presence of a complete and reliable data set with information on stand nutrition and climate for at least one calendar year in the time period 1990 to 1995. These requirements were met by more than 700 stands of Scots pine, Norway spruce or beech, for which data were available for the Recognition project. The PSA sites were also divided into two categories: Category 1 (Figure 4) includes those sites (n=89) which are used for the formulation and calibration of empirical models to describe the spatial relationships between levels (nutrition) and changes (climate variables) of growth-regulating factors and levels and changes of forest growth. Category 2 (n=156) includes the sites used for validation of the empirical models.

The majority (64 out of 89) of the sites and plots included in category 1 of the PSA are selected sites from the Level II network. Out of the total number of Level-II plots dominated by Norway spruce (n=194, status 12/1999), Scots pine (n=226), or beech (n=108) those plots with a mean stand age of 50 years or more, and those for which data on tree foliar macro nutrient levels are available for at least one year within the assessment years 1990–1995/96, 64 plots were selected (Table 2). These plots are characterised by a combination of extreme levels (low, high) of three main factors or state variables, which are supposed to be key variables in regulating tree growth and which are causing or indicating site productivity changes (Figure 5; white circles). The selection of sites was conducted according to the following stratification criteria:

1. **Change in mean seasonal (May–Sept) precipitation sum** between the periods 1961–1990 and 1901–1930 (in %) as variable characterising the atmospheric water supply component of forest stands (together with variable 2);
2. **Change in mean seasonal (May–Sept) air temperature** (in K) between the periods 1961–1990 and 1901–1930; and
3. **Foliar N levels** (mean of assessments during 1990–1995/96), characterising the N nutrition of the trees.

The following criteria were also considered:

1. +/– balanced spatial coverage of plots over Europe (according to species specific distributional ranges);
2. Spatial patterns of nitrogen and sulphur deposition rates (according to available data from recent assessments on the European level).

Table 1. Stands selected for HDI analysis. Category 1: sites used for the formulation and calibration of empirical models. Category 2: sites used for validation of empirical models.

Nr	Country	Site Name	Observation Period	Latitude	Longi-tude	Elevation [m]	Precipi-tation [mm/yr]	Tempe-rature [°C]	Available foliar nutrient concentrations	Climate data	Number of observations	Category
Pinus sylvestris												
1	Finland	Ähtäri 98	1981 - 1996	62.58	24.13	170	605	2.6	N,P,K,Mg,Ca	measured on site	5	2
2	Finland	Alajärvi 108	1981 - 1996	63.02	24.20	195	568	2.1	N,P,K,Mg,Ca	modelled (CRU)	5	2
3	Finland	Halsua 106	1981 - 1997	63.39	24.30	160	542	2.1	N,P,K,Mg,Ca	modelled (CRU)	4	2
4	Finland	Juva 68	1978 - 1996	61.95	27.57	120	629	2.7	N,P,K,Mg,Ca	modelled (CRU)	5	2
5	Finland	Kannonkoski 56	1979 - 1999	62.92	25.61	170	610	2.2	N,P,K,Mg,Ca	modelled (CRU)	5	2
6	Finland	Kolari 135	1982 - 1997	67.25	23.87	190	496	-0.9	N,P,K,Mg,Ca	modelled (CRU)	5	1
7	Finland	Kuorevesi 52	1979 - 1999	62.02	24.81	120	635	3.1	N,P,K,Mg,Ca	modelled (CRU)	5	2
8	Finland	Punkaharju 25	1978 - 1998	61.81	29.33	80	594	3.3	N,P,K,Mg,Ca	modelled (CRU)	6	1
9	Finland	Pylkönmäki 73	1978 - 1996	62.76	24.75	200	613	2.2	N,P,K,Mg,Ca	modelled (CRU)	5	2
10	Finland	Ruokolahti 64	1979 - 1999	61.49	29.07	90	598	3.0	N,P,K,Mg,Ca	modelled (CRU)	4	2
11	Finland	Sodankylä 161	1981 - 1999	67.35	27.02	200	494	-1.2	N,P,K,Mg,Ca	modelled (CRU)	5	2
12	Finland	Sodankylä 184	1984 - 1999	67.34	26.65	188	498	-1.1	N,P,K,Mg,Ca	modelled (CRU)	5	2
13	Finland	Vesijako 26	1978 - 1998	61.39	25.04	115	600	3.4	N,P,K,Mg,Ca	modelled (CRU)	6	2
14	Finland	Vierumäki 157	1982 - 1993	61.11	26.03	140	626	3.5	N,P,K,Mg,Ca	modelled (CRU)	4	2
15	Finland	Vilppula 53	1979 - 1999	62.07	24.48	150	632	3.1	N,P,K,Mg,Ca	modelled (CRU)	5	2
16	Germany	Aalkasten	1964 - 1999	53.26	13.44	88	594	8.3	N,P,K,Mg,Ca	measured on site	36	1
17	Germany	Alt Placht	1964 - 1999	53.17	13.40	65	558	8.5	N,P,K,Mg,Ca	measured on site	36	1
18	Germany	Briescht	1964 - 1999	52.10	14.11	44	557	8.6	N,P,K,Mg,Ca	measured on site	35	1
19	Germany	Gandenitz I	1964 - 1999	53.17	13.43	68	558	8.5	N,P,K,Mg,Ca	measured on site	36	1
20	Germany	Gandenitz II	1964 - 1999	53.16	13.42	65	558	8.5	N,P,K,Mg,Ca	measured on site	36	1

Table 1. continued.

Nr	Country	Site Name	Observation Period	Latitude	Longi-tude	Elevation [m]	Precipi-tation [mm/yr]	Tempe-rature [°C]	Available foliar nutrient concentrations	Climate data	Number of observations	Category
21	Germany	Löpten	1964 - 1999	52.15	13.71	41	557	8.6	N,P,K,Mg,Ca	measured on site	36	1
22	Germany	Pfaffenwinkel	1964 - 1998	50.00	12.33	535	615	5.8	N,P,K,Mg,Ca	measured on site	35	1
23	Germany	Pustert	1964 - 1998	49.21	12.00	480	650	7.2	N,P,K,Mg,Ca	measured on site	23	1
24	Germany	Triebsch	1964 - 1999	52.35	13.79	37	517	8.8	N,P,K,Mg,Ca	measured on site	35	1
25	Sweden	Jädraås	1974 - 2000	60.49	16.30	185	607	3.8	N	measured on site	17	1
26	UK	Black Isle 8/76	1976 - 2000	57.63	-4.24	150	700	8.6	N,P,K,Mg,Ca	measured on site	8	1
27	UK	Rosarie 2 P70	1972 - 1999	57.53	-3.13	320	900	7.6	N,P,K,Mg,Ca	modelled (CRU)	8	2
28	UK	Rosedale 118/70	1970 - 1984	54.35	-0.75	232	760	8.0	N,P,K	modelled (CRU)	5	2
Picea abies												
29	Austria	Dobrowa	1966 - 1999	46.63	14.65	460	978	7.3	N,P,K,Mg,Ca	modelled (CRU)	19	2
30	Austria	Helfenberg	1962 - 1999	48.58	14.15	790	909	6.6	N,P,K,Mg,Ca	measured on site	37	1
31	Austria	Karlstift	1962 - 1999	48.59	14.77	910	884	5.2	N,P,K,Mg,Ca	measured on site	35	1
32	Czech R.	Rajec	1991 - 1999	49.43	16.68	625	690	6.3	N,P,K,Mg,Ca	measured on site	3	1
33	Denmark	Dalgas	1963 - 1978	56.52	9.21	30	717	7.5	N,P,K,Mg,Ca	modelled (CRU)	5	2
34	Denmark	Dalgas	1963 - 1978	56.52	9.21	30	717	7.5	N,P,K,Mg,Ca	modelled (CRU)	4	2
35	Denmark	Djbplantage	1971 - 1987	56.13	9.41	115	703	7.5	N,P,K,Mg,Ca	modelled (CRU)	4	2
36	Denmark	Sdromme	1971 - 1987	55.82	8.87	30	856	7.5	N,P,K,Mg,Ca	modelled (CRU)	4	2
37	Finland	Heinola 113	1978 - 1998	61.17	26.05	115	624	3.6	N,P,K,Mg,Ca	modelled (CRU)	8	2
38	Finland	Kannonkoski 58	1979 - 1999	62.93	25.68	195	609	2.1	N,P,K,Mg,Ca	modelled (CRU)	7	1
39	Finland	Kannonkoski 60	1979 - 1999	62.93	25.67	200	610	2.0	N,P,K,Mg,Ca	modelled (CRU)	5	2
40	Finland	Kemijärvi 194	1985 - 1999	66.85	27.13	272	518	-1.0	N,P,K,Mg,Ca	modelled (CRU)	5	1

Table 1. continued.

Nr	Country	Site Name	Observation Period	Latitude	Longitude	Elevation [m]	Precipitation [mm/yr]	Temperature [°C]	Available foliar nutrient concentrations	Climate data	Number of observations	Category
41	Finland	Moinniemi 49	1979 - 1999	61.72	29.54	115	593	2.7	N,P,K,Mg,Ca	modelled (CRU)	4	2
42	Finland	Patasalo 31	1980 - 1992	61.85	29.39	82	595	3.3	N,P,K,Mg,Ca	modelled (CRU)	5	2
43	Finland	Sahalahti 138	1982 - 1992	61.42	24.41	110	589	3.6	N,P,K,Mg,Ca	modelled (CRU)	4	2
44	Finland	Sodankylä 197	1986 - 1996	67.69	26.23	242	488	-1.6	N,P,K,Mg,Ca	modelled (CRU)	4	2
45	Finland	Sulkava 34	1979 - 1999	61.90	28.28	130	612	2.7	N,P,K,Mg,Ca	modelled (CRU)	5	2
46	France	Bonhomme	1985 - 1999	48.29	6.95	1100	1480	5.0	N,P,K,Mg,Ca	modelled (CRU)	11	2
47	France	Gemaingoutte	1988 - 1999	48.29	6.95	650	993	8.0	N,P,K,Mg,Ca	modelled (CRU)	7	2
48	Germany	Hoeglwald	1983 - 1998	48.29	11.08	540	811	8.2	N,P,K,Mg,Ca	measured on site	8	1
49	Germany	Pfalzgrafenweile	1966 - 1987	48.49	8.50	600	1678	6.6	N,P,K,Mg,Ca	measured on site	5	1
50	Germany	Solling-Fichte	1968 - 1997	51.77	9.58	508	1088	6.4	N,P,K,Mg,Ca	modelled (CRU)	11	1
51	Sweden	Skogaby	1987 - 1999	56.33	13.13	105	1100	7.5	N,P,K,Mg,Ca	measured on site	13	1
Fagus sylvatica												
52	France	Auberive (52)	1969 - 1997	47.79	5.14	456.6667	760	10.0	N,P,K,Mg,Ca	modelled (CRU)	4	1
53	France	Epinal (88)	1969 - 1997	48.17	6.55	423.75	760	10.0	N,P,K,Mg,Ca	modelled (CRU)	4	1
54	France	Souilly (55)	1969 - 1997	49.03	5.32	317.5	760	10.0	N,P,K,Mg,Ca	modelled (CRU)	4	2
55	France	Ste Hélène (88)	1969 - 1997	48.31	6.66	316.6667	760	10.0	N,P,K,Mg,Ca	modelled (CRU)	4	2
56	Germany	Solling-Buche	1968 - 1998	51.76	9.58	504	1088	6.4	N,P,K,Mg,Ca	modelled (CRU)	12	1

Figure 3. Map of sites (n=56) selected for HDI. Tree species: ■ Scots pine (n=28), ▲ Norway spruce (n=23), ● beech (n=5). Filled symbols represent sites included in category 1 (formulation and calibration of empirical models); open symbols represent sites included in category 2 (validation of empirical models).

Figure 4. Map of sites selected for PSA (n=245). Tree species: ■ Scots pine, ▲ Norway spruce, ● beech. Filled symbols represent sites included in category 1; open symbols represent sites included in category 2.

Another group of sites included in the 89 category 1 sites of PSA are those sites that have been included in category 1 of the HDI approach (n=25; grey circles in Figure 5). In most cases these sites do not represent extreme combinations of the three key variables. Their inclusion into the PSA enables the interlinkage of PSA and HDI, which is an important part of the Correlative Approach.

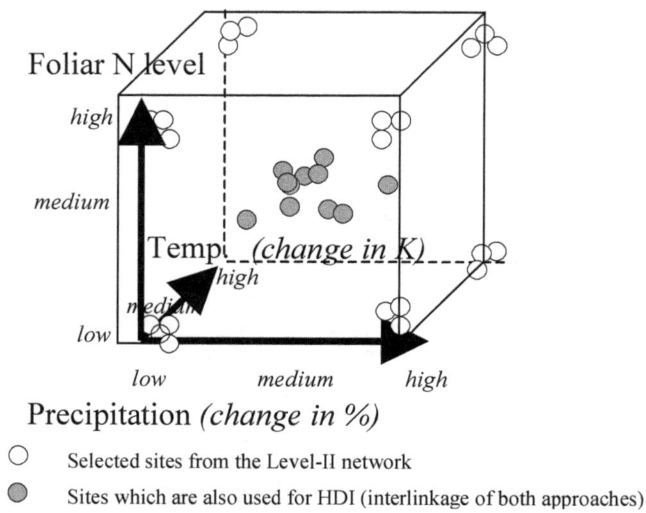

Figure 5. Schematic view of the factorial design used for the selection of suitable stands included in category 1 of the PSA approach (calibration of empirical models).

2.1.6 Discussion

The HDI concept is based on a comprehensive, integrated evaluation of a limited number of forest sites, where stand nutrition has been monitored and recorded for a period of at least 15 years, and up to 40 years. Other environmental variables with relevance for tree growth, such as important meteorological parameters (air temperature, precipitation), have either been monitored directly on site or at nearby meteorological stations. For the first time a detailed insight into the relationships between the changes of important growth factors during the last decades and recent changes of forest growth at the European scale is possible, including the spatial patterns of these relationships. This is achieved by correlating the variation and long-term trends of the different growth-regulating factors with each other and with the year-to-year variation and long-term trends of stand growth at the sites.

However, the HDI concept has to address some inherent problems and limitations:

(1) Spatial distribution of suitable forest stands

Long-term data sets describing changes in tree nutritional status are only available for a limited number (n=56) of suitable forest stands. This is particularly true for beech (n=5; see Table 1). The number of stands where recent growth has been assessed retrospectively by stem analysis (HDI category 1; Table 1) is even smaller (n=25; beech: n=3). Moreover, the suitable stands are not uniformly distributed over Europe (Figure 3). Most stands are situated in Central Europe and Scandinavia, with only a

Table 2. List of stands of the Level-II network (n=64) used for the formulation and calibration of empirical models to describe spatial relationships between levels and changes of growth-regulating factors and levels and changes of forest growth within the PSA approach. Total number of PSA plots: 64 Level-II plots plus 25 HDI category 1 plots (see Table 1 for details of the HDI category 1 plots).

Nr	Country	Level II Plot Nr	Latitude	Longitude	Nr	Country	Level II Plot Nr	Latitude	Longitude	Nr	Country	Level II Plot Nr	Latitude	Longitude
Pinus sylvestris					***Picea abies***					***Fagus sylvatica***				
1	Germany	501	51.76	6.06	24	France	46	45.80	1.82	50	Germany	503	51.69	7.26
2	Germany	1203	52.98	13.64	25	Germany	504	51.84	8.95	51	Germany	703	49.63	7.92
3	Germany	1303	53.64	13.94	26	Germany	506	51.02	8.18	52	Germany	704	50.41	7.73
4	UK	716	53.41	-1.75	27	Germany	917	48.41	10.54	53	Germany	911	48.98	12.88
5	Spain	5	40.85	-3.99	28	Germany	920	50.47	11.35	54	Spain	15	43.00	-1.34
6	Sweden	7301	62.00	14.43	29	Germany	1403	50.49	14.58	55	Sweden	6001	56.30	14.65
7	Sweden	7501	65.95	23.26	30	Germany	1404	51.12	14.58	56	Sweden	6101	56.03	13.92
8	Finland	6	66.36	26.73	31	Denmark	21	55.14	8.89	57	Sweden	6102	56.11	13.52
9	Finland	9	64.97	26.39	32	Sweden	6303	56.78	13.15	58	Poland	125	50.31	18.49
10	Finland	10	61.87	24.21	33	Austria	7302	63.08	16.74	59	Poland	129	49.66	22.49
11	Finland	16	61.77	29.34	34	Austria	17	47.33	12.41	60	Poland	145	51.77	16.21
12	Finland	18	60.70	27.85	35	Austria	19	47.38	11.89	61	Slovakia	202	48.64	19.05
13	Poland	14	51.61	16.92	36	Finland	3	68.00	24.24	62	Slovakia	206	48.63	19.05
14	Poland	16	51.88	15.54	37	Finland	17	61.81	29.32	63	Switzerland	13	46.70	7.76
15	Poland	17	53.88	14.50	38	Finland	21	66.30	29.50	64	Hungary	12	47.66	16.48
16	Poland	46	49.41	21.95	39	Poland	101	54.68	17.99					
17	Poland	60	53.45	19.44	40	Poland	111	49.65	19.43					
18	Poland	63	53.00	19.25	41	Norway	3	58.76	7.57					
19	Poland	75	51.35	15.69	42	Norway	8	60.60	6.52					
20	Poland	77	52.02	21.10	43	Norway	10	61.00	9.23					
21	Poland	82	50.09	18.67	44	Norway	11	60.37	9.74					
22	Poland	91	51.98	22.97	45	Norway	12	62.77	8.89					
23	Hungary	8	46.97	19.55	46	Norway	15	63.29	11.19					
					47	Norway	16	64.65	12.28					
					48	Hungary	2	47.89	19.95					
					49	France	80	47.54	-1.80					

few stands in the UK. Southern Europe and a large part of Eastern Europe completely lack forest stands with long-term assessments of tree nutrition. Consequently the HDI approach can be used to support statements concerning relationships between recent changes of forest growth and recent changes of growth-regulating factors for Central Europe, Scandinavia and for the northern part of the United Kingdom (Scots pine only), but not for the rest of Europe.

Furthermore, most stands for which long-term nutritional data are available are located on control plots of fertilisation experiments. For good reasons these fertilisation experiments were established on particularly poor sites, which often had been subject to severe anthropogenic degradation by intensive land/forest utilisation (such as litter raking, pasture, potash burning, etc). In Central Europe this kind of forest utilisation has been a rather widespread practice (Höhne, 1970; Glatzel, 1991 and 1999; Kirby and Watkins, 1998).

(2) Comparability of data of long-term data sequences and among different sites
In some cases there have been changes in analytical methods during the monitoring period, particularly for data sets comprising a long time period (20–40 years). Any change in methods increases the risk of systematic errors, thus biasing the data sequence. In the Recognition project, most groups (e.g. Prietzel et al., 1997), addressed and quantified these systematic errors by repeated concomitant analyses of reference samples, and corrected the data adequately. Moreover, the Recognition partners assured that any eventual changes in analytical methods had not resulted in systematic errors in their data sets. Thus, the comparability of single data within a long-term data set of a specific site is guaranteed.

The long-term studies included in the HDI approach comprised stands that have been monitored over different time intervals (Table 1). The duration of the investigation periods varied from 15 to 40 years. The investigation periods commenced between 1963 and 1985, and terminated between 1978 and 1999. Moreover, the number of foliar nutrient level assessments for the different sites during the investigation periods varied between 4 and 36. A comparison of the investigated sites regarding their relationships between recent changes in growth and growth-regulating factors must address the differences in the duration and commencement/termination of the investigation periods. A general model, including most possible sites, can only be established on the basis of those calendar years, for which data from all sites are available. Furthermore, the predictive power of the general model is flawed by the circumstance that the various research groups contributing to the Recognition project often have used different sampling and/or analysis methods for foliar analysis. In this context it must be noted again that in the HDI approach of the Recognition project "old" available foliage data of different case studies are compiled, which had been acquired by the various project partners up to three decades ago, without any joint protocol or inter-laboratory comparisons. This circumstance probably reduces the precision of the results and thus the predictive power of the general model, which includes these "old" data obtained by different research groups, resulting in increased residuals and a decreased predictive power. Nevertheless, the various site-specific results of the HDI approach are based on robust, consistent data. The data of selected sites of the European Level II network, which constitute the major part of the sites evaluated in the PSA approach, are generally acquired according to a standard protocol, and the quality and comparability of the data is routinely checked by FIMCI.

(3) Restricted number of potential growth factors included in the assessment
Not all potential growth-regulating factors have been assessed in the various experiments included in the HDI. For example, no information is available about temporal changes in the concentrations of any nutritional or noxious gases and/or aerosols entering the tree directly via foliage (NH_3, SO_2, ozone). Furthermore, in many experiments, the effect of micronutrients (e.g. S, B, Fe, Mn, Zn, Cu) was not analysed. Statistically derived relationships and models based on incomplete data sets are themselves necessarily incomplete. This means that they might not include the most important factors for stand growth on a particular site. This may result in low R^2 values of the formulated models, and a large unexplained variance of the stand growth. This problem is most critical for the integrated models for all HDI sites, since these models include only those variables that have been assessed at all sites.

In contrast to the limited number of plots that are suitable for an evaluation within the HDI concept, the use of data of the forest monitoring plots of the Level II network for the PSA approach resulted in the potential availability of more than 500 Scots pine, Norway spruce, and beech stands, covering most of Europe. In contrast to many sites included in the HDI, the sites included in the PSA generally represent typical forest stands of the region of interest. For most sites a number of various environmental data are available that may be used for a detailed discussion of the site-specific growth trends. Another advantage of the data collective used in the PSA, which mainly consists of plots of the Level II network, is that at all Level II sites the data has been (i) produced using standardised methods, and (ii) subject to a rigorous error assessment, resulting in a high accuracy and precision of the data (EC-UN/ECE, 1998; de Vries et al., 2000). Moreover, some of the Level II stands (e.g. the Solling spruce and beech stands) had already been the subject of intensive studies long before the Level-II network was established. These sites are also included in the HDI, and constitute an ideal link between the HDI and the PSA approaches.

A disadvantage of the sites included in the PSA is the lack of historical data on stand nutrition, climate, and atmospheric deposition. Therefore, in a strict sense, only current spatial relationships between forest growth/recent growth changes on the one hand and the current forest nutritional status on the other hand can be assessed. The temporal development of these interrelations cannot be analysed with the PSA, and a causal analysis of the temporal changes of the dependent variable (growth) and the changes of various site factors is impossible. Even though historical climate and N deposition data (in contrast to historical stand nutritional data) can be estimated using appropriate models (CRU, JRC, EMEP), the applicability of these data is jeopardised by a high degree of uncertainty, particularly when climate or deposition data of large-area (>1000 km²) grids are applied at the stand level (area: <100 ha).

Despite the large number of potentially available plots, the number of forest stands that can actually be studied within the PSA is limited by the number of retrospective growth assessments (stem analysis) that have been conducted within the duration of the Recognition project. These analyses are labour intensive.

Therefore, the number of sites included in PSA where both growth data and data on potential growth factors is available, is limited to 89.

The specific limitations of the HDI and the PSA can be overcome by combining both approaches. There are 25 sites included in the Combined Approach. Such a combination provides a powerful tool to assess past and recent relationships between site fertility, forest nutrition, climate, and forest growth in Europe.

References

Asman, W.A.H., Drukker, B., and Janssen, A.J. 1988. Modelled historical concentrations and depositions of ammonia and ammonium in Europe. Atmospheric Environ. 22: 725–735.

Blake, L., Goulding, K.W.T., Mott, C.J.B, and Johnston, A.E. 1999. Changes in soil chemistry accompanying acidification over more than 100 years under woodland and grass at Rothamsted experimental station, UK. Eur. J. Soil Sci. 50: 401–412.

Braekke, F.H. 1996. Needle analysis and graphical vector analyses of Norway spruce and Scots pine. Trees 11: 23–33.

de Vries, W., Reinds, G.J., van Kerkvoorde, M.S, Hendriks, C.M.A., Leeters, E.E.J.M., Gross, C.P, Voogd, J.G.H. and Vel, E.M. 2000. Intensive Monitoring of Forest Ecosystems in Europe, 2000 Technical Report. EC, UN/ECE 2000, Brussels, Geneva, 191 p.

EC-UN/ECE United Nations Economic Commission for Europe and European Commission 1998. Protection of Forests Against Atmospheric Pollution: European Programme for the Intensive Monitoring of Forest Ecosystems – Basic Documents for the Implementation of the Intensive Monitoring Programme of Forest Ecosystems in Europe. Brussels, European Commission (EC)VI/3908/95-EN. 142 p.

EMEP, 2001. Transboundary acidification and eutrophication and ground level ozone in Europe. Report 1/01, July 2001. EMEP Summary Report 2001. Joint CIAM & CCC & MSC-W report.

Evers, F.H. 1972. Die jahrweisen Fluktuationen der Nährelementkonzentrationen in Fichtennadeln und ihre Bedeutung für die Interpretation nadelanalytischer Befunde. Allg. Forst Jagdz. 143: 68–74.

Evers, F.H. 1986. Die Blatt- und Nadelanalyse als Instrument der Bioindikation. Allg. Forst Zeitschr. 33: 6–9.

Evers, F.H. 1988. Wasser-, nadel- und bodenanalytische Begleituntersuchungen zu den Versuchen mit oben offenen Kammern und dem Luftmeßprogramm auf verschiedenen Standorten beim Edelmannshof/Welzheimer Wald. KfK-PEF-Berichte 351: 245–254.

Fiedler, H.J., Nebe, W. and Hofmann, W. 1973. Forstliche Pflanzenernährung und Düngung. G. Fischer, Stuttgart. 481 p.

Glatzel, G. 1991. The impact of historic land use and modern forestry on nutrient relations of Central European forest ecosystems. Fertilizer Research 27:1–8.

Glatzel, G. 1999. Historic forest use and its possible implications to recently accelerated growth in Central Europe. In: Karjalainen, T., Spiecker, H. and Laroussinie, O. (eds.) Causes and consequences of accelerating tree growth in Europe. EFI Proceedings 27. European Forest Institute. Pp. 65–74.

Guyette, R. P., Henderson, G. S., and Cutter, B. E. 1992. Reconstructing soil pH from Manganese concentrations in tree-rings. Forest Science 38: 727–737.

Guyette, R. P. and Cutter, B. E. 1994. Barium and manganese trends in tree-rings as monitors of sulfur deposition. Water, Air, Soil Poll. 73: 213–223.

Hagemeyer, J. 1997. Die Variabilität der Elementverteilungen im Stammholz von Bäumen – Untersuchungen zur dendroanalytischen Bioindikation. Dissertationes Botanicae 279. 84 p.

Hausser, K., Bilger, W., Evers, F. H., and Werner, H. 1971. Düngungsversuche zu 45- bis 90 jährigen Fichten- und Fichten-Tannenbeständen auf oberem und mittlerem Buntsandstein des Württembergischen Schwarzwalds. Allg. Forst u. Jagd Ztg. 142: 1–11.

Hippeli, P. and Branse, C. 1992. Veränderungen der Nährelementkonzentrationen in den Nadeln mittelalter Kiefernbestände auf pleistozänen Sandstandorten Brandenburgs in den Jahren 1964 bis 1988. Forstw. Cbl. 111: 44–60.

Höhne, H. 1970. Streunutzungsschäden, ihre Entstehung und Verbreitung sowie Möglichkeiten ihrer Behebung. Wiss. Zeitschr. TU Dresden 19: 1047–1054.

Houghton, J.T., Meira Filho, L.G., Callander, B.A., Harris, N., Kattenberg, A. and Maskell, K. (eds.). 1996. Climate change 1995 – The science of climate change. Contribution of Working Group I to the Second Assessment Report of the Intergovernmental Panel on Climate Change. Cambridge. Press Synd. Univ. Cambridge. 572 p.

Hüttl, R.F. 1985. "Neuartige" Waldschäden und Nährelementversorgung von Fichtenbeständen (*Picea abies* Karst.) in Süddeutschland. Freiburger Bodenkundl. Abh. 16, 195 p.

Hüttl, R.F. 1991. Die Nährelementversorgung geschädigter Wälder in Europa und Nordamerika. Freiburger Bodenkundl. Abh. 28, 427 p.

Johann, E., Agnoletti, M., Axelsson, A.L., Bürgi, M., Östlund, L., Rochel, X., Schmidt, U.E., Schuler, A., Skovsgaard, J.P. and Winiwarter, V., 2004. History of secondary Norway spruce forests in Europe. In: Spiecker, H., Hansen, J., Klimo, E., Skovsgaard, J.P., Sterba, H. and Teuffel, K.v. (eds.) Norway Spruce Conversion – Options and Consequences. EFI Research Report 18. Brill, Leiden. Pp. 25–62.

Jones, P.D. and Bradley, R.S. 1992. Climatic variations in the longest instrumental records. In: Bradley, R.S. and Jones, P.D. (eds.) Climate since A.D. 1500. Routledge, London. Pp. 246–268.

Ke, J. and Skelly, J.M. 1990/91. Foliar symptoms on Norway spruce and relationships to magnesium deficiencies. Water, Air, Soil Poll. 54: 75–90.

Kirby, K. J. and Watkins, C. 1998. The Ecological History of European Forests. Wallingford: CAB International. 384 p.

Kreutzer, K. 1972. Über den Einfluß der Streunutzung auf den Stickstoffhaushalt von Kiefernbeständen (*Pinus silvestris* L.). Forstw. Cbl. 91: 263–270.

Levy, G., Brechet, C., and Becker, M. 1996. Element analysis of tree rings in pedunculate oak heartwood: an indicator of historical trends in the soil chemistry, related to atmospheric deposition. Annales des Sciences Forestieres 53: 685–696.

Liu, J. and Trüby, P. 1989. Bodenanalytische Diagnose von K- und Mg-Mangel in Fichtenbeständen (*Picea abies* Karst.). Z. Pflanzenern. Bodenkunde 152: 307–311.

Lundegårdh, H. 1951. Foliar analysis. Hilger and Watts, London. 176 p.

Manly, G. 1974. Central England Temperatures. Quarterly J. Royal Met. Soc. 100: 389–405.

Miller, H.G., Cooper, J.M. and Miller, J.D. 1977. Response of pole-stage Sitka spruce to application of fertiliser nitrogen, phosphorus and potassium in upland Britain. Forestry 65:15–33.

Nebe, W., Gembala, R., and Ilgen, G. 1999. Jahrringchronologische Elementuntersuchungen an Fichte zum Nachweis klassischer sowie neuartiger Waldschäden im Thüringer Wald und Erzgebirge. Beiträge Forstw. Landschaftsökol. 33: 132–138.

New, M., Hulme, M. and Jones, P. 1999. Representing twentieth-century space-time climate variability. Part I: Development of a 1961–90 mean monthly terrestrial climatology. J. Climate 12: 829–856.

New, M., Hulme, M. and Jones, P. 2000. Representing twentieth-century space-time climate variability. Part II: Development of 1901–96 monthly grids of terrestrial surface climate. J. Climate 13: 217–223.

Nohrstedt, H.Ö. and Jakobson, S. 1994. Samband mellan näringshalter och årsbarr. (Relationship between nutrient concentrations in soil and current year-needles). Skogforsk Redegörelse 7: 1–18.

Prietzel, J., Kolb, E. and Rehfuess, K.E. 1997. Langzeituntersuchung ehemals streugenutzter Kiefernökosysteme in der Oberpfalz: Veränderungen von bodenchemischen Eigenschaften und der Nährelementversorgung der Bestände. Forstw. Cbl. 116: 269–290.

Sauter, U. 1991. Zeitliche Variationen des Ernährungszustands nordbayerischer Kiefernbestände. Forstw. Cbl. 110: 13–33.

Skeffington, E. and Wilson, E.J. 1988. Excess nitrogen deposition: Issues for consideration. Environ. Poll. 54: 159–184.

Spiecker, H. 1999. Growth trends in European forests – Do we have sufficient knowledge? In: Karjalainen, T., Spiecker, H., and Laroussinie, O. (eds.): Causes and consequences of accelerating tree growth in Europe. EFI Proceedings 27. European Forest Institute. Pp. 157–169.

Spiecker, H., Hildebrand, E.E., and Evers, F.H. 1992. Ernährung und Zuwachs von Fichten (*Picea abies* [L.] Karst.) in einem langfristig beobachteten Düngungsversuch im Buntsandstein-Schwarzwald. Forst und Holz 47:181–185.

Tamm, C.O. 1964. Determination of nutrient requirements of forest stands, pp 115–170. In: Romberger, J.A. and Mikola, P. (eds.) International Review of forest research. Volume 1. Academic Press New York – London. 404 p.

Trüby, P. 1994. Zum Schwermetallhaushalt von Waldbäumen. Freiburger Bodenkundl. Abh. 33. 297 p.

Wehrmann, J. 1959. Methodische Untersuchung zur Durchführung von Nadelanalysen in Kiefernbeständen. Forstw. Cbl. 78: 77–97.

Wehrmann, J. 1963. Möglichkeiten und Grenzen der Blattanalyse in der Forstwirtschaft. Landwirtsch. Forschung 16: 130–145.

Zech, W., Schneider, B.U., and Röhle, H. 1990. Element composition of leaves and wood of beech (*Fagus sylvatica* L.) on SO_2-polluted sites of the NE-Bavarian mountain. Water, Air, Soil Poll. 54: 97–106.

2.2 Sampling, measurement and analysis methods

Hans-Peter Kahle[1], Heinrich Spiecker[1], Rüdiger Unseld[1],
Pedro-José Pérez-Martínez[1], Jörg Prietzel[2], Karl-Heinz Mellert[2],
Ralf Straussberger[2] and Karl-Eugen Rehfuess[2]

[1]*Institute for Forest Growth, University Freiburg, Germany*
[2]*Chair of Soil Science, Technical University Munich, Germany*

2.2.1 Introduction

Application of harmonised methods in the data generating process (data sampling, measurement and analysis techniques) is an important prerequisite if the data are to be compiled in a unique database and analysed in a uniform way. This is especially true when data from different sources and assessment times are analysed in an integrated study, and when the data originate from assessments where harmonised and commonly accepted standards, which could guarantee a certain quality of the data, do not yet exist. The purpose of this chapter is to give an overview of the sampling, measurement and analysis methods which have been applied in the Recognition project.

The aim of the correlative analyses within this research study is to contribute to a better understanding of the processes involved in the observed changes in growth of forests, and its relations to changes in the environment by means of statistical data analysis methods (Rehfuess et al., 1999). The application and fitting of suitable statistical models is an appropriate way to explore and describe the behaviour of complex adaptive systems (Holland, 1992a and 1992b; Bennett and Chorley, 1978; Grant et al., 1997). Two approaches have been defined for the statistical modelling: (1) the *Historical Development Investigation* (HDI) focuses on the analysis of time series of the variables of interest; and (2) the *Present State Analysis* (PSA) focuses on their cross-sectional and spatial variability (Chapter 2.1). The research design underlying the correlative approaches is that of a comparative observational study (Manly, 1992). The research questions address the long-term growth behaviour of forest trees under the influence of the complex, temporally and spatially varying, and interacting environmental conditions of the real world. Since changes in forest site productivity are hypothesised to be the result of processes which have started already some time ago in the past and which are acting on time scales of up to decades it would not have been a feasible option to set up an experimental treatment design for the clarification of the underlying causal relationships (Likens, 1989 and 1998). Observational studies are characterised by the fact that existing circumstances are used as treatment classifications whereas in designed experiments the investigator assigns treatments to the experimental units in a randomised way. The most important consequence is the difference of the corresponding scientific inference: in designed experiments it is often possible to assign causal relationships between the responses and the treatments; in observational studies it is only possible to identify associative relationships between the responses and the classified treatments (Kuehl, 2000) since the effect of the explanatory variable on the response variable might be confounded (mixed up) with the effect of other variables.

Concerning the database for the empirical studies within the Recognition project, the idea is to utilise and integrate already existing data rather than to conduct new measurements. However, since comparable and actual data on changes in forest growth was not available to the necessary extent, additional growth data were sampled during the lifetime of the research project. Prietzel et al. (Chapter 2.1 in this report) explain how suitable existing data on soil chemistry, foliage nutrition, meteorology, climate and growth were screened and identified, and how selected data were made available for analysis within the project.

2.2.2 Sampling and measurement methods

2.2.2.1 Methods of sampling and chemical analysis of soil and foliage

Already existing data on soil chemistry and foliage nutrition were compiled and analysed within the project (Chapter 2.1). Particular methods of sampling and chemical analysis of soil and foliage vary between the different laboratories; however, in most cases standard methods were applied. In the case of deviating methods these were carefully examined and reviewed in detail. Expert assessments of potential biases and judgements of the relevance for the integrated analyses built the basis for the decision either to include or exclude the data. Moreover, all submitted data were subject to a plausibility check, in which data exceeding preset plausibility ranges were identified by automated queries, and thoroughly checked. The plausibility ranges correspond to those applied by FIMCI (the Forest Intensive Monitoring Coordinating Institute) to check the plausibility of Level II data (Tables 1, 2 and 3).

2.2.2.2 Sampling and assessment of forest growth

The data on forest growth either originated from already existing databases or were collected during the project's lifetime. The methods of sampling and measurement of the newly collected growth data are described in the following sections.

Forest site productivity was not measured directly; instead age-related height growth (site index) was used as an indicator for the productive potential of a forest site (Assmann, 1961). In the HDI, changes in site index over time were used to determine changes in forest site productivity, whereas in the PSA, differences in age-related height growth of trees with different germination dates growing on a similar site were regarded as indicative for changes in site productivity.

Analogous to the physical "distance versus time" function of an object in motion, growth can be defined as the "change in seize versus time" function of a biological entity. Accordingly, as velocity (v) measures the speed of an object in a particular direction, increment (i) is a measure of the rate of change in seize with respect to

Table 1. Plausible values of selected soil parameters according to FIMCI.

Parameter	Unit	Organic layer	Mineral soil	$CaCO_3$-soils
N	mg/g	5.0–35	0.05–10.0	
P	mg/g	0.1–5.0	0.01–5	
organic C	mg/g	150–600	1–200	
pH ($CaCl_2$)		2–6	2–8	6–8
pH (KCl)		2–6	2–8	6–8
pH (H2O)		3–6	3.0–8.5	6.0–8.5
CEC	µmol IE/g	10–1000		
Base saturation		10–100	1–100	

Table 2. Combined plausible values of selected soil parameters according to FIMCI.

Parameter combination	Unit	Range of plausible values
$CaCO_3$, if pH < 6	mg/g	0.00
C/N ratio		3–100
pH, if $CaCO_3$ > 0		6–8
Base saturation, if $CaCO_3$ > 0		90–100
Base saturation, if pH > 6		50–100
Base saturation, if pH < 4		0–50

Table 3. Plausible values for nutrient contents in needles/leaves according to FIMCI.

Parameter	Unit	Range of plausible values
Mass of 100 leaves	g	1–100
Mass of 1000 needles	g	1–100
N	mg/g	5–40
S	mg/g	0.3–10
P	mg/g	0.3–10
Ca	mg/g	1–40
Mg	mg/g	0.1–10
K	mg/g	1–35
Na	µg/g	50–7500
Zn	µg/g	10–500
Mn	µg/g	10–5000
Fe	µg/g	10–1000
Cu	µg/g	1–100
Pb	µg/g	1–100
Al	µg/g	10–1000
B	µg/g	5–150

time. We distinguish (1) the annual increment which measures the instantaneous velocity of growth over one year, and (2) the periodic mean increment which measures the average annual velocity of growth over a period of several years. Change in height increment, Δih, is defined as the rate of change of increment with respect to time. Hence, Δih is a measure of the acceleration (a) of growth ($a = \mathrm{d}v/\mathrm{d}t$). Since we were also analysing changes in height increment with respect to time, another measure needs to be considered: the rate of change of acceleration with respect to time, i.e. the time derivative of the acceleration vector a. This measure formally is called the jerk (j) ($j = \mathrm{d}a/\mathrm{d}t$) (Weisstein, 1999). The following symbols and terms are used for Δih synonymously: $\Delta ih\%$, change in height growth, change in height increment, and height growth acceleration.

Tree height growth can either be determined based on data from repeated height measurements, or retrospectively if height development is reconstructed by means of retrospective analysis. Trends in tree height growth as defined in this study (Chapter 1) are determined by comparing observed with expected growth. Expected growth – or reference growth – usually is derived from some kind of (height growth) model. Differences exist in the way how reference growth is modelled, and, depending on the data used for calibration, how well the model is adapted to a specific forest site. Since level and shape of species specific height growth curves are primarily determined by site factors, model validity is mainly a function of the representativity for the local site conditions of the data used for model calibration. Most commonly applied height growth models, like those used for the construction of yield tables, are considered to be at least of regional validity. For many regions in Europe appropriate site index curves and yield tables for forest management are readily available, whereas this is not the case with local site specific models.

Site specific long-term data from repeated tree height measurements in subsequent stand generations are extremely rare (e.g. Kenk et al., 1991; Untheim, 1996b) and not consistently available for a wide range of different site conditions. The retrospective approach applied here makes use of the fact that the annual diameter and height growth rhythm of trees under seasonal climatic conditions allows reconstruction of diameter and height growth by means of the stem analysis technique.

The aim of the forest growth assessment was to get a precise and unbiased estimate of the average tree height development of dominant trees from pure even-aged stands of the species of interest on the selected sites. The crown class of dominant trees was defined as the target stratum of sampling because height growth of dominant trees is relatively independent of stand density, and is therefore more closely related to site productivity than height growth of trees from other crown classes (Assmann, 1961; Lanner, 1985; Magnussen and Penner, 1996; Clutter et al., 1983; Hägglund, 1981). Furthermore, the probability that trees descend from their dominant crown class is in general far lower than that of trees of subordinate crown classes to ascent (Abetz, 1989; Klädtke, 1990). In order to minimise effects due to non-site factors and to reduce the sampling error, only sample trees from the dominant crown class with similar stem diameter, similar tree height and age were selected. Trees with injuries, or abiotic or biotic damages to an extent which is considered to be not typical for the given site were refused.

To determine changes in tree height growth in the PSA the concept of **cohort comparison** was applied: differences in the average height growth curves between trees with different germination dates on a similar site are used as indicators for changes in forest site productivity over time (Untheim, 1996a). However, differences in height growth between tree cohorts are only indicative for changes in forest site productivity if the following conditions are fulfilled (Spiecker, 1999; Untheim, 1996b):

(i) **site:** the site type of the selected stands has to be the same.
(ii) **stand:** the stands have to be similar with respect to stand establishment, stand composition, and stand structure (indicated e.g. by the shape of the diameter and height distribution of the trees within the stand).
(iii) **sample trees:** in order to obtain an unbiased estimate of changes in the height-age-relation over time the sample trees must be of the same genotype and must at each age represent the same stand component in both cohorts.
(iv) **signal strength:** since tree growth typically fluctuates on a gradual time scale of periods from years to decades a searched for growth trend signal could be obscured or could be disguised by an extraordinary sequence of growth fluctuations especially if the strength of the growth trend signal is low in relation to that of the non-trend variation (Spiecker, 1999; 1995).

The **sites** must be similar with respect to relevant site characteristics such as topographic location, floristic composition and major physical and chemical soil properties. Since gradual changes of a site type at a certain location can occur even within a period of only a few decades (e.g. due to soil acidification) (Spiecker et al., 1992; de Vries et al., 2002; Puhe and Ulrich, 2001; Vanmechelen et al., 1997) it is important that the sample sites to be used for the comparison are chosen in close proximity to each other in order to make sure that not only the site type at the time of sampling but also the environmental conditions governing the processes of soil development at the selected sites are similar.

In practice a similar **stand structure** is mainly the result of similarity in stand establishment, disturbance dynamics and in the management regimes which are applied over the lifetime of the stands (mainly type and intensity of tending and thinning). In addition the stands should be even-aged pure stands, large enough that sample trees from the stand interior are not affected by natural or artificial stand edges. According to the research layout the mean germination dates of the tree cohorts to be selected in a sampling area should differ by approximately 50 (minimum 40) years. The reasons are: (i) a time shift of 40 years covers the most interesting recent time period from 1960 onwards; and (ii) an age overlap of 40 years extends over a period sufficiently long to allow identification of consistent long-term changes in forest site productivity over the time span of several decades. The target ages of the stands to be selected are 50 years for the young, and 90 years for the older stand. The age limits should be between a minimum of 40 years and a maximum of 120 years. These age limits are set in order to exclude juvenile and senescent developmental stages during which trees might be less responsive to environmental stimuli (Kozlowski et al., 1991; Meier-Dinkel and Kleinschmit, 1990).

In order to find suitable pairs of combinations of similar site and stand types in the selected areas, local experts in monitoring forest condition were asked in advance to identify stands in close proximity to one another which fulfil the above mentioned conditions. The suggested stands were critically reviewed in the field by the experienced project field crews before the most suitable stands were finally selected.

Within the stands, **sample trees** were selected with the aim of drawing a representative sample of the crown class of dominant trees. According to earlier experience a sample size of five trees and a minimum number of three replications for each treatment class were defined. In order to reduce the risk of bias the sample trees were centred well around the tree with mean basal area out of the class of dominant trees. A large crown ratio and the occurrence of relatively thick dead branches or large knots in the lower part of the bole were used as indicators of the dominant crown class status of the tree in the past (Abetz, 1989; 1985).

2.2.2.3 Retrospective growth analysis

Since forest site productivity cannot be retrospectively analysed in a direct way, reconstructed changes in site index (age-related height growth of dominant trees) is used in this study as an indirect measure of past development of site productivity (Spiecker et al., 1994; Spiecker, 1999).

Data on tree growth were collected by applying the stem analysis technique (Abetz, 1985; Ferretti et al., 2002). The stem analysis technique involves retrospective analysis of the annual radial growth rates by measuring the width of the annual growth rings on a series of cross-sections and of the annual or periodic height growth rates by determining the length of the internodes on the main stem axis (Roloff, 2001). The field and laboratory assessments were conducted according to the standard protocols applied at the Institute for Forest Growth, University Freiburg.

Breast height (1.3 m) was marked on the north-side of the tree before felling. On the felled log, the shortest distance between the breast height mark and the tip of the tree was measured. This measurement plus 1.3 m was taken as the tree height. In the case of oblique trees this figure would systematically overestimate the tree height in comparison with methods usually used to measure heights of standing trees. Since oblique trees were refused from the sample, no bias of this type does affect the data.

For the Norway spruce and Scots pine samples, the height growth rates were retrospectively assessed by sequentially measuring the distance of each annual height shoot tip or whorl of primary branches to the stem base. Norway spruce and Scots pine are monocyclic species in which branch and stem elongation occur annually, and each annual shoot constitutes one growth unit. Within the green crown of the coniferous species, whorls in most cases can easily be identified. They are in general marked by the insertion of primary branches and by the occurrence of bud scars on the stem surface (Roloff, 2001). In both species lateral branching usually occurs only one year after bud formation (Gruber, 1990). Since the occurrences of thick dead branches and knots were used as criteria for the selection of sample trees, whorls could in most cases also easily be found in the dead branch zone of the lower stem. Already in the field the number of height growth years was cross-checked with the number of tree-rings on the cross-sections taken from predefined heights. For European beech height growth was reconstructed on a periodic basis by determining the cambial age on a series of cross-sections from predefined heights along the stem axis. From each sample tree an additional cross-sectional segment from the root-stem junction was taken. The number of tree-rings determined on this segment is used to estimate the absolute tree age.

Parallel sided cross-sections with a thickness of about 5 cm were taken from the following standard heights:

if tree height ≤ 20 m: 1.3 m, 11.5 m, 14.0 m, 16.5 m, 19.0 m
if tree height > 20 m and ≤ 30 m: 1.3 m, 11.5 m, 16.5 m, 21.5 m, 24.0 m, 26.5 m, 29.0 m
if tree height > 30 m: 1.3 m, 11.5 m, 21.5 m, 24.0 m, 26.5 m, 29.0 m, 31.5 m ...

These heights were measured from the breast height mark and refer to the upper side of the cross-section. Since the cross-sections should ideally be taken from the middle of an internode, the optimum location for sampling was found by looking for the next nearest place where no branch disturbs the radial growth pattern on the cross-sectional surface.

The upper side of the air-dried cross-sections was polished with increasingly fine sanding paper. The upper surface was then scanned with a digital camera, and the annual radial increments were measured to the nearest 1/100 mm with the semi-automated image analysis system developed at the Institute for Forest Growth. On the cross-sections from breast height the annual radial increments were measured along eight equidistant radii (45°). On all other cross-sections, radial increments were measured along four 90°-radii (Weise, 1987). The correct dating of the individual radial measurement series was visually controlled during the measurement process by comparing the series graphically with a correctly dated suitable reference series. For each year the radial increment in a given stem height was calculated as the quadratic mean of the eight (for the breast height cross-sections) or four (for the other cross-sections) measurements for each cross-section. In the radial increment data quality control, measurement and dating errors were checked by applying standard dendrochronological cross-dating methods (Briffa and Jones, 1990; Schweingruber, 1988). The dating of the annual height growth measurements was then re-checked based on the quality checked cambial age data.

After the stem analyses were conducted and the retrospective height and diameter growth data of the sample trees were available, changes in the competitive status of the spruce and pine sample trees in the past were assessed by tracking the development of the height-to-diameter-ratio (h/d-ratio: tree height divided by tree diameter in breast height). The h/d-ratio development curve first describes basic stem geometry (taper). However, changes of its level and shape over time are indicative of the sample trees' competitive status and its development (Abetz, 1976). In even-aged mono-species spruce or pine stands a tree which belonged to a subordinate crown class in the past and which later on successfully ascent to the dominant crown class typically would have a larger h/d-ratio at young developmental stage and show a steeper decrease in h/d-ratio with increasing age than an ever-dominant tree. Referencing height growth of an ever-dominant crown class tree retrospectively to such a "social climber" would result in an overestimation of the height growth change during the phase of reduced height growth (Magnussen and Penner, 1996) and an underestimation during the ascending phase of the climber. Hence, the risk of such a bias can be evaluated by this method. Since in literature no such relation between the h/d-ratio and the competitive status and its changes has been established for beech, this type of analysis was conducted only for the coniferous species.

2.2.3 Data analysis methods

2.2.3.1 Statistical methods for the Historical Development Investigation

The specific aim of the Historical Development Investigation (HDI) is to improve the insight into the effects of changes in tree nutrition, weather, climate, and deposition on changes in tree growth over time by means of the statistical analysis of time series data of the respective variables of interest. The underlying concept, data sources, and data requirements are introduced and described in detail in Chapter 2.1.

The particular statistical methods applied in the HDI are analyses of multiple linear regression models, either of the classical linear model type or of the generalised linear model type. In the classical linear model the errors in the error term are assumed to be normally and independently distributed with a mean of zero and a constant variance. With time series data the assumption of independency is often violated. In case of autocorrelated errors the generalised linear model is a more appropriate model type because it allows the error covariance matrix to be structured according to the stochastic process in the error variable. The underlying stochastic process needs to be specified and modelled, for example, by an autoregressive process (Fahrmeir and Hamerle, 1984). Since the more complex models are richer in parameters to be estimated the requirements concerning size and balance of the input data matrix are higher also. The *proc autoreg* and *proc mixed* procedures of the SAS® system were used for conducting the statistical analysis within the HDI approach (Chapter 4.1).

2.2.3.2 Statistical methods for the Present State Analysis

The present state analysis (PSA) aims to improve the insight into the effects of tree nutrition, nitrogen deposition, and changes in air temperature and precipitation on changes in tree growth by means of the statistical analysis of the spatially scattered data.

In the PSA the relationships between growth and environmental predictor variables are analysed using multiple linear regression analysis (ordinary least squares) and geostatistical methods (Wackernagel, 1998) to explicitly include spatial information.

The statistical t-tests for the significance of correlations and regression models and for significant differences in means and in regression slopes follow Zar (1974). The test of the null hypothesis of coincidence of k population regressions is based on an F-test also suggested by Zar (1974: 234–235).

Inference in regression models that combine time series and cross-sectional data

The estimation of relationships that combine multiple observations over time on a set of cross-sectional units is faced with the problem to specify a model that will adequately allow for differences in behavior over cross-sectional units (Judge et al., 1980). The specific model that is applied in the context of the PSA is the variable-intercept model (Hsiao, 1988). Basic assumption of this model is that, conditional on the observed explanatory variables, the variable intercepts capture individual specific effects, whereas the regression parameters for the exogenous variables are assumed to be constant across cross-sectional units. This particular model type was preferred against more complex alternative models because of its straightforward and logical structure. In general the model can be written as:

$$y_{it} = \beta_{1i} + \sum_{k=2}^{K} \beta_k x_{kit} + e_{it}$$

where $i = 1,2, \ldots, N$ refers to cross-sectional units (individuals), and $t = 1,2, \ldots, T$ refers to a given time. Thus, y_{it} is the value of the dependent variable for individual i at time t, and x_{kit} is the value of the kth non-stochastic explanatory variable for individual i at time t. The stochastic term e_{it} is assumed to have mean zero $E[e_{it}] = 0$, and constant covariance, $E[e_{it}^2] = \sigma_e^2$. The intercept β_{1i} captures differences in behaviour between individuals, whereas the slope coefficients β_k are assumed to be constant. More specific the intercept for the ith individual is formulated as $\beta_{1i} = \overline{\beta_1} + \mu_i$, where $\overline{\beta_1}$ is the "mean intercept", and μ_i represents the difference from this mean for the ith individual. The coefficients are estimated by using the fixed or random effects estimator in the context of general linear models. A Hausman test (Hausman, 1978) is computed to show which of both models is the correct specification. The *proc TSCS* of the GAUSS® software was used for estimating the pooled time-series cross-section regression models within the PSA approach.

Geostatistical analysis

The following steps were taken in the geostatistical analyses (Wackernagel, 1998): (1) the data were explored for spatial and multivariate structure and checked for outliers; (2) the interpretation of the spatial structure and the detected associations between the variables are built into a variogram model which is fitted to the data; and (3) based on the model of the spatial variation values of the regionalised variable at locations different from the sample points were estimated.

The ISATIS® (Geovariances, 2000) software package was used for conducting the statistical analysis within the PSA approach. Peculiarities of the applied geostatistical methods are described in Goovaerts (1997) and Webster and Oliver (2001).

References

Abetz, P. 1976. Beiträge zum Baumwachstum: Der h/d-Wert – mehr als ein Schlankheitsgrad! Forst- und Holzwirt 31: 389–393.

Abetz, P. 1985. Ein Vorschlag zur Durchführung von Wachstumsanalysen im Rahmen der Ursachenforschung von Waldschäden in Südwestdeutschland. AFJZ 156: 177–187.

Abetz, P. 1989. Zu den Ursachen des „Umsetzens" von (Z-)Bäumen. AFZ 44: 1334–1337.

Assmann, E. 1961. Waldertragskunde: Organische Produktion, Struktur, Zuwachs und Ertrag von Waldbeständen. Bayerischer Landwirtschaftsverlag, München. 490 p.

Bennett, R.J. and Chorley, R.J. 1978. Environmental Systems: Philosophy, Analysis and Control. Methuen and Co, London. 624 p.

Briffa, K.R. and Jones, J.D. 1990. Basic chronology statistics and assessment Methods of Dendrochronology: Applications in the Environmental Sciences. In: Cook, E.R. and Kairiukstis, L.A. (eds.), Methods of Dendrochronology: Applications in the Environmental Sciences. Kluwer Academic Publishers, Dordrecht. Pp. 137–152.

Clutter, J.L., Fortson, J.C., Pienaar, L.V., Brister, G.H. and Bailey, R.L. 1983. Timber Management: A Quantitative Approach. John Wiley and Sons, New York: 333 p

de Vries, W., Reinds, G.J., Dobben, H.v., Zwart, D.d., Aamlid, D., Neville, P., Posch, M., Auée, J., Voogd, J.C.H. and Vel, E.M. 2002. Intensive Monitoring of Forest Ecosystems in Europe – Technical Report 2002. UN/ECE, EC, Forest Intensive Monitoring Coordinating Institute (FIMCI). Geneva, Technical Report: 175 p.

Fahrmeir, L. and Hamerle, A. 1984. Multivariate statistische Verfahren. De Gruyter, Berlin.

Ferretti, M., Innes, J.L., Jalkanen, R.E., Saurer, M., Schäffer, J., Spiecker, H. and Wilpert, K.v., 2002. Air pollution and environmental chemistry – what role for tree-ring studies? Dendrochronologia 20: 159–174.

Geovariances, 2000. ISATIS® geostatistical package, Fontainebleau.

Goovaerts, P. 1997. Geostatistics for Natural Resources Evaluation. Oxford University Press, Oxford. 483 p.

Grant, W.E., Pedersen, E.K. and Marin, S. 1997. Ecology and Natural Resource Management – Systems Analysis and Simulation. John Wiley and Sons, New York. 373 p.

Gruber,F. 1990. Verzweigungssystem, Benadelung und Nadelfall der Fichte (*Picea abies*). Birkhäuser, Basel. 136 p.

Hägglund, B. 1981. Evaluation of forest site productivity. Forestry Abstracts 42: 515–527.

Hausman,J.A. 1978. Specification tests in econometrics. Econometrica 46: 1251–1271.

Holland, J.H. 1992a. Adaptation in Natural and Artificial Systems. MIT Press, Cambridge. 211 p.

Holland, J.H. 1992b. Complex Adaptive Systems. Daedalus 121: 17–30.

Hsiao, C. 1988. Analysis of Panel Data. Cambridge University Press, Cambridge, Reprint. 246 p.

Judge, G.G., Griffiths, W.E., Hill, R.C. and Lee, T.C. 1980. The Theory and Practice of Econometrics. John Wiley and Sons, New York. 793 p.

Kenk, G., Spiecker, H. and Diener, G. 1991. Referenzdaten zum Waldwachstum. KfK-PEF 82: 59.

Klädtke, J. 1990. Umsetzungsprozesse unter besonderer Berücksichtigung Z-Baum-bezogener Auslesedurchforstung. AFJZ 161: 29–36.

Kozlowski, T.T., Kramer, P.J. and Pallardy, S.G. 1991. The Physiological Ecology of Woody Plants. Academic Press, San Diego. 657 p.

Kuehl, R.O. 2000. Design of Experiments: Statistical Principles of Research Design and Analysis. Duxbury Press, Pacific Grove. 666 p.

Lanner, R.M. 1985. On the insensitivity of height growth to spacing. Forest Ecology and Management 13: 143–148.

Likens, G.E. (ed.), 1989. Long-Term Studies in Ecology – Approaches and Alternatives. Springer, New York. 214 p.

Likens, G.E. 1998. Limitations to Intellectual Progress in Ecosystem Science. In: Pace, M.L. and Groffman, P.M. (eds.). Successes, Limitations, and Frontiers in Ecosystem Science. Springer, New York. Pp. 247–271.

Magnussen, S. and Penner, M. 1996. Recovering time trends in dominant height from stem analysis. Canadian Journal of Forest Research 26: 9–22.

Manly, B.F.J. 1992. The Design and Analysis of Research Studies. Cambridge University Press, Cambridge. 353 p.

Meier-Dinkel, A. and Kleinschmit, J. 1990. Aging in tree species: present knowledge. In: Rodriguez, R., Tames, R.S. and Durzan, D.J. (eds.), Plant Aging: Basic and Applied Approaches. Plenum Press, New York. Pp. 51–63.

Puhe, J. and Ulrich, B. 2001. Global Climate Change and Human Impacts on Forest Ecosystems: Postglacial Development, Present Situation, and Future Trends in Central Europe. Springer, Berlin. 592 p.

Rehfuess, K.E., Agren, G.I., Andersson, F., Cannell, M.G.R., Friend, A.D., Hunter, I., Kahle, H.P., Prietzel, J. and Spiecker, H. 1999. Relationships between recent changes of growth and nutrition of Norway spruce, Scots pine, and European beech forests in Europe – RECOGNITION. European Forest Institute Working Paper 19. 94 p.

Roloff, A. 2001. Baumkronen – Verständnis und praktische Bedeutung eines komplexen Naturphänomens. Eugen Ulmer, Stuttgart. 164 p.

Schweingruber, F.H. 1988. Tree Rings: Basics and Applications of Dendrochronology. Kluwer Academic Publishers, Dordrecht. 276 p.

Spiecker, H. 1995. Growth dynamics in a changing environment – long-term observations. Plant and Soil 168–169: 555–561.

Spiecker, H., 1999. Forest growth. In: Rehfuess, K. E., Agren, G. I., Andersson, F., Cannell, M. G. R., Friend, A. D., Hunter, I., Kahle, H. P., Prietzel, J., and Spiecker, H. Relationships between recent changes of growth and nutrition of Norway spruce, Scots pine, and European beech forests in Europe – RECOGNITION. European Forest Institute Working Paper 19. Pp. 8–17.

Spiecker, H., Hildebrand, E.E. and Evers, F.H. 1992. Ernährung und Zuwachs von Fichten (*Picea abies* [L.] Karst.) in einem langfristig beobachteten Düngungsversuch im Buntsandstein-Schwarzwald. Forst und Holz 47: 181–185.

Spiecker, H., Mielikäinen, K., Köhl, M. and Untheim, H. 1994. Growth trends of European Forests - Has site productivity changed? European Forest Institute Working Paper 4. 68 p.

Untheim, H. 1996a. Has site productivity changed? A case study in the Eastern Swabian Alb, Germany. In: Spiecker, H., Mielikäinen, K., Köhl, M. and Skovsgaard, J.P. (eds.). Growth Trends of European Forests – Studies From 12 Countries. Springer, Berlin: Pp. 133–147.

Untheim, H. 1996b. Zur Veränderung der Produktivität von Waldstandorten: Untersuchungen zum Höhen- und Volumenwachstum von Fichte (*Picea abies* [L.] Karst.) und Buche (*Fagus sylvatica* L.) auf Standorteinheiten der Ostalb und des Flächenschwarzwaldes. Mitt. FVA Baden-Württemberg 198: 239 + Appendix.

Vanmechelen, L., Groenemans, R. and Ranst, E.v. 1997. Forest Soil Condition in Europe - Results of a Large-Scale Soil Survey. Forest Soil Co-ordinating Centre. Geneva and Brussels, United Nations Economic Commission for Europe (UN/ECE); European Commission (EC): 198 p.

Wackernagel, H. 1998. Multivariate Geostatistics. Springer, Berlin. 291 p.

Webster, R. and Oliver, M.A. 2001. Geostatistics for Environmental Scientists. John Wiley and Sons, Chichester. 271 p.

Weise, U. 1987. Beiträge zur Methodik und Genauigkeit von Stammscheiben-, Bohrkern- und Stammanalysen an Fichte und Tanne. Mitt. FVA Baden-Württemberg 130: 133.

Weisstein, E.W. 1999. MathWorld – A Wolfram Web Resource, http://mathworld.wolfram.com.

Zar, J.H. 1974. Biostatistical Analysis. Prentice-Hall, Englewood Cliffs. 620 p.

3. Material and methods for the Modelling Approach

3.1 Description of the models applied in the Modelling Approach

Göran I. Ågren[1], Oleg Chertov[2], Seppo Kellomäki[3], Alexander Komarov[2] and Marcel van Oijen[4]

[1] *Swedish University of Agricultural Sciences, Uppsala, Sweden*
[2] *European Forest Institute*
[3] *University of Joensuu, Joensuu, Finland*
[4] *Centre for Ecology and Hydrology, Edinburgh Research Station, UK*

3.1.1 Background

Vast amounts of data are available to describe the states and changes in European forests, as described in Chapter 2.1. There is also an abundance of data and knowledge about plants and soils at several levels of resolution. From all this information we need to select what is relevant for understanding of recent growth changes in European forests and for prediction of its consequences for the future. Our understanding of forest ecosystem processes is not such that there exists one optimal model. Instead we have to face the problem that by increasing the level of resolution and thereby also the possibilities to represent various aspects, we also increase the requirement for additional information about the system under study. This additional information is not always available and has to be estimated under uncertainty. For example, "What were the soil conditions when a stand was planted?" Such uncertainty leads to errors that may or may not be reproduced in predictions. For this reason, simpler models may be more robust (Rastetter et al., 1992; Ågren, 1996). To circumvent this problem we selected a suite of models to represent different scales – both in time and space as well as resolution in model processes – with the idea that when the models agree in their predictions, the results should be robust with respect to the way the details are described and when the models disagree, we should be able to understand the causes of the differences. The models differ considerably in how they are organised and in the emphasis placed on different processes. For this reason the descriptions of the models take quite different forms.

Table 1. Overview of models used in Recognition.

	EFIMOD	*EFM*	*FinnFor*	*Q*
Model type	Stochastic, spatially-distributed, individual-based	Deterministic, non-spatial, pool-based	Stochastic, non-spatial, pool-based	Deterministic, non-spatial, pool-based
# State variables	33	50	34	36
# Parameters	119	261	249	44
Time step	monthly	<hourly	hourly	monthly
Input variables	Radiation, T_{max}, T_{min}, precipitation, N-deposition	CO_2, radiation, T_{max}, T_{min}, precipitation, RH, wind, N-deposition	CO_2, radiation, T_{max}, T_{min}, precipitation, RH, wind, N-deposition	T_{max}, T_{min}, N-deposition
Ecosystem fluxes	C, N	C, N, water	C, N, water	C, N
C-input submodel	NPP = potential NPP * f([N], radiation)	NPP = Photosynthesis – Respiration	NPP = Photosynthesis – Respiration	NPP = N-productivity * (N–$[N]_{min}$ * biomass)
Number of tree C pools	5 for every tree	10	5	5
Number of tree N pools	5 for every tree	10	5	5
Number of soil pools (C, N, SOM, microbes, ...)	8	10	8	24
Number of surface litter pools	3 for every tree species plus deadwood	3	5 for every tree species	0

3.1.2 Models

Based on the competence in the project four models were selected:

EFIMOD – developed by the *European Forest Institute*, Joensuu, Finland;
EFM – developed by the *Natural Environment Research Council*, Centre for Ecology and Hydrology, Edinburgh, United Kingdom;
FinnFor – developed by the *University of Joensuu*, Faculty of Forest Sciences, Joensuu, Finland; and

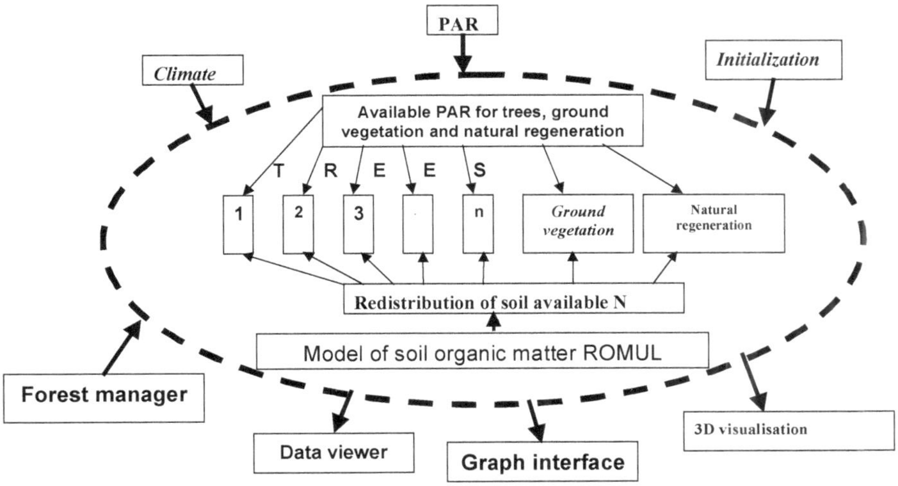

Figure 1. Flow chart of the EFIMOD model.

Q – developed by the *Swedish University of Agricultural Science Uppsala*, Department of Ecology and Environmental Research, Uppsala, Sweden.

A brief overview of the model characteristics is given in Table 1. Due to the limited space for the description of the models, the following descriptions will focus on the main characteristics of the models. Processes not discussed are in general given the simplest possible formulation.

3.1.3 EFIMOD

The EFIMOD model (Chertov et al., 1999b; Komarov et al., 2003) represents a realisation of the concept of a 'single plant ecosystem' (SPE) (Chertov, 1983a). EFIMOD simulates the biological cycle of carbon and nitrogen for every tree with a hypothetical soil pot in a stand. The principal feature of the model is that tree growth is mainly driven by available soil nitrogen with a fixed species-specific potential for utilisation and a climate-dependent net primary production (NPP) value determined by photosynthesis. This allows for a dynamic calculation of ecosystem 'tree-soil' feedback and effects on carbon and nitrogen in forest ecosystems. The species-specific ecological parameters are 'maximum biological productivity of leaves/needles' and 'specific consumption of nitrogen'. The tree model calculates biomass growth of five compartments: (1) the total biomass, (2) leaf biomass; (2) branch biomass, (3) stem biomass, (4) coarse root biomass and (5) fine root biomass. The soil submodel (ROMUL) (Chertov et al., 2001), simultaneously allows for a calculation of the rate of tree litter and soil organic matter mineralisation and humification with the corresponding carbon dioxide and nitrogen release for plant growth (Figure 1).

The model is based on the following assumptions:

- it is a spatial-individual-based model – i.e. the growth of each tree is simulated as a function of light competition, which depends on the tree's position within the stand, and the amount of nitrogen available for this tree in its area of nutrition;
- the model of tree growth is the simplest one that allows accounting for the carbon and nitrogen turnover;
- simulation of a large ensemble of interacting units allows the rough individual models to be smoothed out, and the main properties of the population's dynamics to be reflected – i.e. the standard procedure of Monte-Carlo simulations may be used;
- a set of input parameters is based on ordinary and traditional experimental data.

The model simulates pure and mixed stands, which are growing on different soils and in different climates.

A single plant occupies a certain space in the above-ground environment and the soil. The SPE can be treated as an elementary cell of forest ecosystems, and the ecosystem itself can be represented as SPE groups, where trees are situated in 'individual soil pots'. The pot size depends on the root mass and rate of nutrient consumption. The pot size (the area of nutrition) changes with time as the tree grows. The concept of the SPE unites the ideas of 'phytogenic field' (an area strongly influenced and controlled by a plant) of Uranov (1965) and of 'tessera' (a single tree's soil) by Karpachevsky (1981). Moreover, 'the ecological field theory', which has been recently formulated (Wu et al., 1985; Mou et al., 1993), closely corresponds to the concept of the SPE.

The EFIMOD model is not a gap or a process model as such. It can be defined as a 'combined (hybrid)' model, because the major biological cycles (production and decomposition processes) of every SPE in a stand are included. The model uses a simple approach for the calculation of tree productivity from a set of tree-specific ecological parameters. The model avoids eco-physiological processes and uses for calculation of potential growth the parameters of biological productivity and main dendrometric and population characteristics.

The driving forces for description of the tree production are (a) the maximum biological productivity of leaves/needles, a_{max}, [annual biological production (biomass increment) per kg of leaf/needle mass], and (b) growth determined by the avaliable nitrogen in a soil. The maximum biological productivity of leaves/needles reflects potential net primary production (NPP) of a tree in a defined climate with no limitations due to crown shading, soil moisture and soil nutrients. The value of a_{max} has been determined as a function of growing degree-days for Scots pine and Norway spruce using data of Alexeev (1975), Chertov (1983b), Kazimirov and Morozova (1973), Kazimirov et al. (1977 and 1978) and Kostin (1997). The growth determined by the available nitrogen is calculated using the value of 'specific consumption of nitrogen' (kg nitrogen per kg of biomass increment) reflecting species-specific nitrogen consumption. The actual growth is either the maximum growth reduced by factors taking into account water stress and light conditions, or the growth allowed by available soil nitrogen (Liebig's law of the minimum). Tree age determines the allocation of biomass increments and these proportions do not change within tree age state (actually pregenerative, generative and post generative age periods). Fixed concentrations of nitrogen is assumed for the increase of biomass.

The main features of the soil submodel (ROMUL) can be outlined as follows. The main assumption is that there is a succession of complexes of soil organisms-substrates in the process of SOM decomposition corresponding to the concepts of 'raw humus', 'moder' and 'mull' existing in forest pedology (Wilde, 1958; Duchaufour, 1961). We assume that the biomass of the soil organisms represents a negligible part of all decomposed matter and has a high rate of decomposition. We do not consider it as a separate pool. The number and species composition of the decomposer organisms depend on the biochemical properties of the organic debris, and the hydrological and thermal conditions. Thus, it is possible to calculate decomposition coefficients for the communities as a function of temperature, moisture and the biochemical properties of litter. The organic debris is divided into two main groups: (1) above-ground litter falling on the soil surface; and (2) below-ground litter consisting of root debris. Using ROMUL it is also possible to take into account above-ground and below-ground cohorts of litter that have different chemical properties. The number of cohorts is not restricted.

The rate of the nutrient release due to mineralisation corresponds to the rate of the organic matter mineralisation. The only exception is the nitrogen kinetics in the organic layers (forest floor and peat). Because of its high rate of consumption by soil micro-organisms, the gross rate of nitrogen mineralisation is lower than the same rate of carbon mineralisation in the organic layers. In mineral horizons the rate of nitrogen release corresponds to the rate of carbon mineralisation when C/N ratio < 8 (Alexandrova, 1970). The dynamics of nitrogen, including the nitrogen available for plants, has been described and evaluated using ROMUL by Chertov (1985) and Chertov et al. (2001).

A community of *Bacteria* and *Arthropoda* produces humus with a C/N ratio 15, and a community of *Lumbricidae* forms humus with a C/N ratio 8. These organisms use 20% of the nitrogen in undecomposed plant debris for their biomass, and 80% of nitrogen goes to humus formation. These considerations are used for the calculations of the humification rate. The nutrients of the consumed matter are fully mineralised and available for plants after one year because of the short life cycles of *Arthropoda* and *Lumbricidae*. We assume that nitrogen and other elements released as a result of mineralisation are fully consumed by plants. If excess nitrogen remains, it is immobilised in the soil. Leaching, deposition of nitrogen, nitrogen fixation, and denitrification are not taken into account in the ROMUL submodel, but can be included in the ecosystem model. Other elements (phosphorus, potassium, calcium and magnesium) can also be included.

The validation of the model, with various initial stand densities on different soils, shows a realistic picture of tree growth, stand self-thinning and soil changes. Convergence of leaf/needle mass has been observed for various initial stand densities (Chertov et al., 1999b). The model has also been applied for the evaluation of the rate of soil restoration on industrial barrens (Chertov et al., 1999a) and for the evaluation of different silvicultural strategies in Central European Russia (Komarov et al., 2002; Chertov et al., 2005). The basic EFIMOD version was slightly modified for the Recognition project in relation to the format of output parameters to be compatible with other models' outputs.

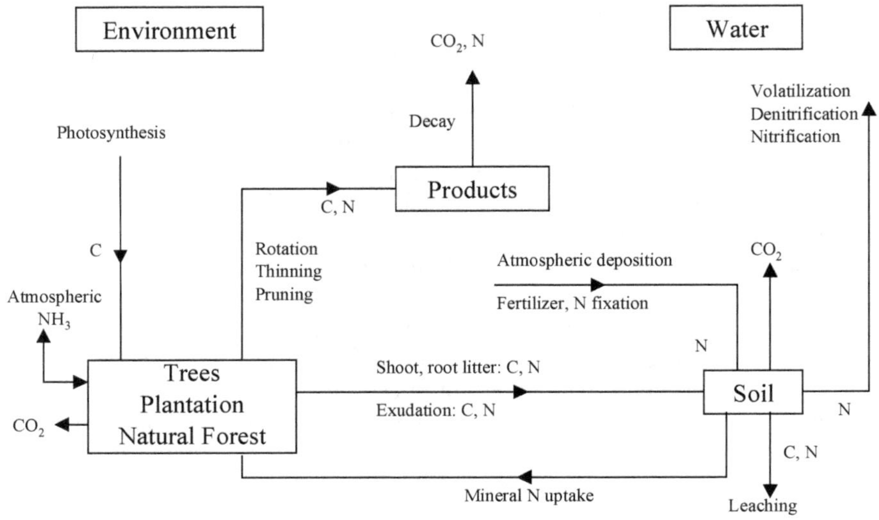

Figure 2. Outline of the EFM model.

3.1.4 EFM

The Edinburgh Forest Model (EFM) is mechanistic, dynamic, and deterministic. Published descriptions of the model can be found in Thornley (1991) and Thornley and Cannell (1992). The model can be used to simulate plantation forests, natural forests, or natural forests subject to thinning or pruning. Figure 2 illustrates the ecosystem components represented in the model. The model accounts for the pools and fluxes of carbon (C), nitrogen (N) and water in the forest ecosystem. There are submodels for trees (Figure 3), soil (Figure 4), and water (Figure 5).

The input and output fluxes to a box (state variable) are obtained using some assumed mathematical equations. For example, photosynthesis in Figures 2 and 3, which is an input to the foliage C substrate box (labile carbohydrates), is computed using a hyperbolic relationship and assuming that the rate of photosynthesis depends upon light, CO_2, air temperature, and nitrogen (Cannell and Thornley, 1998). The processes represented in Figure 2 (and in Figures 3 and 4) and their equations can be regarded as empirical relationships, which supply the mechanisms used to interpret and provide an understanding of the top-level diagram in Figure 2. The empirical leaf photosynthesis equation represents one of the contributing mechanisms when viewed from the ecosystem/whole-plant level.

Many of the elements of the EFM are the same or very similar to those of a grassland ecosystem model, for which full details are given in Thornley (1998a). Plant dry matter is regarded as being structural (e.g. cellulose) or storage (e.g. mono- and di-saccharides, starch, fructans for C substrates; amino acids, nitrate, some labile proteins for N substrates). The sources and sinks have different environmental dependencies, and this assumption partially decouples the sources

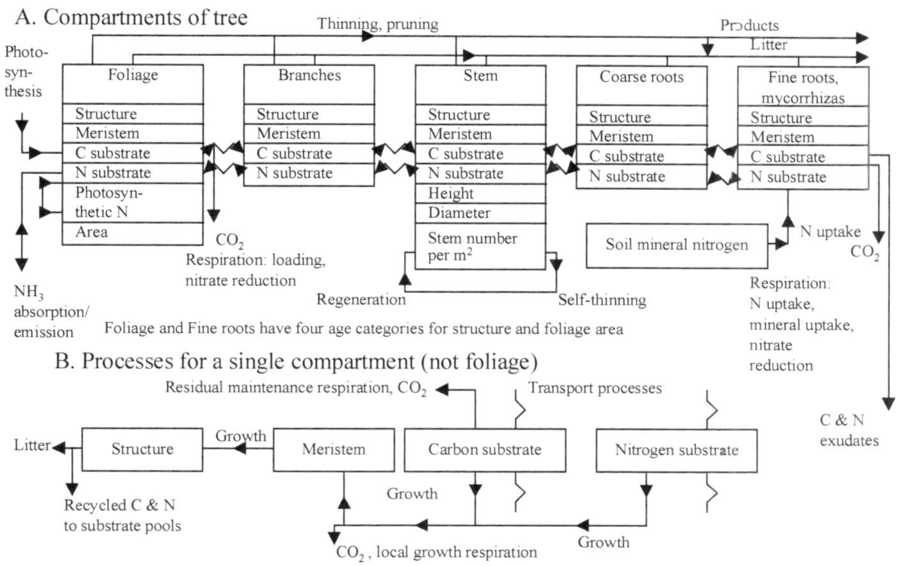

Figure 3. The tree submodel in the EFM model.

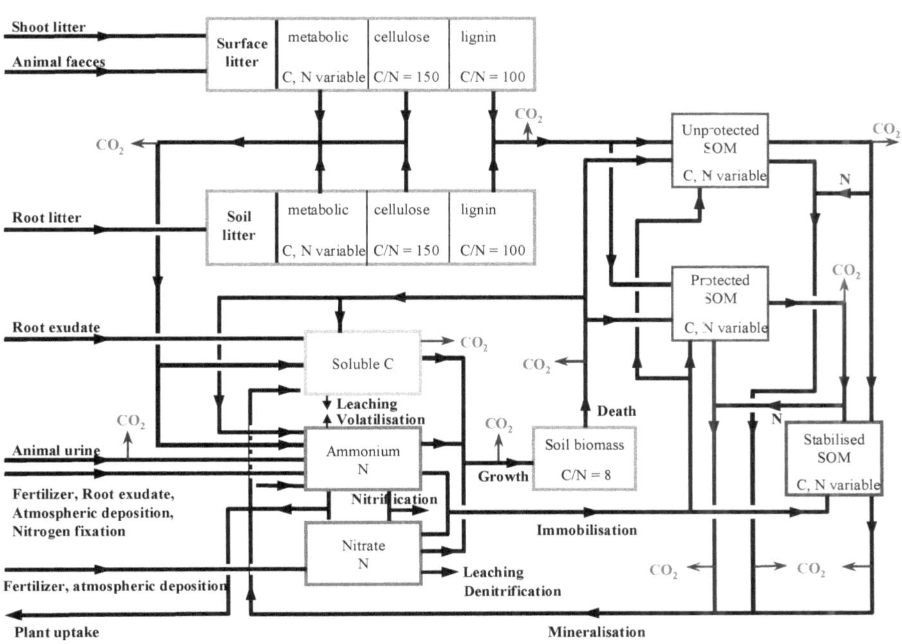

Figure 4. The soil submodel of the EFM model.

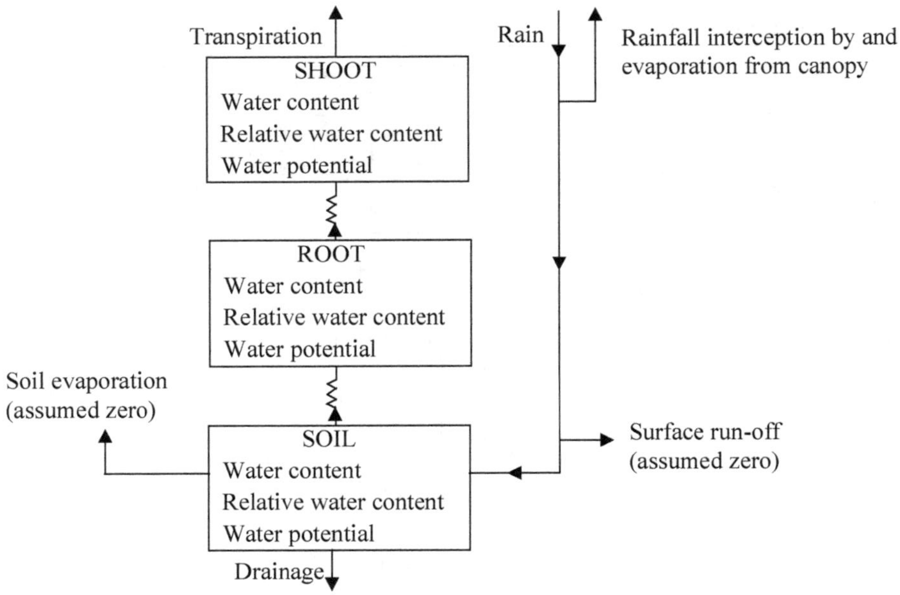

Figure 5. The water submodel of the EFM model.

from the sinks. It permits a more meaningful calculation of growth, and also allows realistic dynamics of the within-plant allocation patterns (Thornley, 1998b). Foliage and fine root structural materials are assigned an age structure so that tissue ageing and turnover can be represented. Stem height: diameter ratio is simulated as a variable (Thornley, 1999).

The soil submodel is standard in most respects (for details see Thornley, 1998a). The present formulation departs from most other models by representing ammonium N and nitrate N pools separately. This allows loss processes, which act on these different substrates, to be more realistically accounted for – for example, volatilisation and nitrification occur from the ammonium pool, whereas nitrate is the substrate for denitrification and leaching. Also, plants and microbes do not respond equally to the two substrates.

A soluble carbon pool represents C substrates such as carbohydrates and organic acids in the soil. This pool drives C leaching, microbial growth, and non-symbiotic N fixation. Plant water potential and its components are also represented. Soil water potential and soil hydraulic conductivity are calculated from soil relative water content. The plant cells are elastic. Osmotic effects are included. Water fluxes are driven by water potential gradients.

Some recent applications of the EFM are described by Thornley and Cannell (1996), Cannell et al. (1996), and Thornley and Cannell (2000).

For the purposes of the Recognition project, the EFM was modified in some ways. Because the project required the model to be run for many sites across Europe, using many different environmental scenarios, calculation speed and input facilities needed to be improved. The model was modified to accept time series of changing atmospheric CO_2 and nitrogen deposition. Furthermore, the model was modified to run with relatively short (< 10 years) time series of its weather variables (radiation, temperature, rain, wind, humidity), provided that the appropriate values are given for the so-called 'climate-change coefficients'. These coefficients quantify the yearly rate of change of the weather variables and are discussed in more detail in Chapter 3.2.

3.1.5 FinnFor

FinnFor is a model for forest ecosystem dynamics controlled by the physiological and ecological performance of trees under the control of selected environmental factors (Kellomäki and Väisänen, 1997). Photosynthesis and respiration are core physiological processes driving the growth and development of trees. Photosynthesis is a function of temperature (T), carbon dioxide (CO_2), water (H_2O), nitrogen (N), radiation (R) and foliage; i.e. PhN = f(LAI)×f(R)× f(H_2O)×f(N)×f(T,CO_2,), where LAI is the leaf area index. Photosynthesis is converted to the mass of tree organs and the dimensions of stem of trees occupying the site (Figure 6). These calculations are done for tree cohorts, which are described in terms of species and number of trees, diameter, height and age. Each cohort is represented by an object tree, whose performance is modelled in terms of gross photosynthesis and respiration in order to obtain the amount of photosynthates available annually for the mass growth of object trees. The model has been parameterized for Scots pine, Norway spruce, and silver birch (Kellomäki and Väisänen 1997, Matala et al. 2003).

In the calculations, the radiation above the canopy is divided in short-wave (400–2500 nm) and long-wave (> 2500 nm). The short-wave radiation falling on the canopy is profiled through the canopy, and the direct and diffuse radiation incidental on different layers of tree crown is calculated. Direct and diffuse average gap probabilities are calculated in order to obtain the transmitted fraction of direct and diffuse radiation on the layers of tree crowns, which are ellipsoids specific for different tree species.

Hydrological processes in the model represent e.g. the interception of water in canopy, throughfall onto soil surface and infiltration of water into soil profile. Soil moisture and temperatures are interlinked in such a way that the volumetric water content depends on the freezing temperature and evaporation of water from the soil (Kellomäki and Väisänen 1996). Similarly, the water in the soil influences soil temperature through its heat content and heat capacity as related to the physical properties of the soil. Rainfall is treated also in the form of snow, which accumulates on soil surface and affects soil frost and the soil temperature. In spring time the snow melt will substantially recharge the water in soil. The hydrological processes work on an hourly basis.

The model for the decomposition of litter (dead organic material from any mass compartment of trees) and humus (soil organic matter) with mineralization of nitrogen utilizes the algorithm developed by Chertov and Komarov (1997). Decomposition rates of different types of litter and soil organic matter are determined by soil temperature and the moisture, and nitrogen and ash content of the litter. The moisture of the litter is a linear function of that of the mineral topsoil.

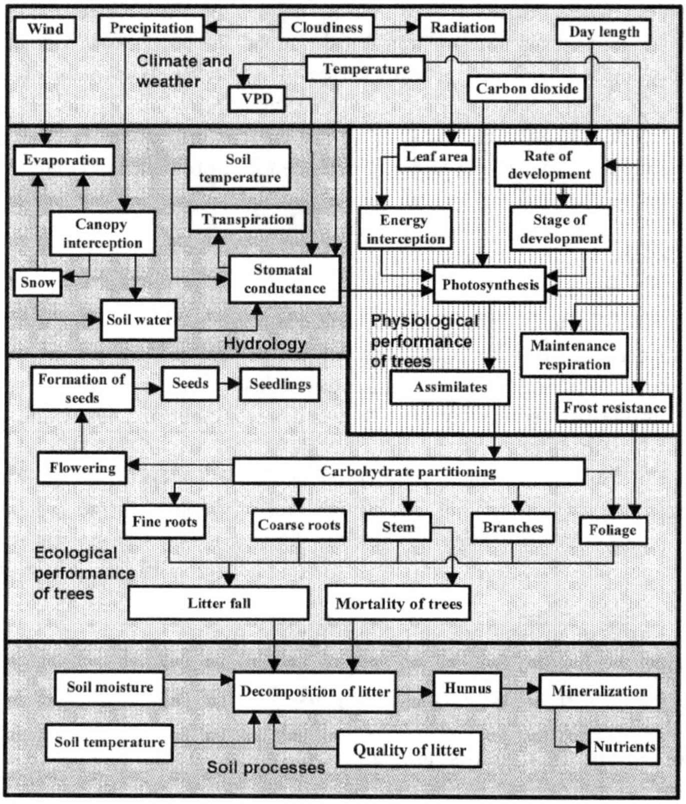

Figure 6. Outlines of the effects of climate on the soil processes and the metabolism of plants and the consequent dynamics of the forest ecosystem in the FinnFor model.

Nitrogen is released through the decomposition of soil organic matter, and the degree of retention of nitrogen in the decomposing material is a function of nitrogen content in the litter and in the forest floor. The decomposition of the soil organic matter works on the monthly basis.

The photosynthetic rate of the object trees is calculated on the basis of the biochemical model developed by Farquhar et al. (1980). The photosynthetic rate is updated on an hourly basis utilizing solar radiation, air temperature and humidity, and CO_2 concentration, which affect the stomatal conductance and the biochemical fixation rate of carbon. The photosynthetic rate is also affected by the soil moisture and temperature through the stomatal conduction. Furthermore, the nitrogen content of foliage affects the photosynthetic rate of different tree species. Seasonality of physiological performance of coniferous trees is introduced through the stage of development, which controls the seasonal activity of trees, e.g. the sensitivity of photosynthetic response to environmental factors (Pelkonen and Hari 1980). In the case of silver birch, the build-up of foliage in spring and senescence of foliage in autumn are modelled as presented by Kellomäki et al. (2001).

The gross photosynthesis over a year provides the total amount of photosynthates available for the maintenance and growth of the object trees and their organs (foliage, branches, stem, coarse root, fine roots). The allocation between the organs is based on the allometry between mass of the organs and total mass of the object tree. Photosynthates available for the organs are converted to the mass by removing the maintenance and growth respiration and translating the remaining photosynthates into dry mass, which is added to the existing mass of the organs. At the same time, the litter losses from the organs are calculated in order to update the mass.

The mass (M(t), kg) and mass growth (ΔM(t), kg) of stems are used to update the diameter (d, cm) and height (h(t), m) of object trees

$$\Delta h(t) = c \times \frac{\Delta M(t)^{d}}{M(t)^{e}} \Rightarrow h(t) = h_o + \sum_{t=1}^{n} \Delta h(t) \qquad (1)$$

where c, d and e are species specific parameters and h_0 the initial height of a tree and n the number of years simulated. Based on Marklund (1988), the mass of a stem is a function of its height and diameter

$$\ln M(d,h) = \beta_o + \beta_1 \times \frac{d}{d+\gamma} + \beta_2 \times h + \beta_3 \times \ln h \qquad (2)$$

where β_0, β_1, β_2, β_3 and λ are species-specific parameters. Using the height from Eq. (1), the diameter can be solved from Eq. (2).

The inverse relationship between the stand density and the mean volume of stems volume is used to drive the mortality of trees, but this is supplemented by the survival of object trees (Hynynen et al. 2001). At the beginning of each simulation step, the survival probability is calculated for each cohort based on (i) the within-stand competition model and (ii) the life-span of trees. At the end of each simulation step, the total stocking of trees is checked in order to ensure that the stocking falls below the self-thinning line determining the maximum allowable number of trees in a stand. Thereafter, the ratio between maximum tree number and predicted number is calculated. If this ratio is less than one, the predicted number in each cohort will be multiplied by the ratio to decrease the survival and increase the mortality, until the simulated number of trees equals to the maximum allowable number.

3.1.6 Q

The Q model is actually a family of ecosystem models that use the continuous-quality theory to describe soil organic matter turnover and the nitrogen productivity to describe plant growth (Ågren and Bosatta, 1998). The Q model has been developed with the aim of analysing the carbon and nitrogen cycles and their relation to forest stand development (Rolff and Ågren, 1999). The model is based on the interaction between a soil subsystem and a plant subsystem, each being described in terms of their content of carbon and nitrogen, respectively, Fig. 7. The model is named after one of its central variables, substrate quality (q).

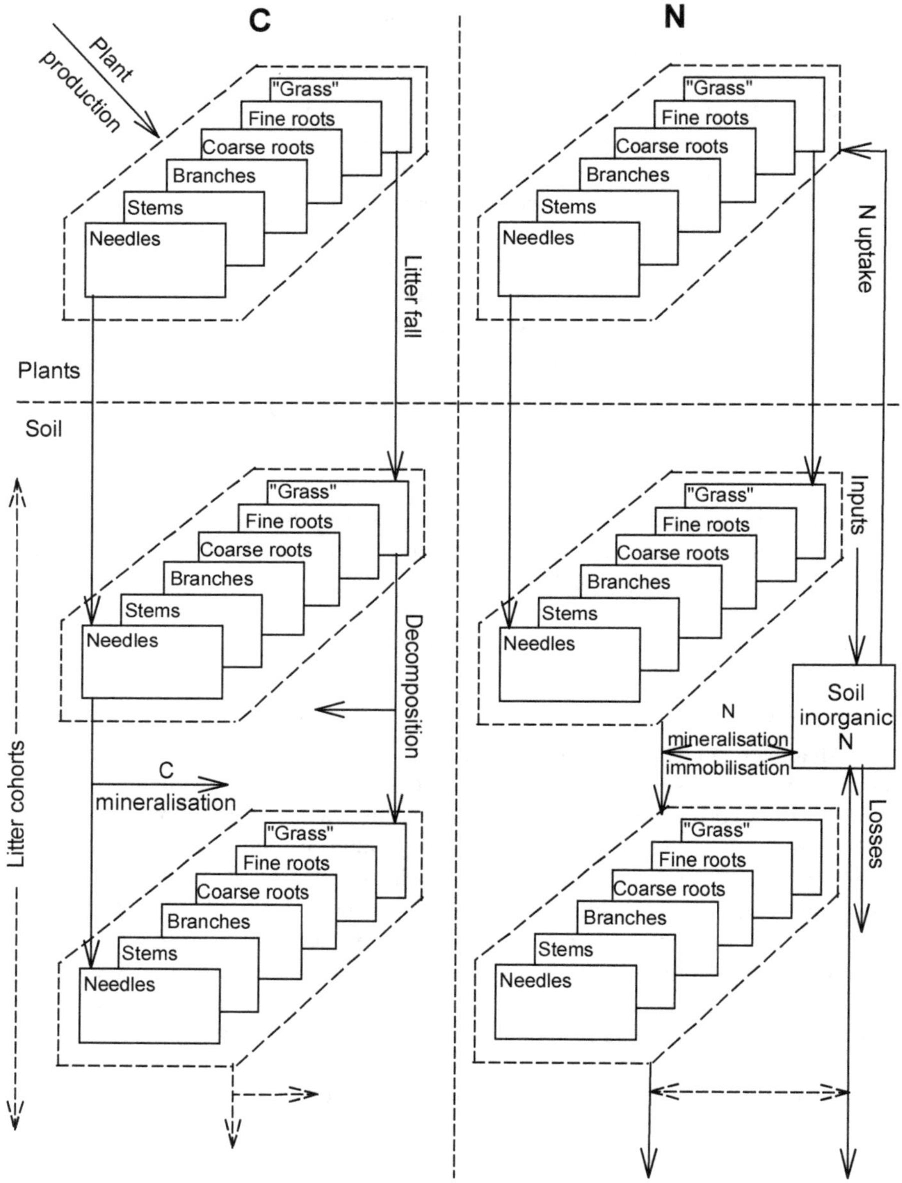

Figure 7. Schematic view of the Q model.

The model belongs to the 'green slime' class of models, i.e. individuals are not distinguished in the model. The plants are represented in the model by different tree components and a 'grass layer' covering all vegetation other than the dominating tree layer (Figure 7). For each plant component in the model there is one carbon and one nitrogen state variable. Normally, the grass plays a negligible role in the model; it is only after clear-fellings that the grass, absorbing mineralised nitrogen that otherwise would have leached out of the system, is important as a conservation mechanism for nitrogen. The organic matter in the soil is divided into litter cohorts, thus keeping track of both the source of the material as well as the time of its appearance in the soil. The main aim of the model is to study long-term development of forests, and the time step is one month. Monthly average air temperature is the only climatic driver and nitrogen deposition is included as an external variable. Temperature only affects soil processes as it has been found that nitrogen productivity is insensitive to temperature (Ladanai and Ågren 2004; Salih et al., 2005). Carbon dioxide concentration could easily be included but due to lack of experiments from which parameters can be estimated, no effects of CO_2 were included.

Growth of needles is the key process and it is described by the nitrogen productivity:

$$\frac{dW_{needle}}{dt} = \left(a - bW_{needle}\right) N_{needle}$$

where W_{needle} is the biomass of needles and N_{needle} is the amount of nitrogen in the total needle biomass. a and b are empirical parameters. Growth of other woody tree components is described with a similar equation, but where W_{needle} replaces N_{needle}. Fine-root growth includes also a function of needle nitrogen concentration that decreases fine-root growth as the needle nitrogen concentration increases. The upper limit to needle biomass, a/b, is never attained because of needle mortality and the limitation from available nitrogen, which keeps N_{needle} down.

Litter fall occurs once per year and is calculated as a fraction of the biomass components. It is assumed that there is no litter formation from stem wood, except that which might be left after harvesting operations.

Each litter cohort of a given litter type is treated individually. During decomposition a litter cohort changes in two ways: (1) its quality decreases; and (2) its mass decreases.

The substrate quality (q) changes when decomposers assimilate the substrate. The rate of change of the quality depends on the change in quality of a carbon atom being assimilated into decomposer biomass $\eta_1(q)$ (the model used in this project is $\eta_1(q) = \eta_{11}q$) and the growth rate of decomposers per unit carbon $u(q,T)$ (the model used in this project is $u(q) = (u_0 + aT)q^{\beta}$, where u_0, a, and β are parameters). The linear temperature (T) response function, rather than more conventional exponential ones is based on empirical data and can also capture variability in soil water availability because the negative correlation between temperature and soil water (Andrén and Paustian 1987)). The rate of change is then calculated from the following equation:

$$\frac{dq}{dt} = -f_C \eta_1(q) u(q,T)$$

where f_C is the carbon concentration in the decomposers.

The fraction of a cohort remaining at a given time (g) is calculated from the quality that the cohort has reached from the following relation

$$\frac{dg}{dq}=\frac{1-e(q)}{\eta_1(q)e(q)}g(q).$$

Decomposer efficiency or production-to-assimilation ratio $e(q)$ is a critical function as it partitions the carbon flow into biomass and respiration (the model used in this project is $e(q) = e_0$, where e_0 is a parameter). Carbon mineralisation is calculated as the difference between soil carbon this month and soil carbon previous month.

Nitrogen mineralisation/immobilisation is calculated from the difference in nitrogen content in a litter cohort between the beginning and end of a month. The nitrogen concentration in a litter cohort r depends on the quality that it has reached according to

$$\frac{dr}{dq}=-\left(\frac{f_N}{f_C}-1\right)\frac{1}{\eta_1(q)}$$

where f_N is the nitrogen concentration in the decomposers. As decomposition progresses, the nitrogen concentration approaches more and more that of the decomposers, f_N/f_C.

The nitrogen uptake by the trees is proportional to the amount of available nitrogen, mineralisation plus inputs, and a function describing the extent to which tree roots occupy the soil. The grass takes up whatever nitrogen is left. There is, however, an upper limit on how much nitrogen the grass can take up. Nitrogen losses occur when the vegetation system lacks capacity to utilise all the nitrogen that is available from mineralisation and inputs.

A previous version of the Q model has been used to analyse responses of pine forests to climate change (Ryan et al., 1996a and 1996b). Temperature effects on soil organic matter decomposition (Ågren, 2000; Ågren and Bosatta, 2002) have also been analysed and large-scale effects of biomass harvesting on forest soil carbon stores predicted (Ågren and Hyvönen, 2003). One major change in the model for application within Recognition is the adoption of monthly time steps. The way that soil organic matter is handled has also been modified in order to simplify the adoption of the model to many different sites.

3.1.7 Comparison of the models

The models differ considerably in how they are structured around the driving variables. Table 2 summarises how the four major environmental variables for Recognition (nitrogen, CO_2, temperature, and precipitation) appears in the models.

Two of the models, EFIMOD and FinnFor, have no loss mechanisms for nitrogen, which should make them more sensitive to nitrogen inputs as all incoming nitrogen is retained in the system, whereas in the other two models, EFM and Q, losses will limit growth responses under high nitrogen loads. The importance of fixed versus floating nitrogen concentrations in woody tree components is probably minor, but the floating nitrogen concentration in needles is of major importance in Q.

Carbon dioxide appears explicitly in only two of the models, EFM and FinnFor. The extent to which it is reasonable to exclude CO_2 (EFIMOD and Q) on the grounds that the changes in $[CO_2]$ has been, and will be over the next 80 years, too small to have any significant effect, or if its inclusion, mainly based on results from short-term laboratory studies, exaggerates CO_2 effects must still be considered an unresolved question.

Temperature is important for nitrogen mineralisation in all models. However, the inclusion of temperature in tree growth varies considerably from no effect in Q over a balanced effect on photosynthesis and respiration in EFM and FinnFor to a strict growth limitation in EFIMOD. It is possible that the four models in practise do not differ as much as might appear from the formal descriptions. Since temperature affects both photosynthesis and respiration in EFM and FinnFor, the net effect is less than the response of either process on its own. In EFIMOD, the growth restriction from temperature goes hand in hand with restriction from nitrogen mineralisation. The overall result could be that these three models might respond similarly to Q with respect to temperature.

Because of its high temporal variability, water is a demanding variable to include in models. This is explicitly recognised in EFM and FinnFor whereas EFIMOD and Q use simplifications to circumvent this demand.

References

Ågren, G.I. 1996. Nitrogen productivity or photosynthesis minus respiration to calculate plant growth. Oikos 76: 529–535.

Ågren, G.I. 2000. Temperature dependence of old soil organic matter. Ambio 29: 55.

Ågren, G.I. and Bosatta, E. 1998. Theoretical Ecosystem Ecology – Understanding Element Cycles, Cambridge University Press.

Ågren, G.I. and Bosatta, E. 2002. Reconciling differences in predictions of temperature response of soil organic matter. Soil Biology and Biochemistry 34: 129–132.

Ågren, G.I and Hyvönen, R. 2003. Changes in carbon stores in Swedish forest soils due to increased biomass harvest and increased temperatures analysed with a semi-empirical model. Forest Ecology and Management 174: 25–37.

Alexandrova, L.N. 1970. Sources of humic substances in soil. Trans. Leningrad Agric. Inst., 142: 5–25 (in Russian).

Alexeev, V.A. 1975. Svetovoy Rezhim Lesa. Leningrad: Nauka. 228 p. (in Russian).

Andrén, O. and Paustian, K. 1987. Barley straw decomposition in the field: A comparison of models. Ecology 68: 1190–1200.

Cannell, M.G.R., Thornley, J.H.M., Mobbs, D.C. and Friend, A.D. 1996. UK conifer forests may be growing faster in response to increased N deposition, atmospheric CO_2 and temperature. Forestry 71: 277–296.

Cannell, M.G.R. and Thornley, J.H.M. 1998. Temperature and CO_2 responses of leaf and canopy photosynthesis: a clarification using the non-rectangular hyperbola model of photosynthesis. Annals of Botany 82: 883–892.

Chertov, O.G. 1983a. The mathematical model of single plant ecosystem. J. Obshchey Biol. (J. Fundamental Biology, Moscow) 44: 406–414. (in Russian with English abstract).

Chertov, O.G. 1983b. Quantitative approach for evaluation of ecological parameters of species. Scots pine (*Pinus sylvestris* L., *Pinaceae*) as an example. Bot. J. (Leningrad). 68: 1318–1324. (in Russian with English abstract).

Chertov, O.G. 1985. Simulation model of forest litter and floor mineralization and humification. J. Obshchei Biol., Moscow (J. of General Biol.) 46: 794–804. (in Russian with English summary).

Chertov O.G., and Komarov A.S. 1997. SOMM – a model of soil organic matter dynamics. Ecological Modelling 94: 177–189.

Chertov, O.G., Komarov, A.S. and Tsiplianovsky, A.M. 1999a. The simulation of soil organic matter and nitrogen accumulation in Scots pine plantations on bare parent material using the combined forest model EFIMOD. Plant and Soil 213: 31–41.

Chertov, O.G., Komarov, A.S. and Tsiplianovsky, A.M. 1999b. A combined simulation model of Scots pine, Norway spruce and Silver birch ecosystems in the European boreal zone. Forest Ecology and Management 116: 189–206.

Chertov, O.G. Komarov, A.S., Nadporozhskaya, M.A., Bykhovets, S.A. and Zudin, S.L. 2001. ROMUL – a model of forest soil organic matter dynamics as a substantial tool for forest ecosystem modelling. Ecological Modelling 138 (1–3): 289–308.

Chertov, O., Komarov, A., Mikhailov, A., Andrienko, G., Andrienko, N. and Gatalsky. P. 2005. Geovisualisation of forest simulation modelling results: a case study of carbon sequestration and biodiversity. Computers and Electronics in Agriculture 49: 175–191.

Duchaufour, Ph. 1961. Précis de Pédologie. Évolution de Sol (Russian translation: 1970, Moscow, Mir). 591 p.

Farquhar. G.D., von Caemmerer, S. and Berry, J.A. 1980. A biochemical model of photosynthetic assimilation in leaves of C_3 species. Planta 149: 67–90.

Hynynen, J., Ojansuu, R., Hökkä, H., Salminen, H., Siipilehto, J., Haapala, P. 2001. Models for predicting stand development in MELA system. Finnish Forest Research Institute. Research Papers 835: 1–116.

Karpachevsky, L.O. 1981. Forest and forest soils. Lesnaya Prom., Moscow. 264 p. (in Russian).

Kazimirov, N.I. and Morozova, R.M. 1973. Biologicheskiy Krugovorot Veshchestv v Elnikah Karelii. Nauka, Leningrad. 176 p. (in Russian).

Kazimirov, N.I., Volkov, A.D., Ziabchenko, S.S., Ivanchikov, A.A. and Morozova, R.M. 1977. Obmen Veshchestv i Energii v Sosnovyh Lesah Evropeiskogo Severa. Nauka, Leningrad. 304 p. (in Russian).

Kazimirov, N.I., Morozova, R.M. and Kulikova, V.K. 1978. Organicheskaya Massa i Potoki Veshchestsv v Berezniakah Sredney Taigi. Nauka, Leningrad. 216 p. (in Russian).

Kellomäki, S., and Väisänen, H., 1996. Model computations on the effect of rising temperature on soil moisture and water availability in forest ecosystems dominated by Scots pine in the boreal zone in Finland. Climate Change 32: 423–445.

Kellomäki, S. and Väisänen, H. 1997. Modelling the dynamics of the boreal forest ecosystems for climatic change studies in the boreal conditions. Ecological Modelling 97(1,2): 121–140.

Kellomäki, S., Rouvinen, I., Peltola, H. and Strandman, H. 2001. Density of foliage mass and area in the boreal forest cover over Finland with applications to estimate the emissions of monoterpene and isoprene. Atmospheric Environment 35: 1491–1503.

Komarov, A., Chertov, O., Andrienko, G., Andrienko, N., Mikhailov, A. and Gatalsky, P. 2002. DESCARTES & EFIMOD: An Integrated System for Simulation Modelling and Exploration Data Analysis for Decision Support in Sustainable Forestry. A. Rizzoli and A. Jakeman (Eds.) Integrated Assessment and Decision Support, Proceedings IEMSS 2002, 24–27 June 2002, Lugano, Switzerland. Volume 1: 239–244.

Komarov, A., Chertov, O., Zudin, S., Nadporozhskaya, M., Mikhailov, A., Bykhovets, S., Zudina, E. and Zoubkova, E. 2003. EFIMOD 2 – A model of growth and elements cycling in boreal forest ecosystems. Ecological Modelling 170 (2–3): 373–392.

Kostin, N.V. 1997. Geographic variability of productivity of Norway spruce stands in European forest zone. PhD Thesis. St. Petersburg Forest Technical Academy.

Ladanai, S. and Ågren, G.I. 2004. Temperature sensitivity of nitrogen productivity for Scots pine and Norway spruce. Trees 18: 312–319.

Marklund, L.G. 1988. Biomass functions for pine, spruce and birch in Sweden. Department of Forest Survey, Swedish University of Agricultural Sciences, Report 45.

Matala, J., Hynynen, J., Miina, J., Ojansuu, R. Peltola, H., Sievänen, R., Väisänen, H. and Kellomäki, S. 2003. Comparison of a physiological model and a statistical model for prediction of growth and yield in boreal forests. Ecological Modeling 161: 95–116.

Mou, P., Mitchell, R.J. and Jones, R.H. 1993. Ecological field theory model: a mechanistic approach to simulate plant-plant interactions in south-eastern forest ecosystems. Canadian J. For. Res. 23(2): 2180–2193.

Pelkonen, P. and Hari, P. 1980. The dependence of the springtime recovery of CO_2 uptake in Scots pine on temperature and internal factors. Flora 169: 398–404.

Rastetter, E.B., King, A.W. Cosby, B.J., Homberg, G.M., O'Neill, R.V. and Hobbie, J.E. 1992. Aggregating fine-scale ecological knowledge to model coarser-scale attributes of ecosystems. Ecological Applications 2: 55–70.

Rolff, C. and Ågren, G.I. 1999. Predicting effects of different harvesting intensities with a model of nitrogen limited forest growth. Ecological Modelling 118: 193–211.

Ryan, M.G., Aber, J.D. Ågren, G.I., Friend, A.D., Linder, S., McMurtrie, R.E., Parton, W.J., Raison, R.J. and Rastetter, E.B. 1996a. Comparing models of ecosystem function for temperate conifer forests. I. Model description and validation. In: Breymeyer, A., Hall, D.O., Melillo, J.M. and Ågren, G.I. (eds.) Global Change: Effects on Forests and Grasslands. J.Wiley. Pp. 313–362.

Ryan, M.G., Aber, J.D. Ågren, G.I., Friend, A.D., McMurtrie, R.E., Parton, W.J. and Rastetter, E.B. 1996b. Comparing models of ecosystem function for temperate conifer forests. II. Predictions. In: Breymeyer, A., Hall, D.O., Melillo, J.M. and Ågren, G.I. (eds.) Global Change: Effects on Forests and Grasslands. J.Wiley. Pp. 363–387.

Salih, N., Ågren, G.I. and Hallbäcken, L. 2005. Modeling response of N addition on C and N allocation in Scandinavian Norway spruce stands. Ecosystems 8: 373–381.

Thornley, J.H.M. 1991. A transport-resistance model of forest growth and partitioning. Annals of Botany 68: 211–226.

Thornley, J.H.M. 1998a. Grassland Dynamics: An Ecosystem Simulation Model, CAB International, Wallingford, Oxon OX10 8DE UK. xii + 241 p.

Thornley, J.H.M. 1998b. Modelling shoot: root relations: the only way forward? Annals of Botany 81: 165–171.

Thornley, J.H.M. 1999. Modelling stem height and diameter growth in plants. Annals of Botany 84: 195–205.

Thornley, J.H.M. and Cannell, M.G.R. 1992. Nitrogen relations in a forest plantation-soil organic matter ecosystem model. Annals of Botany 70: 137–151.

Thornley, J.H.M. and Cannell, M.G.R. 1996. Temperate forest responses to carbon dioxide temperature and nitrogen: a model analysis. Plant Cell and Environment 19: 1331–1348.

Thornley, J.H.M. and Cannell, M.G.R. 2000. Managing forests for wood yield and carbon storage: a theoretical study. Tree Physiology 20: 477–484.

Uranov, A.A. 1965. Phytogenic field. In: Problems of Soviet botany. Vol. 1. Proc. 3rd All-Union Botanical Congress. Nauka, Moscow; Leningrad. Pp. 251–254 (in Russian).

Wilde, S.A. 1958. Forest Soils. John Wiley and Sons. New York.

Wu, H., Sharpe, P.J.H., Walker, J. and Penridge, L.K. 1985. Ecological field theory: a spatial analysis of resource interference among plants. Ecological Modelling, 29: 215–243.

3.2 Methodology for the application of process-based models to analyse changes in European forest growth

Marcel van Oijen[1], Göran I. Ågren[2], Oleg Chertov[3], Seppo Kellomäki[4], Alex Komarov[3], Deena Mobbs[1] and Maureen Murray[1]

[1]*Centre for Ecology and Hydrology, Edinburgh Research Station, UK*
[2]*Swedish University of Agricultural Sciences, Uppsala, Sweden*
[3]*European Forest Institute*
[4]*University of Joensuu, Joensuu, Finland*

3.2.1 Introduction

Recognition was set up to answer two questions: (1) What has caused many European forests to grow faster during the 20th century (Spiecker et al., 1996)? and (2) How will the trends continue in the 21st century? Process-based modelling was chosen as one of the tools to help answer these questions, because it directly addresses the role of the different factors thought to have been responsible for the growth changes, namely changes in N-deposition, climate and atmospheric CO_2. Moreover, process-based models attempt to represent the complexity of the forest ecosystems and thereby help identify the key underlying physiological, phenological and morphological processes that determine the response of forests to environmental change. Therefore, process-based modelling is the best tool available for predicting forest growth under conditions that have never occurred before in nature and rarely in field-scale experiments, such as elevated CO_2 and warming. However, the use of models comes at a cost: a very large amount of information is required to define the variables (climate, atmospheric composition, soil conditions) as well as system parameters.

The four process-based models that have been used in Recognition are EFIMOD, EFM, FinnFor and Q. These models all simulate flows of resources (C, N, water) into and out of the forest system, but in very different ways. The structure of the four models is described in detail Chapter 3.1. Here, we will describe how the models were supplied with input data, how they were parameterised, tested and applied. We shall end by discussing the usefulness of process-based modelling to answer questions concerning past and future forest growth. The actual answers to these questions – i.e. the modelling results themselves – are presented in Chapter 4.4.

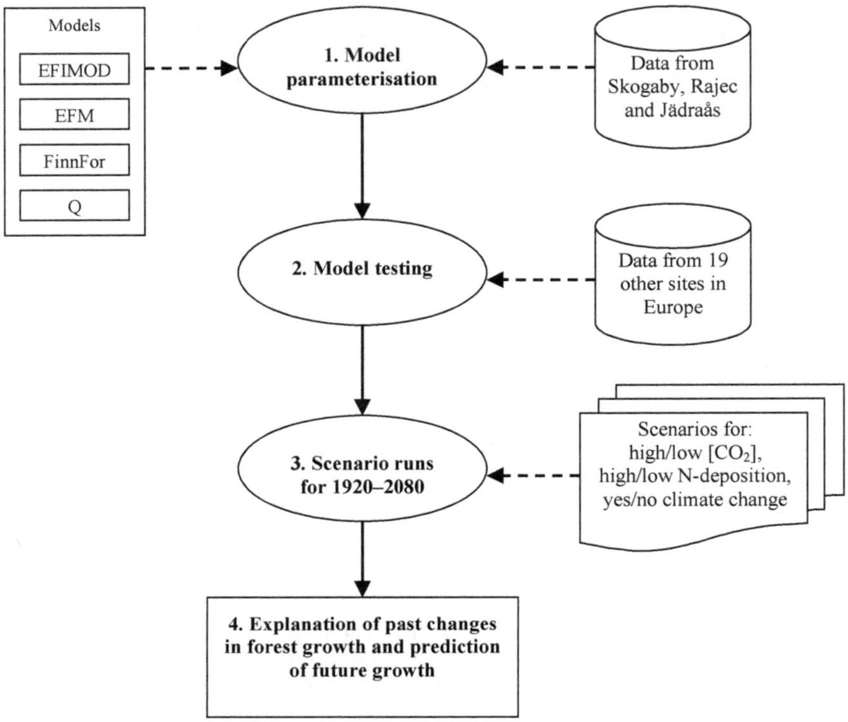

Figure 1. Flowchart of process-based modelling in Recognition.

3.2.2 The modelling approach

An overview of the approach is shown in Figure 1. The approach differs from some earlier projects in that the four models in Recognition are parameterised, provided with input, applied and analysed with common methods. The models differ and therefore give different answers to the questions posed in Recognition, so a detailed comparison of the models is essential to identify key areas of uncertainty in our knowledge of forest ecosystems. Previous model comparisons often consisted of extensive descriptions of the detailed structure of the models that were compared (Ryan et al., 1996; Sonntag, 1997; Constable and Friend, 2000; Cramer and Field, 1999). Comprehensive lists have been produced of submodels, model components and algorithms. We will not present such lists here. Instead, our model comparison focuses on differences in model behaviour and uses this information to derive the key differences in model structure, which require further study (compare Medlyn et al., 2000).

Each of the four models used in the project (EFIMOD, EFM, FinnFor, Q) was used to simulate the growth of two tree species: Norway spruce and Scots pine. Identical model versions were used for both species with differences between species expressed in the tree-defining parameter values.

Forest growth was simulated for 22 sites (13 Scots pine and 9 Norway spruce), which were selected on the basis of availability of tree and environment data (Table 1). The sites differed greatly in growing conditions, with average temperature over the period 1975–1990 varying from –0.6 to 10.0 °C, rainfall from 403 to 1169 mm yr^{-1}, and soil carbon and nitrogen contents varying by nearly an order of magnitude (Table 1). Besides the process-based modelling approach described here, growing conditions and forest growth on most of these sites was studied by statistical means as well, and more information on this can be found in Chapters 2.1 and 2.2.

Process-based modelling studies need two types of sites. The first type is used for model parameterisation, and must have long-term detailed data for both environment (climate, soil, N deposition) and trees. The data must be detailed because parameters have to be quantified for all processes represented in the models. The data must be long-term because the models are applied to a full forest rotation period of 80 years, and the parameter values derived from the data should lead to reliable simulations over the whole period. For Norway spruce, we used sites Skogaby and Rajec for this purpose (see Table 1 and Chapter 2.1). For Scots pine, we used Jädraås (see Table 1 and Chapter 2.1). The second type of site is used for model validation and application. This type of site also has good information on environmental conditions, but the data on tree growth may be less detailed than those for the parameterisation sites. For this purpose, 19 other sites were selected (Table 1).

For all sites, information was needed on atmospheric CO_2, N-deposition, climate and forest management. Management regimes – i.e. planting densities and thinning practices – were chosen for each site to conform to regional forestry practice.

3.2.3 Construction of environmental scenarios

3.2.3.1 Atmospheric CO_2 concentration

The Intergovernmental Panel for Climate Change (IPCC) has developed a range of scenarios, IS92a–f, of future greenhouse gas and aerosol precursor emissions based on assumptions concerning population and economic growth, land use, technological changes, energy availability and fuel mix during the period 1990 to 2100. These emissions have been used to calculate future atmospheric concentrations of CO_2 and other atmospheric particles. In Recognition, we used projections of CO_2 consistent with the mid-range IPCC emission scenario, IS92a (Figure 2), calculated with the Bern model (Siegenthaler and Joos, 1992; Joos et al., 1996).

3.2.3.2 N deposition

For each of the 22 sites, the time course of yearly total N-deposition for the years 1920–2080 was estimated using three sources of information (de Vries and Reinds, personal communication 2001). First, literature information suggested very low levels of N-deposition (< 3 kg N ha^{-1} yr^{-1}) across Europe in the year 1900 (Galloway, 1985). Second, data and calculations by the Co-operative Programme for Monitoring and Evaluation of the Long-Range Transmission of Air Pollutants in Europe (EMEP) show increasing N-deposition values during most of the 20th

Table 1. Sites used for process-based modeling in project RECOGNITION. Lat, Lon = latitude and longitude in decimal degrees. Temp = average temperature; Rain = average yearly precipitation; Rad = average daily global radiation; N-deposition = average yearly total N-deposition. All averages are for the years 1975-1990. Weather variables taken from the JRC-Ispra database. N-deposition based on EMEP-data (de Vries and Reinds, pers. comm. 2001).

	Country	Lat. (°)	Long. (°)	Altitude (m)	Soil C (kg C ha^{-1})	Soil N (kg N ha^{-1})	N-deposition (kg N ha^{-1} yr^{-1})	Temp. (°C)	Rain (mm yr^{-1})	Rad. (MJ m^{-2} d^{-1})	Wind (m s^{-1})
Picea abies											
Kemijärvi	Finland	66.85	27.13	272	100 027	3 791	2.3	-0.1	403	7.1	2.5
Kannonkoski	Finland	62.93	25.68	195	78 524	2 949	4.0	3.5	614	8.0	2.1
Skogaby [(1) (2)]	Sweden	56.55	13.22	105	91 337	3 999	12.7	7.5	1 169	8.6	2.4
Solling	Germany	51.77	9.58	508	131 000	6 067	27.8	9.5	559	9.7	3.1
Rajec [(1) (2)]	Czech R.	49.40	16.63	625	82 840	3 358	21.8	7.1	613	10.3	3.5
Karlstift	Austria	48.59	14.77	910	244 012	10 385	21.8	7.7	455	10.7	2.5
Helfenberg	Austria	48.58	14.15	790	176 038	7 878	21.8	8.8	612	10.7	2.5
Pfalzgrafenweiler	Germany	48.49	8.50	600	90 688	3 944	22.0	9.3	653	10.8	2.7
Höglwald [(2)]	Germany	48.29	11.08	540	88 398	4 852	30.3	8.8	870	11.3	2.8
Pinus sylvestris											
Kolari	Finland	67.25	23.87	190	61 517	2 058	2.0	-0.6	450	7.2	2.1
Punkaharju	Finland	61.81	29.33	80	65 705	2 248	4.6	3.6	575	7.8	2.5
Jädraås [(1) (2)]	Sweden	60.82	16.49	190	37 483	1 100	4.5	3.4	677	9.2	2.5
Black Isle 8/76	UK	57.63	–4.24	150	310 000	10 333	6.2	8.7	646	8.2	4.9
Aalkasten	Germany	53.26	13.44	88	52 477	3 454	18.9	9.5	513	10.0	3.6
Alt Placht	Germany	53.17	13.40	65	47 180	2 103	21.0	9.5	513	10.0	3.6
Gandenitz I	Germany	53.17	13.43	68	51 197	2 276	21.0	9.5	513	10.0	3.6
Gandenitz II	Germany	53.16	13.42	65	53 925	2 207	21.0	9.5	513	10.0	3.6
Triebsch	Germany	52.35	13.79	37	57 575	2 388	21.0	10.0	534	10.0	3.7
Löpten	Germany	52.15	13.71	41	38 457	1 542	21.0	10.0	534	10.0	3.7
Briescht	Germany	52.10	14.11	44	61 812	2 260	24.4	10.0	535	10.0	3.7
Pfaffenwinkel	Germany	50.00	12.33	535	89 329	4 394	21.9	8.0	460	10.5	2.5
Pustert	Germany	49.21	12.00	480	39 932	2 014	21.9	9.3	580	10.5	2.1

[(1)] Sites used in model calibration
[(2)] Weather data measured on-site

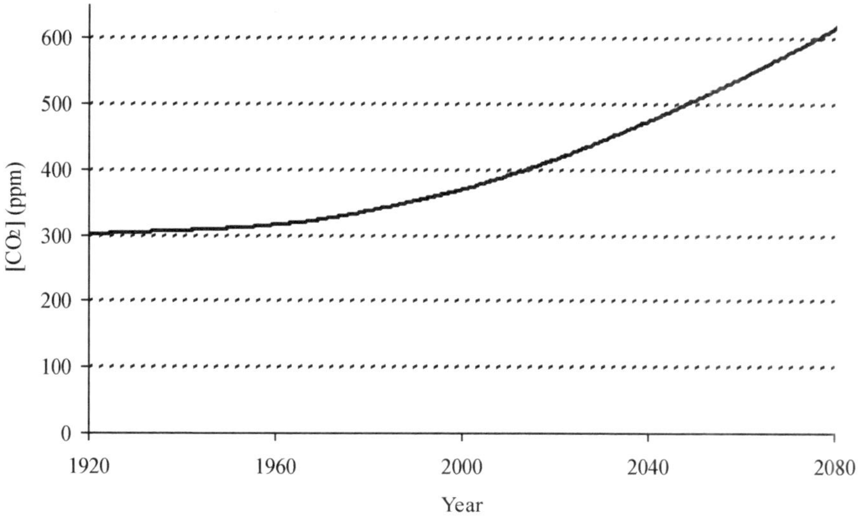

Figure 2. Time course of atmospheric CO_2 concentration as derived from the IS92a scenario by means of the Bern model (Siegenthaler and Joos, 1992; Joos et al., 1996).

century with maxima reached around 1990. Third, the 1999 Gothenburg Protocol to Abate Acidification, Eutrophication and Ground-level Ozone sets emission ceilings for 2010 for NO_X, ammonia and other pollutants. Hence site-specific further reductions of N-deposition until the year 2010 were assumed and deposition was assumed to remain constant thereafter. The default scenarios of N-deposition for each of the 22 sites were based on these assumptions, and are shown in Figure 3. Simulations were carried out with these default scenarios but also with constant N-deposition rates to assess the sensitivity of European forests to deposition. This approach using different deposition levels provided an estimate of the consequences of uncertainty regarding N-deposition.

3.2.3.3 Climate

To drive the Recognition models, time series with daily values for a total of six weather input variables were required as follows (Chapter 3.1): minimum air temperature (T_{min}), maximum air temperature (T_{max}), precipitation (Prec), wind speed (Wind), atmospheric humidity (RH) and global radiation (Rad). These weather variables had been measured, for periods exceeding eight years, on only four out of the 22 sites, Skogaby, Rajec, Höglwald and Jädraås. For the remaining 18 sites, weather data were derived from the appropriate grid cells in the MARS-project Meteorological Database (http://mars.jrc.it/). The MARS database contains daily meteorological data spatially interpolated on a 50 × 50 km grid, and covers the period 1975–2001 for most of its grid cells.

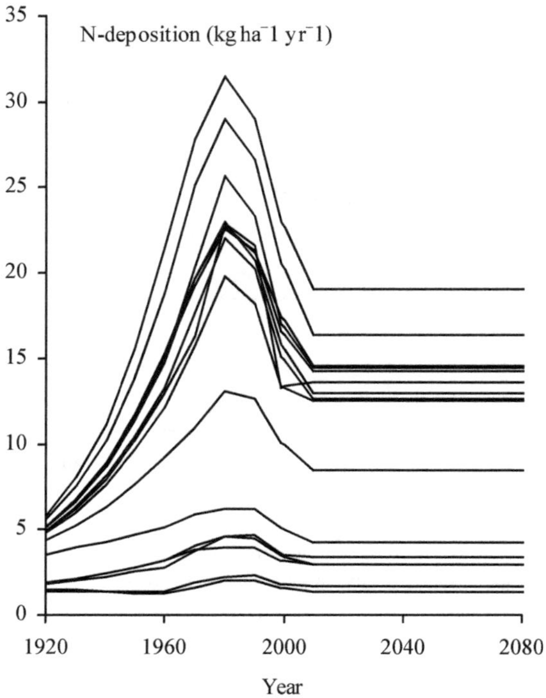

Figure 3. Time courses of atmospheric nitrogen deposition, at 22 sites across Europe (Table 1) as used by the process-based models in Recognition.

As the process-based modelling in Recognition covers the period 1920–2080, the short time series with on-site and MARS data needed to be extended. We extended the time series back to the year 1920 using the Climatic Research Unit (CRU) dataset (New et al, 2000) and forward to 2080 by means of weather data produced with the HadCM3 General Circulation Model (Viner, personal communication 2000; Hulme et al., 1999). The CRU dataset is a 0.5 degrees latitude-longitude gridded dataset of monthly terrestrial surface climate for the period of 1901–1995. The dataset comprises the weather variables used in Recognition except radiation and wind speed. The spatial coverage includes all of Europe. CRU weather variables were mostly derived by direct interpolation from station observations (New et al., 2000). The HadCM3 weather data were taken from the called HadCM3GGA1 experiments. The data include all six weather variables and cover the whole period 1920–2080, but they have lower spatial resolution than the CRU data: 3.75 ° longitude × 2.5 ° latitude, which corresponds to grid cells in Europe of 200–300 km W–E by 278 km N–S.

For each of the 22 sites in the project, we calculated the average past and future rates of change in the six weather variables, by means of linear regression of the CRU and HadCM3 data on time. We refer to the slopes of the linear regressions as

'Climate-Change Coefficients' (CCCs). CCCs were calculated for each month of the year separately. The CRU data were used to calculate CCCs for T_{min}, T_{max}, Prec and RH over the period 1901–1995, and the HadCM3 data were used to calculate the CCCs for Wind and Rad over the same period, and for all six variables over the period 1988–2099. This procedure led to a total of 22 sites × 12 months × 6 weather variables = 1584 CCCs for the years 1901–1995, and the same number for 1988–2099. Latitudinal and seasonal patterns in CCCs are shown in Figures 5 and 6. For the period 1901–1995, most sites showed positive CCC-values for minimum and maximum temperature and wind speed, and negative values for radiation and humidity. Changes in precipitation were more mixed. However, all the changes seemed relatively small when compared to CCCs for the period 1988–2099, when temperature in particular is expected to increase more rapidly. The expected changes in temperature are quite evenly distributed over the year, but the summer months in particular may be brighter and drier (Figure 6).

Finally, the short site-specific time series of weather data that had been derived from on-site measurements and MARS data were extended, separately for each month of the year, back to 1920 and forward to 2080 using the CCCs. The method conserved year-to-year and day-to-day variability present in the on-site/MARS data but did affect month-to-month weather variability whenever CCCs from consecutive months differed for a specific site and weather variable.

3.2.4 Model calibration and testing

The models were parameterised using data for Norway spruce from Skogaby and Rajec and data for Scots pine from Jädraås (Table 1). For the most part, these datasets did not contain measurements of parameters (i.e. model-specific constants), but instead contained time series of growth variables such as tree height and diameter, and needle nitrogen concentration. Therefore model parameterisation was carried out by calibration: setting plant and soil parameters to values that make simulated growth correspond to measurements. The datasets for the sites Skogaby and Jädraås were used because they comprised measurements taken from multiple treatments, with different levels of fertilisation and irrigation. This ensured that the sensitivity of the models to environmental change was optimised to some extent. Because all four models were calibrated on the same datasets, their results for these sites were similar. Figure 4A shows this for simulated and measured height growth in Skogaby; model Q is absent because it does not simulate height growth as such.

The models were tested by running them for all 22 sites, using the actual planting dates and densities, thinning regimes and weather conditions for the sites. Height growth was the only tree growth variable measured on the majority of the sites, so model testing consisted of comparing the final simulated heights with measured heights (Figure 4B). Model Q was excluded from this test because it did not simulate height growth. The test showed that the three remaining models were able to account for 18–51% of variation in height growth across the studied sites. More extensive tests of all four models, on other data than those gathered in project Recognition, have been published elsewhere (Ågren and Bosatta 1998; Arah et al., 1997; Chertov et al., 1999; Ryan et al., 1996).

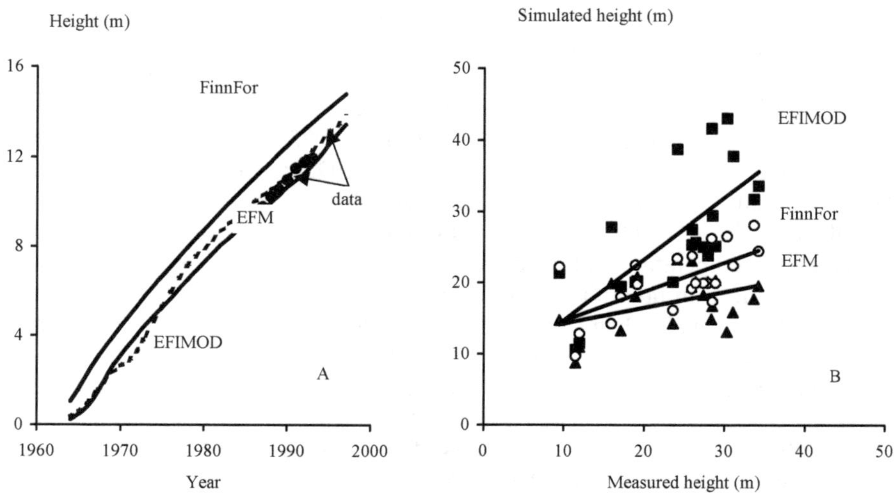

Figure 4. A. Calibration of process-based models on data. Dots indicate measurements of height growth in Skogaby, Sweden; lines are results from modelling after calibrating model parameters on the data; B. Testing of process-based models against data. Dots indicate simulated final tree heights at 22 sites across Europe (Table 1) vs the corresponding measured heights. Lines are linear regressions: r^2(EFIMOD) = 0.51, r^2(EFM) = 0.18, r^2(FinnFor) = 0.42.

3.2.5 Model application

After the four models were parameterised for Norway spruce and Scots pine, they were run using the environmental scenarios described above and variants thereof. For each of the 22 sites, simulations were run for three different time frames: 1920–2000, 1960–2040 and 2000–2080. For each combination of time frame and site, 16 runs were done characterised by the presence or absence of thinning and of changes in CO_2, nitrogen deposition and climate. This gives a total of 1056 runs per model, with half that number for the models EFIMOD and Q, which did not include CO_2 as a factor (Chapter 3.1).

The four models used differ in their structure. Precisely which structural differences among the models are responsible for any differences in behaviour? In other words, how were the differences in simulated growth caused by differences in model formulations of underlying processes and feedbacks? A common protocol was used to answer these questions. This involved calculating acquisition rates and use-efficiencies for the different tree growth resources (light, nitrogen and water; Bartelink et al., 1997; Reich and Schoette, 1988) with each of the different models, and assessing to what extent changes in growth are effected through these indicators of resource use.

Results of the simulations, including the analysis of rates of acquisition and use of resources, are reported in Chapter 4.4, and compared with the statistical approaches used in Recognition Chapters 5.1.

3.2.6 Discussion

3.2.6.1 The models

In Recognition, four different process-based models were used: EFM, Q, FinnFor and EFIMOD. These models are highly different in their structure. The EFM model has the most detailed description of tree physiology and represents ammonium and nitrate separately. The Q model has the most advanced formulation of decomposition processes and is the only one to include other vegetation than the dominant tree crop. The FinnFor model has the most detailed accounting for climate conditions and is also the only one to include regeneration processes. EFIMOD is a spatial-individual model. In the context of Recognition, the presence of these differences was seen as an advantage, as it allowed us to investigate the consequences of different structural assumptions.

The four models differ not only in the set of ecosystem processes that they simulate, but also in their overall feedback structure. An important example of differences in feedback structure is how the models deal with different levels of nitrogen availability. Only the Q model is based on Ingestad's "nitrogen productivity" concept, which implies that carbon relations, including growth rate, will adapt to the N-concentration in the plants. Compared to Q, the other models attribute a larger role to photosynthesis as the supplier of carbon-resources for growth. The consequence is that nitrogen-use efficiency (growth per unit of nitrogen taken up) is expected to be less sensitive to environmental change in Q than in the other models. This example of structural differences between the models is one of many where we as yet have insufficient empirical evidence to decide on the best approach. We propose that our approach to the analysis of structural differences between ecosystem models, which consists of quantifying resource acquisition rates as well as resource use efficiencies, and the responses of these indicators to environmental change, is a useful first step in identifying the key areas of disagreement in current quantitative ecology.

3.2.6.2 Uncertainty analysis

In global change research, attention has focused on the uncertainty in the environmental drivers (climate, atmospheric composition). This seems a natural thing to do: the primary change is environmental, and tree physiology may not adapt quickly, so why not focus on uncertainty in environmental scenarios? The climate change community strongly recommends that impact studies are based on more than one climate change scenario ("ensemble simulations") (IPPC-TGCIA, 1999). Pollution research also stresses the importance of accounting for different possible emission scenarios. So uncertainty regarding the environment tends to be emphasised. However, uncertainty regarding ecosystem processes is not negligible. This is apparent from the large structural differences that exist between current ecosystem models; especially regarding the way they simulate plant physiology. In Recognition, we have addressed the environmental uncertainties, by using a range of sites and environmental scenarios, but we have concentrated more than is usually done in impact studies on the ecosystem uncertainties, by comparing growth simulations from four different models and analysing the causes for differences in their behaviour in terms of their simulations of the underlying physiological processes (Chapter 4.4).

3.2.6.3 Choice of sites

The number of sites used for process-based modelling was kept small (22). This fact, and the uneven distribution of the sites across Europe (Table 1), implies that the modelling results cannot be interpreted as being representative for all growing conditions in Europe, or as providing an estimate of average rates of forest growth change. However, the focus on a small number of sites, each being characterised by good availability of growth data and environmental data, allowed us to carry out very detailed model- and site-specific analyses of simulation results and uncertainties. Lack of breadth was compensated for by gain in depth.

3.2.6.4 Construction of environmental scenarios: Nitrogen deposition and importance of other nutrients

For none of the 22 sites, complete and accurate time series of on-site measured nitrogen deposition were available. We therefore used a simple method of producing site-specific time series, for the years 1920–2080, which centred on data provided by EMEP (de Vries and Reinds, pers. comm. 2001). One source of uncertainty in this approach is that the data provided by EMEP represent open field deposition, and deposition on forests can differ. For example, Lövblad et al. (1995) reported that deposition on forests in southern Sweden was about twice as high as on open field. However, such conversion factors cannot be applied generally. Forests differ from open fields especially regarding their relatively efficient capture of aerosols, and Lövblad's example applies to Swedish conditions where aerosol transport is more dominant than elsewhere (Fowler, pers. comm. 2001).

None of the process-based models in Recognition includes nutrients other than nitrogen. The absence of other elements than N precludes a direct comparison with the results from the correlative approach concerning P, Ca, Mg, and K (see Chapters 2.2 and 4.1). However, available information on non-N nutrient dynamics, especially concerning their role in the physiology of the trees, is poor, and did not permit reliable inclusion in the models. The absence of other minerals than N from the models does mean that the process-based modellers need the results of the correlative approach (Chapters 4.1, 4.2 and 4.3) to identify areas in Europe where the results of the process-based modelling are of limited use (e.g. areas where P or K is limiting).

3.2.6.5 Construction of environmental scenarios: Weather

Our method of constructing site-specific time series of weather variables for the years 1920–2080, centred around short time series (usually 1988–1995) of measurements taken on-site or from the high spatial resolution JRC MARS weather database. The data were extrapolated to the full 160-year time range by applying

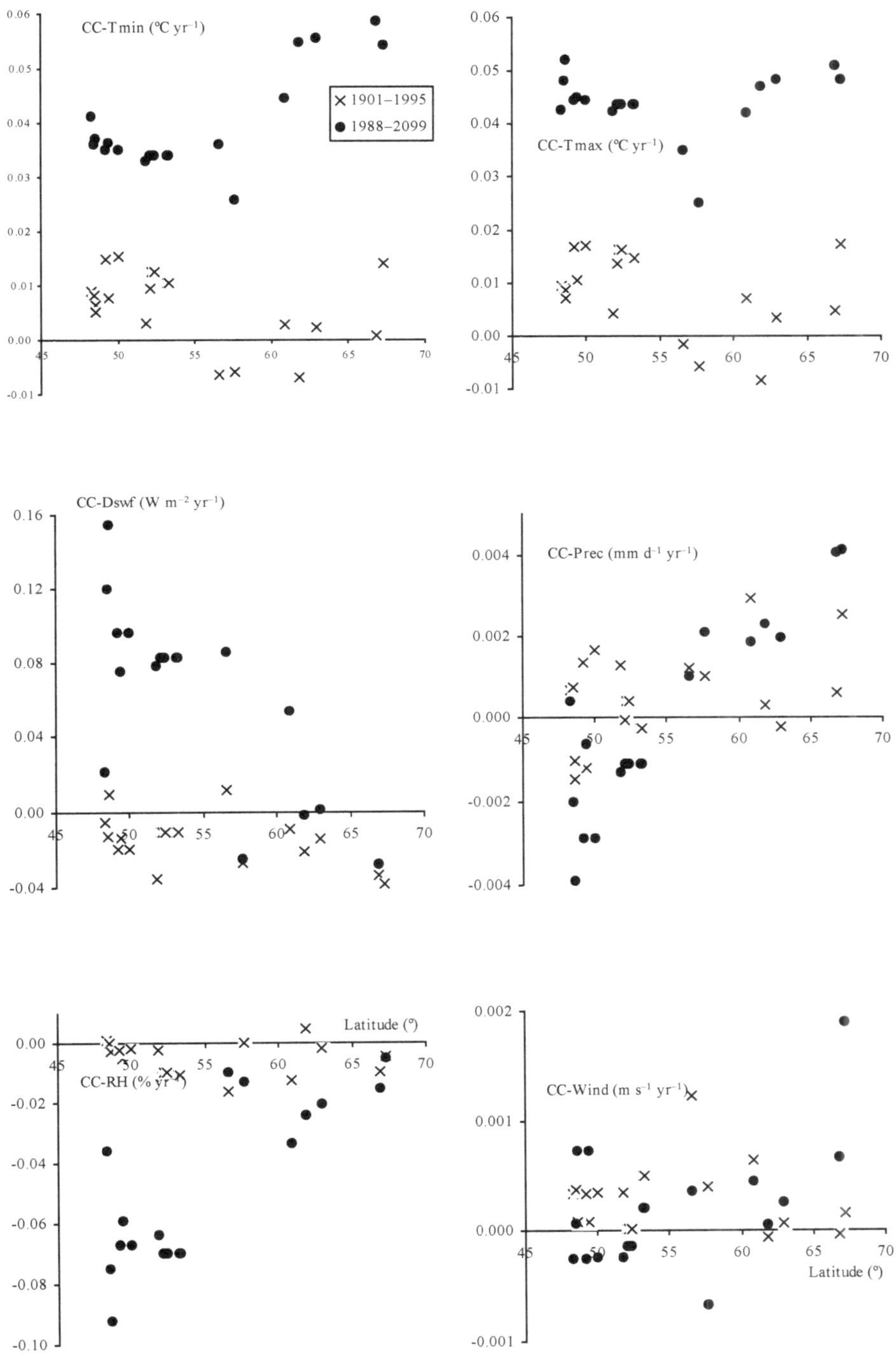

Figure 5. Climate change coefficients (CCCs) for past and future weather. CCCs are the average rates of change of the six weather variables used for process-based modelling: Daily minimum and maximum temperature (Tmin, Tmax), Daily short wave irradiance (Dswf), Precipitation (Prec), Relative humidity (RH) and wind speed (Wind). Each dot represents the average of the twelve monthly CCCs for one of the 22 sites (Table 1), vs. latitude of the site.

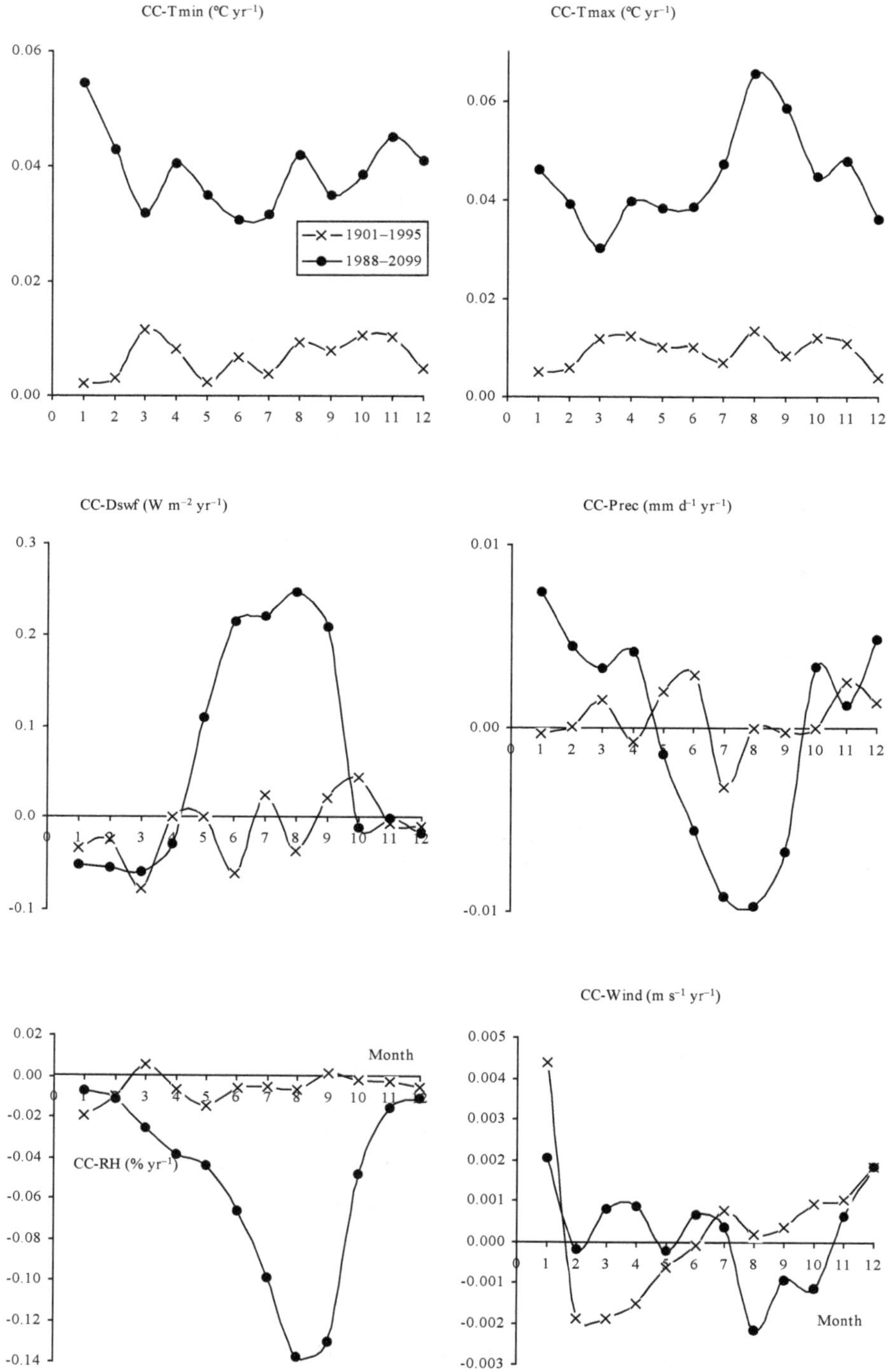

Figure 6. Climate change coefficients (CCCs) for past and future weather. CCCs are shown for each of the six weather variables used for process-based modelling. Each dot represents the average over the 22 sites (Table 1) of the twelve monthly CCCs.

month-specific climate change coefficients derived from sources with low spatial resolution (CRU-database of measurements 1901–1995, HadCM3-GCM calculations 1988–2099). This method of extrapolation ensures that the day-to-day and year-to-year weather variability in the measured short time series is maintained over the whole time range, whereas month-to-month (and inter-seasonal) variability is allowed to change. The method thus accounts for the strengths and limitations of our present understanding of climate change: our capability of predicting changing frequencies of extreme events (single days or short series of days with extreme weather) is still limited (Easterling et al., 2000), but we do have some confidence in predictions of differences in the magnitude of climate change among seasons. A good discussion of the use of scenario data and downscaling issues is provided by Carter et al. (IPCC-TGCIA, 1999).

3.2.6.6 Conclusions

We have sketched the modelling approach in Recognition. We conclude that the two project questions, i.e. explaining past forest growth and predicting future growth, can be answered with the methods outlined in this chapter and applied in the project. Key physiological responses, i.e. changes in rates of acquisition and efficiency of use of resources can be identified. The results of the application of the method are reported elsewhere (Chapter 4.4), as is a comparison with other, statistical approaches (Chapter 5.1).

References

Ågren, G.I and Bosatta, E. 1998. Theoretical Ecosystem Ecology – Understanding Element Cycles, Cambridge University Press.

Arah, J.R.M., Thornley, J.H.M., Poulton, P.R. and Richter, D.D. 1997. Simulating trends in soil organic carbon in long-term experiments using the ITE (Edinburgh) Forest and Hurley Pasture ecosystem models. Geoderma 81: 61–74.

Bartelink, H.H., Kramer, K. and Mohren, G.M.J. 1997. Applicability of the radiation-use efficiency concept for simulating growth of forest stands. Agricultural and Forest Meteorology 88: 169–179.

Carter, T.R., Hulme, M. and Viner, D. (Eds.) 1999. Representing Uncertainty in Climate Change Scenarios and Impact Studies. ECLAT-2 Workshop Report No 1, Climatic Research Unit, UEA, Norwich, UK.

Chertov, O.G., Komarov, A.S. and Tsiplianovsky, A.M. 1999. A combined simulation model of Scots pine, Norway spruce and Silver birch ecosystems in the European boreal zone. Forest Ecology and Management 116: 189–206.

Constable, J.V.H. and Friend, A.L. 2000. Suitability of process-based tree growth models for addressing tree response to climate change. Environmental Pollution 110:47–59.

Cramer, W. and Field, C.B. (eds.) 1999. The Potsdam NPP Model Intercomparison. Global Change Biology 5 (Supplement 1): iv + 76 p.

Easterling, D.R., Meehl, G.A., Parmesan, C., Changnon, S.A., Karl, T.R. and Mearns, L.O. 2000. Climate extremes: observations, modeling, and impacts. Science 289: 2068–2074.

Galloway, J.N. 1985. The deposition of sulphur and nitrogen from the remote atmosphere. In: Biogeochemical Cycling of Sulphur and Nitrogen in the Remote Atmosphere. Galloway, J.N., Charlson, R.J., Andrew, M.O. and Rodhe, H. (eds.). NATO ASI Series D. Reidel, Dordrecht. Pp. 143–175.

IPCC-TGCIA 1999. Guidelines on the Use of Scenario Data for Climate Impact and Adaptation Assessment. Version 1. Prepared by Carter, T.R., Hulme, M. and Lal, M. Intergovernmental Panel on Climate Change. Task Group on Scenarios for Climate IMpact Assessment. 69 p.

Joos, F., Bruno, M., Fink, R., Siegenthaler, U., Stocker, T.F., Le Quere, C. and Sarmiento, J.L. 1996. An efficient and accurate representation of complex oceanic and biospheric models of anthropogenic carbon uptake. Tellus 48B: 397–416.

Lövblad, G., Kindblom, K., Grennfelt, P., Hultberg, H. and Westling, O. 1995. Deposition of acidifying substances in Sweden. Ecological Bulletins 44: 17–34.

Medlyn, B.E., McMurtrie, R.E., Dewar, R.C. and Jeffreys, M.P. 2000. Soil processes dominate the long-term response of forest net primary productivity to increased temperature and atmospheric CO_2 concentration. Can. J. For. Res. 30: 873–888.

New, M., Hulme, M. and Jones, P. 2000. Representing twentieth-century space-time climate variability. Part II: Development of 1901–96 monthly grids of terrestrial surface climate. Journal of Climate 13: 2217–2238.

Reich, P.B. and Schoetle, A.W. 1988. Role of phosphorus and nitrogen in photosynthetic and whole plant carbon gain and nutrient-use efficiency in eastern white pine. Oecologia 77: 25–33.

Ryan, M.G., McMurtrie, R.E., Ågren, G.I., Hunt, E.R., Aber, J.D., Friend, A.D., Rastetter, E.B. and Pulliam, W.M. 1996. Comparing models of ecosystem function for temperate conifer forests. II. Simulations of the effect of climate change. In: Breymeyer, A.I., Hall, D.O., Melillo, J.M. and Ågren, G.I. (Eds), Global Change Effects on Coniferous Forests and Grasslands. John Wiley and Sons Ltd, Chichester, U.K. Scientific Committee on Problems of the Environment (SCOPE) No. 56: 363–387.

Ryan, M.G., Aber, J.D. Ågren, G.I., Friend, A.D., Linder, S., McMurtrie, R.E., Parton, W.J., Raison, R.J. & Rastetter, E.B. 1996. Comparing models of ecosystem function for temperate conifer forests. I. Model description and validation. In. Breymeyer, A, Hall, D.O., Melillo, J.M. & Ågren, G.I. (eds.) Global Change: Effects on Forests and Grasslands, J.Wiley. Pp. 313–362.

Siegenthaler, U. and Joos, F. 1992. Use of a simple model for studying oceanic tracer distribution and the global carbon cycle. Tellus 44B: 186–207.

Sonntag, M. 1997. Comparison of forest models at the leaf and canopy scale. In: Mohren, G.M.J., Kramer, K. and Sabaté, S. (eds.). Impacts of Global Change on Tree Physiology and Forest Ecosystems. Kluwer Academic Publishers. Dordrecht, The Netherlands. Pp 299–306.

Spiecker, H., Mielikäinen,K., Köh, M. and Skovsgaard, J.P. (eds.) 1996. Growth Trends in European Forests. EFI Research Report 5. Springer. 372 p.

4. Results

4.1 Historical development of nutrition and climate and their relationships with tree growth for three European tree species

Karl-Heinz Mellert[1], Jörg Prietzel[1], Ralf Straussberger[1], Brigitte Dorsch[3], Hans-Peter Kahle[2], Pedro-José Pérez-Martínez[2], Karl-Eugen Rehfuess[1] and Heinrich Spiecker[2]

[1]*Chair of Soil Science, Technical University Munich, Germany*
[2]*Institute for Forest Growth, University Freiburg, Germany*
[3]*Ludwig-Maximilians-University, Munich, Germany*

4.1.1 Introduction

The task of the Historical Development Investigation (HDI) was to investigate whether there are long-term changes of the nutritional status and the climatic conditions of representative forest stands which potentially could explain the increase of growth. The plots used in the HDI have been monitored for several decades with regard to growth and nutrition. The climatic conditions for these plots have also been measured or can be reconstructed. These stands are in general control plots of fertilization experiments and are located mainly in Central Europe and in Finland. The monitoring periods vary from 15 to 40 years. For further details of the sites see Section 2.1.3.1 of Chapter 2.1.

In this chapter the following specific hypotheses were tested for a subset of sample sites within the HDI. The hypotheses are that for distinct regions in Europe temporal trends can be identified for: (i) the height growth of dominant trees; (ii) for climatic variables; and (iii) for the supply of the repective stands with major nutrients. The relationships between the detected trends of height growth, the climatic conditions, and of the nutritional status of stands were analysed in order to identify the driving factors for tree growth acceleration.

4.1.1 Model formulation

4.1.2.1 General models

In the Historical Development Investigation (HDI) concept, plot data from 56 long-term forest nutrition and growth studies in six European countries were analysed and compared with each other (see Figure 3 in Chapter 2.1). The basic requirements for the suitability of a plot to be included in the HDI are described in detail in Chapter 2.1.

The following specific hypotheses were tested within the HDI:

1. For distinct regions in Europe, there are temporal trends for: (a) the height increment of dominant trees; (b) the supply of these stands with major nutrients; and (c) climatic variables.

2. The identified trends in height increment are related to systematic changes of the stand's nutritional status and climate conditions.

The specific aim of the HDI, however, is to improve the insight into the complex relationships between temporal changes of tree growth, nutrition and of atmospheric environmental conditions by means of statistical analyses of time series of data concerning all respective variables of interest. Our second hypothesis for this investigation requests that the identified temporal trends of referenced height increment – regardless whether they are positive or negative – can be explained statistically by systematic changes of the stand's nutritional status and/or of climatic conditions either for individual stands or for the whole population of sites under study. Hypothesis 2 thus results in the formulation of the regression equation described in the following section.

4.1.2.2 Regression analysis

The regression models that were applied to the data set were derived from hypothesis 2. Due to the limited amount of data regression models which include the whole set of macronutrients as predictors were thought to be over-parameterised. Therefore only the results of the analysis derived from data including nitrogen as the only nutrient variable are given in this report. Dorsch (2002) tested whether the water availability for the pine stand based on Thornthwaite's water balance (Thornthwaite and Mather, 1955) would improve the portion of explained variability compared to models involving precipitation as factor for water availability. Water balance data did not increase the explanation of growth variance. Consequently, we decided to use the original and more conclusive temperature and precipitation variables when formulating the final regression model. It was hypothesized that there may be interactions between stand nutrition and climate (e.g. the supply of N may interact with water supply), and therefore we tested regression models that included interaction terms between stand nutrition and climate. However, there was no significant interaction between these factors. The lack of any significant interaction may be due to the limited size of the evaluated data set.

The SAS procedures *proc autoreg* and *proc mixed* were run with the formulations.

Regression Model 1:

$$rih = \alpha + \beta 1 \times prec4\_6 + \beta 2 \times prec\_1\_5 + \beta 3 \times temp4\_6 + \beta 4 \times temp\_1\_5 + \beta 5 \times n\_fol + \beta \times n\_1\_5 + \varepsilon$$

The response variable is:

- referenced height increment *(rih)*: measured annual height increment as percentage of the reference value according to the refering yield table

The predictor variables for Regression Model 1 are:

- *prec4_6*: precipitation sum for months April to June (current year);
- *prec_1_5*: 5-year backward running average (linear degressive) of the

precipitation sum for months May to September of the previous five years: (precipitation in year 1×0.34) + (precipitation in year 2×0.27) + (precipitation in year 3×0.2) + (precipitation in year 4×0.13) + (precipitation in year 5×0.06);
- *temp4_6*: average temperature for months April to June;
- *temp_1_5*: 5-year backward running average (linear degressive) of the average temperature for months May to September (for weighting see *prec_1_5*);
- *n_fol*: foliar N concentration of current-year needles; and
- *n_1_5*: 5-year backward running average (linear degressive) of *n_fol* (for weighting see *prec_1_5*).

The analysis carried out with the described regression model showed that variables that include retrospective information of the previous five years are significant in describing changes of referenced height increment. Regression models involving all macro nutrients (results not given in this report) were specifically used to compare the relative influence of stand N nutrition on referenced height growth with the relative influence of the supply with other macronutrients (P, K, Mg, Ca). The supply of the investigated stands with other macronutrients generally showed a poorer correlation with referenced height increment (Dorsch, 2002) than N nutrition. This result agrees well with the conclusions drawn in section 4.1.4.2.

The applied mixed linear model analyses to what extent the referenced height growth of the studied Scots pine stands is affected by time-lagged effects concerning the N nutritional status and the climate conditions that they had experienced during the previous five years. Compared to the simple linear model, the mixed linear model is in general more appropriate for analysing temporal data because it is possible to:

- adapt the model to the temporal structure of the data by selecting suitable covariance structures for the error term; and
- add random effects to the model – these effects address the heterogeneity among the sites. Random effects were modelled in two ways: (Model A) as a different intercept for each site; and (Model B) as the random factor 'water storage capacity of the soil'.

Moreover, the data included into the mixed linear model do not have to be balanced – i.e. the time series of the sites are allowed to have different lengths.

Auto-correlation was used to analyse the covariance structure of the error term – i.e. the residual of a given year is correlated with the residual of the previous year. The statistical significance of the particular covariance structure and the random effect was tested with a likelihood quotient (LQ) test. For this analysis we used the *proc mixed* procedure of SAS.

The regression model was applied in two ways:

Model A: to all available sites within an analytical unit (e.g. all Scots pine sites with sufficient observations); and

Model B: in order to examine the possible causes of tree growth acceleration, Regression Model 1 was also applied to a restricted data set comprising only sites that show increased height growth.

4.1.2.3 Criteria for a nutrient driven tree growth acceleration

The following criteria for the identification of a **nutrient**-driven tree growth acceleration were formulated rigorously:

The probability of a nitrogen-driven tree growth acceleration is high if:
1. a sustained increase in N nutrition can be observed; and
2. the stands have insufficient N supply at the start of the time series; and
3. the nutrition with other macro-nutrients is classified at least sufficient – in cases where the nutrition with P or base cations falls below the threshold of sufficient nutrition, growth will probably be limited by these factors.

If criteria 1 and 2 are fulfilled an increase in N nutrition is indicated by positive trends in:

1a. foliar N concentration and *constant or increased* needle/leaf mass (100-needle/leaf weight); or
1b. foliar N content (product of foliar N concentration and needle/leaf mass).

The case where all of the above mentioned criteria (1a, 1b, 2, 3) are fulfilled, is the strongest indication for an N-driven growth acceleration. However, an N-driven growth acceleration can also be assumed if at least one of the criteria (1a) or (1b) in combination with both of the criteria 2 and 3 are realised. Cases, where an increase in foliar N concentration is associated with a reduction of needle mass are excluded from this consideration, because in this case the increased foliar N concentration is rather caused by concentration effects due to needle unit dry mass reduction than by an improvement in N nutrition.

The analysis of trends regarding the nutritional status of the investigated stands with various nutrients is based on the combined assessment of: (i) foliar nutrient concentrations in half-year-old needles/leaves classified according to the evaluation system of Heinsdorf (1999); (ii) the mass of 100 needles/leaves (100-needle/leaf weight); and (iii) foliar nutrient contents (product of 100-needle/leaf weight and the foliar nutrient concentration). The evaluation system of Heinsdorf (1999) is particularly suitable for our approach. Heinsdorf classifies five levels of nutritional status: Level 1 – extreme deficiency; Level 2 – insufficient supply; Level 3 – sufficient to optimal supply; Level 4 – luxurious supply (i.e. there is an excess nutrient supply); Level > 4 very luxurious supply (i.e. the supply of the nutrient may be potentially toxic). The nutritional classes in the Heinsdorf system are not equidistant with respect to the foliar nutrient concentrations. The scaling of this system allows a physiologically-based judgement of the nutritional status of trees with respect to growth. Moreover, it allows a direct comparison of the supply of trees with different nutrients.

For comparisons of the nutritional status among different stands, their average nutrient levels in the year 1985 were used as a reference. This year was selected because it provided a maximum number of observations on various sites for a relatively early period. In case there was no nutrient data available for that year, the average nutrient level in that year was estimated by linear regression based on the available period of foliar nutrient monitoring. For graphical analyses the investigated stands are sorted according to their nitrogen nutritional status, since in most cases nitrogen is the most determining nutrient for foliage, height and volume growth of trees, whereas the supply with other macronutrients is limiting tree growth only in rare cases.

Micronutrients were excluded from the analysis for two reasons: (1) a lot of missing values in the database of these parameters would have resulted in an unbalanced statistical design – the fact that the database is at the lowermost edge of the requirements for suitable statistical procedures concerning the number of observations even for the macronutrients warn against the inclusion of an incomplete micronutrient data set; and (2) our knowledge about the significance of micronutrients for tree growth is still poor.

4.1.2.4 Criteria for a climate driven tree growth acceleration

The explorative analysis of time series of climate data is similar to that of the nutritional data. The regional distribution of climate trends is also addressed, and sites where a climate-driven increase of tree growth is likely to have occurred, are identified.

4.1.3 Temporal variation of climate variables

4.1.3.1 General remarks

The analysis of the climate of the past decades indicates that climate is changing in Europe (e.g. Schönwiese et al., 1993). As an example, air temperature anomalies show positive trends in large parts of Europe. These changes should also be reflected in the database of the HDI sites. As the specific task of HDI is to identify temporal trends of climate and stand nutrition, and since the availability of nutrient data is limited in our approach (temporal resolution: one observation per year), the analysis of climate data has to be adapted accordingly.

In contrast to meteorological surveys on climate change, which are based on large climate data sets for long reference periods, we have to consider changes of climate parameters especially within the period of foliar nutrient monitoring – which varied from site to site, but were all within the second half of the 20th century. The specific period of foliar nutrient monitoring varies widely for the different sites and there was often a large deviation from conventional climatological reference periods (e.g. 1931–1960 or 1961–1990). Therefore, it was not useful to apply standard procedures for the characterisation of climate anomalies within HDI. Consequently, for each climate parameter to be tested, the value of the fitted linear regression for the first year of the foliar nutrient measurement period was used as a reference.

Tables 1 to 4 show the mean linear trends of the average temperature and precipitation according to: (1) the CRU (see Chapter 3.1) data set; and (2) the individual measurements at the meteorological stations. Average temperature and precipitation were calculated for: (a) the whole year; (b) the growing season (defined as the period from May to September); and (c) the spring season (defined as the period from April to June). The trends are considered for the entire period from 1951 to 1998, and also for the sub-periods of foliar nutrient monitoring. This allows comparison of the temporal development during the shorter period of foliar

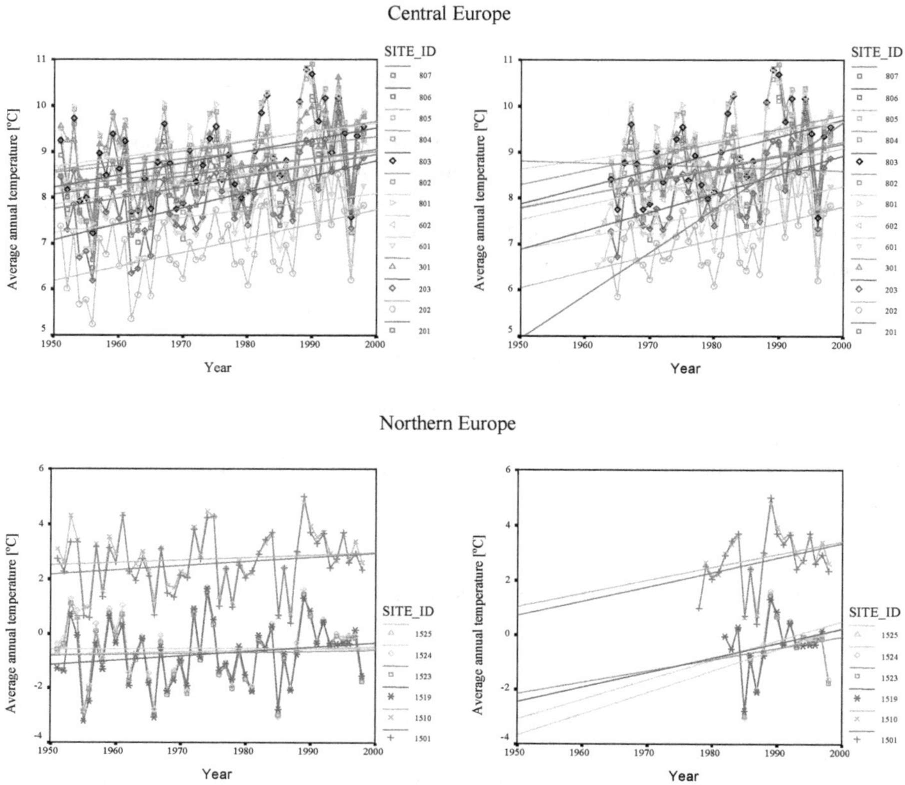

Figure 1. Time series of average annual temperature according to CRU data at those HDI sites, for which height increment was analysed. Presented are annual data and linear trends for each investigated site. The graphs on the left show annual data and linear trends for each investigated site for the 1951–1998 period. The graphs on the right show data and trends for the period of foliar monitoring.

nutrient monitoring with the long-term tendency during the second half of the 20th century. This approach also allows trends of measured and modelled data to be compared. The computed trends refer to changes within 10-year periods and in addition represent mean trends over the entire period of foliar nutrient monitoring. According to their slopes the trends were classified into seven classes on a logarithmic scale: (1) <0.625%; (2) 0.625–<1.25%; (3) 1.25–<2.5%; (4) 2.5–<5%; (5) 5–<10%; (6) 10–<20%; (7) ≥20%.

4.1.3.2 Air temperature

At most sites the linear trend of the average annual air temperature is positive for both periods of comparison, and for modelled and measured values (Tables 1 and 2).

Table 1. Linear trends of temperature at HDI sites in Central Europe (CE) and the UK for: the period 1951–1998 (CRU 51–98); the period of foliar nutrient monitoring according to the CRU data (CRU obs per); and data measured at the closest meteorological station (measured). The table is sorted by region and latitude of the sites. Trends larger than ± 0.625%/10 years ↗ or ↘; ± 1.25%/10 years ↗↗ or ↘↘; ± 2.5%/10 years ↗↗↗ or ↘↘↘; ± 5%/10 years ▲ or ▽; ± 10%/10 years ▲▲ or ▽▽; ± 20%/10 years ▲▲▲ or ▽▽▽. Significance level of linear regression p <0.05: cell colour dark grey; p <0.1: cell colour light grey.

			Average annual temperature			Average temperature May–Sept.			Average temperature April–June		
Site name	Site ID	Region	CRU 51–99	CRU obs per	measured	CRU 51–99	CRU obs per	measured	CRU 51–99	CRU obs per	measured
Dobrowa	603	CE	↗↗↗	▲	↗	↗↗	↗↗↗	↗↗	↗↗	↗↗↗	↔
Epinal (88)	1309	CE	↗↗↗	↗↗↗		↗↗	↗↗↗		↔	↗↗↗	
Solling-Buche	1002	CE	↗	↗↗↗		↔	↗		↔	↗↗	
Bonhomme	1326	CE	↗↗↗	▲		↗↗	↗↗↗		↗	▲	
Gemaingoutte	1327	CE	↗↗↗	↘↘↘		↗↗	↔		↗	↗↗↗	
Höglwald	201	CE	↗↗↗	▲▲	▲▲	↗↗	▲	↗↗↗	↗	▲▲	▲
Ste Hélène (88)	1323	CE	↗↗	↗↗↗		↗↗	↗↗↗		↗	↗↗↗	
Pfalzgrafenweiler	301	CE	↗↗	↔	↗↗	↗↗	↗↗	↔	↔	↘↘	↘
Helfenberg	601	CE	↗↗	↗↗↗	↗↗↗	↗	↗↗	↗↗↗	↗	↗↗	↗↗↗
Karlstift	602	CE	↗↗↗	↗↗↗	↘↘	↗↗	↗↗	↔	↗↗	↗↗	↔
Souilly (55)	1321	CE	↗↗	↗↗↗		↗	↗↗		↔	↗↗↗	
Pustert	203	CE	↗↗↗	▲	↗↗↗	↗↗	↗↗↗	↗	↗↗	↗↗↗	↗
Rajec	1801	CE	↗↗↗	▽▽		↗↗	↘↘↘		↗↗	▲▲	
Pfaffenwinkel	202	CE	↗↗↗	▲	↗↗	↗↗	↗↗	↗	↗↗	↗↗↗	↔
Auberive (52)	1301	CE	↗↗	↗↗↗		↗	↗↗		↔	↗↗↗	
Solling-Fichte	1001	CE	↗	↗↗↗		↔	↗		↔	↗↗	
Brieschtt	802	CE	↗↗↗	↗↗↗	↗↗	↗	↗↗	↗	↗↗	↗↗	↗
Löpten	801	CE	↗↗	↗↗	↗↗	↗	↗	↗	↗	↗↗	↗
Triebsch	803	CE	↗↗↗	↗↗↗	↗↗↗	↗↗	↗↗	↗↗	↗↗	↗↗↗	↗↗↗
Gandenitz II	805	CE	↗↗	↗↗↗	↗↗↗	↗	↗	↗↗	↗	↗↗	↗↗
Gandenitz I	806	CE	↗↗	↗↗↗	↗↗↗	↗	↗	↗↗	↗	↗↗	↗↗
Alt Placht	804	CE	↗↗	↗↗↗	↗↗↗	↗	↗	↗↗	↗	↗↗	↗↗
Aalkasten	807	CE	↗↗	↗↗↗	↗↗	↗	↗	↔	↗↗	↗↗	↗↗↗
Rosedale 118/70	2205	UK	↗↗	↘	▽	↗	↔	↘↘↘	↗	↔	▽
Rosarie 2 P70	2201	UK	↔	↗↗↗	▲	↔	↗↗	↘↘	↔	↗↗	↗↗
Black Isle 8/76	2203	UK	↔	▲	▲	↔	↗↗	↗↗↗	↔	↗↗	▲

Table 2. Linear trends of temperature at HDI sites in Northern Europe (NE) for: the period 1951–1998 (CRU 51–98); the period of foliar nutrient monitoring according to the CRU data (CRU obs per); and data measured at the closest meteorological station (measured). Trends larger than ± 0.625%/10 years ↗ or ↘; ± 1.25%/10 years ↗↗ or ↘↘; ± 2.5%/10 years ↗↗↗ or ↘↘↘; ± 5%/10 years ▲ or ▽; ± 10%/10 years ▲▲ or ▽▽; ± 20%/10 years ▲▲▲ or ▽▽▽. Significance level of linear regression $p < 0.05$: cell colour dark grey; $p < 0.1$: cell colour light grey.

			Average annual temperature			Average temperature May–Sept.			Average temperature April–June		
Site name	Site ID	Region	CRU 51–98	CRU obs per	measured	CRU 51–98	CRU obs per	measured	CRU 51–98	CRU obs per	measured
Sdromme	2040	NE	↗↗↗	▽		↗	↘		↗↗	⇔	
Djbplantage	2042	NE	↗↗↗	▽		↗	↘↘		↗	⇔	
Dalgas	2032	NE	↗↗↗	▲▲		↗↗	↗↗		↗↗	⇔	
Dalgas	2033	NE	↗↗↗	▲▲		↗↗	↗↗		↗↗	⇔	
Skogaby	402	NE	⇔	↘↘↘		⇔	↗↗		⇔	▽	
Jädraås	401	NE	↗↗↗	▲		⇔	⇔		⇔	↘	
Vierumäki 157	1521	NE	↗↗↗	▲▲▲		⇔	↘↘		↗	↗	
Heinola 113	1518	NE	↗↗↗	▲▲		⇔	↗		↗	↘↘	
Vesijako 26	1502	NE	↗↗	▲▲		⇔	⇔		⇔	↘↘↘	
Sahalahti 138	1520	NE	↗	▲▲▲		⇔	↗↗		⇔	↗↗↗	
Ruokolahti 64	1512	NE	▲	▲▲		⇔	⇔		↗↗	↘↘	
Moinniemi 49	1506	NE	▲	▲▲		⇔	⇔		↗↗	↘↘	
Punkaharju 25	1501	NE	▲	▲▲▲		⇔	⇔		↗↗	↘↘	
Patasalo 31	1503	NE	▲	▲▲▲		⇔	↗↗↗		↗↗	↗↗↗	
Sulkava 34	1505	NE	▲	▲▲		⇔	⇔		↗↗	↘↘↘	
Juva 68	1513	NE	↗↗↗	▲▲▲		⇔	⇔		↗↗	⇔	
Kuorevesi 52	1507	NE	↗	▲▲		⇔	↘		⇔	▽	
Vilppula 53	1508	NE	↗↗	▲▲		⇔	↘		⇔	↘↘↘	
Ähtäri 98	1515	NE	↗↗↗	▲▲▲		⇔	↘		↗	⇔	
Pylkönmäki 73	1514	NE	↗↗↗	▲▲▲		⇔	↘↘		↗	↘↘	
Kannonkoski 56	1509	NE	↗↗↗	▲▲		⇔	↘		↗	▽	
Kannonkoski 60	1511	NE	↗↗↗	▲▲		⇔	↘		↗	▽	
Kannonkoski 58	1510	NE	↗↗↗	▲▲		⇔	↘		↗	▽	
Alajärvi 108	1517	NE	▲	▲▲▲		⇔	⇔		↗↗	⇔	
Halsua 106	1516	NE	▲	▲▲▲		⇔	↗↗		↗↗	↘↘↘	
Kemijärvi 194	1524	NE	↘↘	▲▲▲		↘	↘↘		↘	▽▽▽	
Kolari 135	1519	NE	▲▲	▲▲▲		↗	↗		↗↗↗	▽▽	
Sodankylä 184	1523	NE	▲	▲▲▲		↘	↘↘		↘	▽▽▽	
Sodankylä 161	1522	NE	↗↗↗	▲▲▲		↘	⇔		↘	▽▽	
Sodankylä 197	1525	NE	▲	▲▲▲		⇔	▽		⇔	▽▽▽	

Height analyses were carried out at 12 sites for Scots pine (site 202 Pfaffenwinkel, site 203 Pustert, site 801 Löpten, site 802 Brieschtt, site 803 Triebsch, site 804 Alt Placht, site 805 Gandenitz II, site 806 Gandenitz I, site 807 Aalkasten, site 1501 Punkaharju 25, site 1519 Kolari 135 and site 1523 Sodankylä 184), and seven sites for Norway spruce (site 201 Höglwald, site 301 Pfalzgrafenweiler, site 601 Helfenberg, site 602 Karlstift, site 1510 Kannonkoski 58, site 1524 Kemijärvi 194 and site 1525 Sodankylä 197). For these sites the trend for temperature was generally positive (Figure 1).

The average temperature in the growing season (May–September) and in spring (April–June) generally increased in UK and Central Europe (CE) (Table 1), whereas in Northern Europe (NE) (Table 2) there is no obvious trend.

A comparison of the temperature development during the 1951–1998 period and during the period of foliar nutrient monitoring indicates that in NE the period of foliar nutrient monitoring does not represent the dominant trend during the second half of the 20th century. General conclusions on temperature effects should take this into account.

The general temperature trends during the period of foliar nutrient monitoring at the HDI sites are summarised as follows:

- the average annual temperature increased at almost all sites;
- in CE and the UK the temperature showed positive trends at nearly all sites during spring and the growing season; and
- in NE the temperature of the growing season and spring did not change much.

This pattern suggests that a growth acceleration due to an increase in temperature is most likely to occur at HDI sites in CE, provided that the water supply is sufficient. An increase in air temperature connected with reduced water availability would lead to a growth decline rather than to accelerated growth. In NE any warming seems to have taken place primarily in the dormant season, and consequently the effect on tree growth might be expected to be small.

4.1.3.3 Precipitation

In CE there is no uniform trend of annual precipitation. Significant changes are rare (Table 3) and most of the significant trends are negative. Figure 2 indicates that the long-term changes in annual precipitation in CE are small compared to the inter-annual fluctuations of precipitation. For the period of foliar nutrient monitoring, there are negative trends of precipitation during the growing season and during spring at most sites in CE. In contrast, there are positive trends in precipitation at all three UK sites during the growing season and spring.

At most sites in NE, there are positive trends in annual precipitation for the CRU 51–98 data (Table 4). However, for the period of foliar nutrient monitoring, precipitation seems to have declined at many sites. In NE the observed trends of precipitation during the growing season and during spring are very heterogeneous (Table 4). Even in neighbouring regions (e.g. Denmark and southern Sweden) the trends differ strongly between different sites. All types of changes occurring in NE are represented in the sites chosen for height analysis (Figure 2).

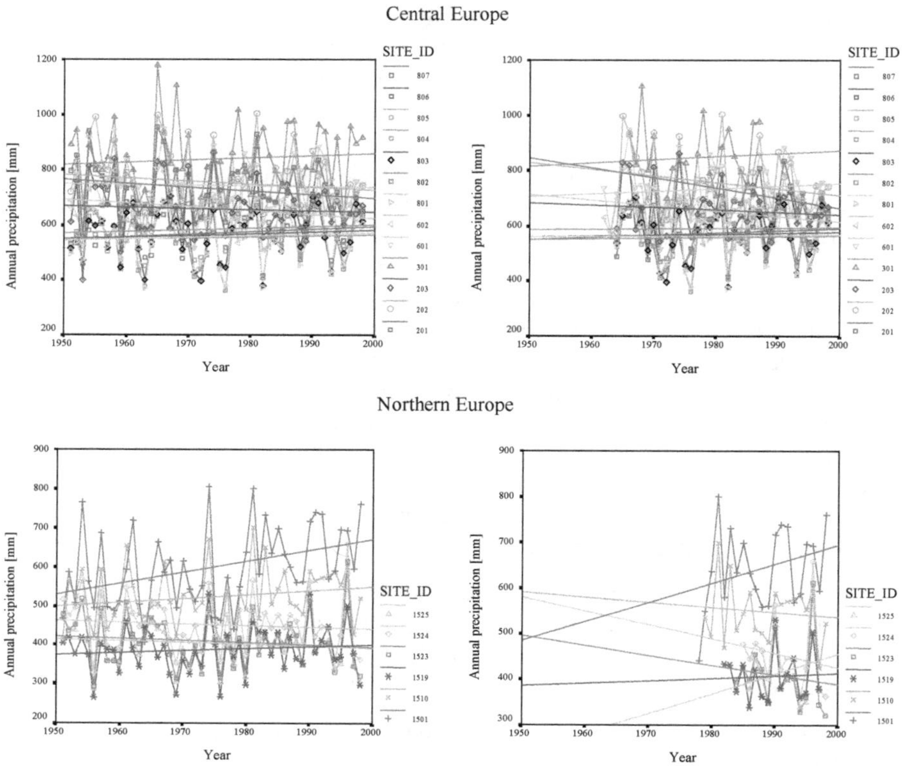

Figure 2. Time series of annual precipitation according to CRU data at those HDI sites for which height increment was analysed. The graphs on the left show annual data and linear trends for each investigated site for the 1951–1998 period. The graphs on the right show data and trends for the period of foliar monitoring.

It is quite clear that for many of the sites there are considerable deviations between the trends as calculated for the CRU 51–98 data and for the period of foliar nutrient monitoring.

The pattern for the period of foliar nutrient monitoring deviates from the general pattern for the CRU 51–98 data. The following findings for this period should be emphasised:

- in general annual precipitation did not show any long-term trend in CE;
- annual precipitation decreased at many sites in NE;
- slight decreases of precipitation during spring and the growing season were observed at many sites in CE; and
- the spatio-temporal pattern of precipitation during spring and the growing season in NE is inconsistent.

Table 3. Linear trends of precipitation at HDI sites in Central Europe (CE) and the UK for: the period 1951–1998 (CRU 51–98); the period of foliar nutrient monitoring according to the CRU data (CRU obs per); and data measured at the closest meteorological station (measured). Table is sorted by region and latitude of the sites. Trends larger than ± 0.625%/10 years ↗ or ↘; ± 1.25%/10 years ↗↗ or ↘↘; ± 2.5%/10 years ↗↗↗ or ↘↘↘; ± 5%/10 years ▲ or ▽; ± 10%/10 years ▲▲ or ▽▽; ± 20%/10 years ▲▲▲ or ▽▽▽. Significance level of linear regression p <0.05: cell colour dark grey; p <0.1: cell colour light grey.

			Annual precipitation			Precipitation May–September			Precipitation April–June		
Site name	Site ID	Region	CRU 51–98	CRU obs per	measured	CRU 51–98	CRU obs per	measured	CRU 51–98	CRU obs per	measured
Dobrowa	603	CE	↘↘	↔	↘↘	↘↘	↘↘	↘↘↘	↘↘↘	↔	↘↘↘
Auberive (52)	1301	CE	↔	↗↗		↘↘	↗↗		↗↗↗	↗↗	
Epinal (88)	1309	CE	↗	↘		↘	↘↘↘		↗↗↗	↘↘↘	
Bonhomme	1326	CE	↗↗	▽▽		↗	▽▽		↗↗↗	▽▽▽	
Gemaingoutte	1327	CE	↗↗	▽		↗	↘		↗↗↗	▽▽▽	
Höglwald	201	CE	↘↘	↘↘↘	↗↗	↘↘↘	▽▽	▽	↘↘↘	▽▽	▽▽
Ste Hélène	1323	CE	↗↗	↗↗↗		↔	↘↘		↗↗↗	↘↘↘	
Pfalzgrafenweiler	301	CE	↗	↗↗	▲	↔	↘↘	↘↘↘	↗↗	↔	↗↗↗
Helfenberg	601	CE	↔	↗	↗↗	↘↘	↔	↘↘	↔	↘	▽
Karlstift	602	CE	↘↘	↘↘↘	↗↗	↘↘↘	↘↘↘	↘↘	↘↘↘	↘↘↘	▽
Souilly (55)	1321	CE	↗	↗↗		↘	↘		↗↗↗	↘↘	
Pustert	203	CE	↘	↘	↔	↘↘	↘	↘↘	↘↘↘	▽	↔
Rajec	1801	CE	↘↘	▽▽		↘↘	▽▽▽		↘↘	▽▽▽	
Pfaffenwinkel	202	CE	↘↘	↘↘↘	↔	↘↘↘	↘↘↘	↘↘↘	↘↘↘	▽	↘↘
Solling-Buche	1002	CE	↗↗	↗↗↗		↘↘	↘↘		↔	▽	
Solling-Fichte	1001	CE	↗↗	↗↗↗		↘↘	↘↘		↔	▽▽	
Briescht	802	CE	↗	↔	↗	↘	↘	↔	↗↗↗	↘	↗↗↗
Löpten	801	CE	↗	↔	↗	↘	↘↘	↔	↗↗	↘↘↘	↗↗↗
Triebsch	803	CE	↗	↔	↔	↘	↔	↗	↗↗↗	↔	↗
Gandenitz II	805	CE	↗	↔	↘↘↘	↔	↔	▽	↔	↘↘	↘↘↘
Gandenitz I	806	CE	↗	↔	↘↘↘	↔	↔	▽	↔	↘↘	↘↘↘
Alt Placht	804	CE	↗	↔	↘↘↘	↔	↔	▽	↔	↘↘	↘↘↘
Aalkasten	807	CE	↗↗	↗↗	↘	↔	↗↗↗	↘↘↘	↗	↘↘	↘↘↘
Rosedale	2205	UK	↘	▲▲	▲▲▲	↘↘↘	▲	▲▲	↔	▲	▲▲
Rosarie 2 P70	2201	UK	↘	↗↗	▽	↘↘	↗↗	↗↗	↗↗↗	▲	▲▲
Black Isle 8/76	2203	UK	↗↗↗	▲	↗↗	↘	↗↗	↗	↗↗↗	▲▲	▲▲

Table 4. Linear trends of precipitation at HDI sites in Northern Europe (NE) for: the period 1951–1998 (CRU 51–98); the period of foliar monitoring according to the CRU data; and data measured at the closest weather station (measured). Table is sorted by region and latitude of the sites. Trends larger than ± 0.625%/10 years ↗ or ↘; ± 1.25%/10 years ↗↗ or ↘↘; ± 2.5%/10 years ↗↗↗ or mmm; ± 5%/10 years ▲ or ▽; ± 10%/10 years pp or ▽▽; ± 20%/10 years ▲▲▲ or ▽▽▽. Significance level of linear regression $p < 0.05$: cell colour dark grey; $p < 0.1$: cell colour light grey.

Site name	Site ID	Region	Annual precipitation		Precipitation May–September		Precipitation April–June	
			CRU 51–98	CRU obs per measured	CRU 51–98	CRU obs per measured	CRU 51–98	CRU obs per measured
Sdromme	2040	NE	⇔	▲▲	↘↘	▲▲	↗	▲▲
Djbplantage	2042	NE	↘↘	▲▲	↘↘	▲▲	⇔	▲▲
Dalgas	2032	NE	⇔	▽	↘↘	▽▽▽	↗↗	▽▽▽
Dalgas	2033	NE	⇔	▽	↘↘	▽▽▽	↗↗	▽▽▽
Skogaby	402	NE	↗↗↗	▲▲	⇔	↘↘↘	↗↗	▽▽▽
Jädraås	401	NE	↗↗	↘↘	⇔	↘	▲	▲
Vierumäki 157	1521	NE	⇔	▽	↗↗	▲▲	↗↗	↗↗↗
Heinola 113	1518	NE	⇔	↘↘↘	↗↗	▽	↗↗	↗↗
Vesijako 26	1502	NE	↗↗	⇔	↗↗	↘↘	↗↗	↗↗↗
Sahalahti 138	1520	NE	↗↗	▲▲	↗↗	▲▲▲	↗↗↗	▲▲▲
Ruokolahti 64	1512	NE	↗↗↗	↘	↗↗	↘↘	▲	▽▽
Moinniemi 49	1506	NE	↗↗↗	⇔	↗↗↗	↗↗	▲	▽▽
Punkaharju 25	1501	NE	▲	▲	↗↗↗	▲	▲	↘↘
Patasalo 31	1503	NE	▲	↗	↗↗↗	▲▲▲	▲	▽▽
Sulkava 34	1505	NE	↗↗↗	↘↘	↗↗	↘↘↘	▲	▽▽
Juva 68	1513	NE	↗↗	↘↘	↗↗	↘↘↘	↗↗↗	↘↘
Kuorevesi 52	1507	NE	↗↗↗	↘↘	↗↗	↘↘↘	↗↗↗	⇔
Vilppula 53	1508	NE	↗↗↗	↘↘↘	↗↗	↘↘↘	↗↗↗	↘↘
Ähtäri 98	1515	NE	↗↗	↘↘↘	↗↗	⇔	↗↗↗	▽▽
Pylkönmäki 73	1514	NE	↗↗	▲	↗	▲	↗↗	▲
Kannonkoski	1509	NE	↗↗	↘↘	↗	↘↘↘	↗↗↗	↘↘↘
Kannonkoski	1511	NE	↗↗	↘↘	↗	↘↘↘	↗↗↗	↘↘↘
Kannonkoski	1510	NE	↗↗	↘↘	↗	↘↘↘	↗↗↗	↘↘↘
Alajärvi 108	1517	NE	↗↗	↘	↗	↗↗↗	↗↗	▽▽
Halsua 106	1516	NE	↗↗	▽	⇔	↘↘	↗↗	▽▽
Kemijärvi 194	1524	NE	↘	▽	↘↘↘	▽	↗	▲▲
Kolari 135	1519	NE	↗↗	↗↗	⇔	↗	↗↗	↗↗
Sodankylä 184	1523	NE	↘↘	↘↘↘	↘↘↘	↘↘↘	↗	▲▲
Sodankylä 161	1522	NE	⇔	↘↘↘	↘↘	▽	↗↗	↗↗
Sodankylä 197	1525	NE	↘	▲▲	↘↘↘	▲▲▲	⇔	▲▲

Consequently the expectation at present is that if only temporal trends of precipitation changes are considered, decreases of growth during the period of foliar nutrient monitoring at the HDI sites are more likely to occur than growth increases.

4.1.4 Scots pine (*Pinus sylvestris*)

4.1.4.1 General remarks

For Scots pine, data from 28 European sites were suitable for evaluation within the HDI. The sites are located in Sweden (1), Finland (15), UK (3), and Germany (9). However, data regarding changes of tree growth were only available for 12 of these sites. These sites are situated in Finland (site 1501 Punkaharju 25, site 1519 Kolari 135, site 1523 Sodankylä 184), and Germany (site 202 Pfaffenwinkel, site 203 Pustert, site 801 Löpten, site 802 Briescht, site 803 Triebsch, site 804 Alt Placht, site 805 Gandenitz II, site 806 Gandenitz I, site 807 Aalkasten).

4.1.4.2 Temporal variation and trends of stand nutrition

Initial nutritional status of sample stands

According to the foliar N concentrations, seven of the 28 investigated Scots pine stands suffered from extreme N deficiency in 1985 (Figures 3 and 4). For another 17 stands the N nutrition was insufficient, and only four stands were adequately supplied with N.

The numbers of stands in each N supply class in CE, the UK and NE are shown in Table 5.

Foliar P levels indicate that the P supply of the Scots pine stands was sufficient to luxurious in all cases (Figure 3). The supply of Ca, Mg, and K was sufficient to optimal, luxurious or very luxurious in most cases. Only in three cases for Ca, and one case for K, was there an insufficient supply or extreme deficiency.

Trends of stand nutrition

Table 6 shows the linear trends of various foliar parameters of Scots pine relative to the status at the beginning of the period of foliar nutrient monitoring (see also Figures 5, 6 and 7). The trends are distinguished into seven classes on a log scale (see Chapter 4.1.2.1). For four of these classes (rates of change per 10 years: $<0.625\%$, $0.625\leq1.25\%$, $1.25\leq2.5\%$, $2.5\leq5\%$) the trends are relatively weak. Three classes represent stronger trends (rates of change per 10 years: $5\leq10\%$, $10\leq20\%$, $>20\%$). The entries are sorted by geographical latitude (south to north) to facilitate the visualisation of spatial patterns.

In the following paragraphs, the trends of various foliage parameters are described and discussed with respect to the affiliation of the investigated sites to different European regions (Central Europe – CE; United Kingdom – UK; Northern Europe – NE).

The foliage unit dry mass expressed as 100-needle-weight is a widely used variable in forest nutrition studies. It is used in the analysis of time series to address

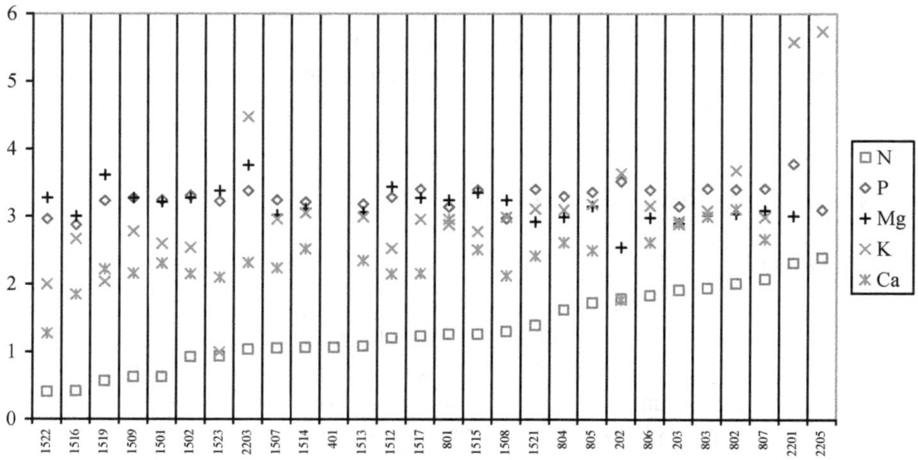

Figure 3. Nutritional status of the investigated Scots pine stands (reference year 1985). Levels according to Heinsdorf (1999): Level 1 = extreme deficiency; 2 = insufficient supply; 3 = sufficient to optimal supply; 4 = luxurious supply; >4 = very luxurious supply.

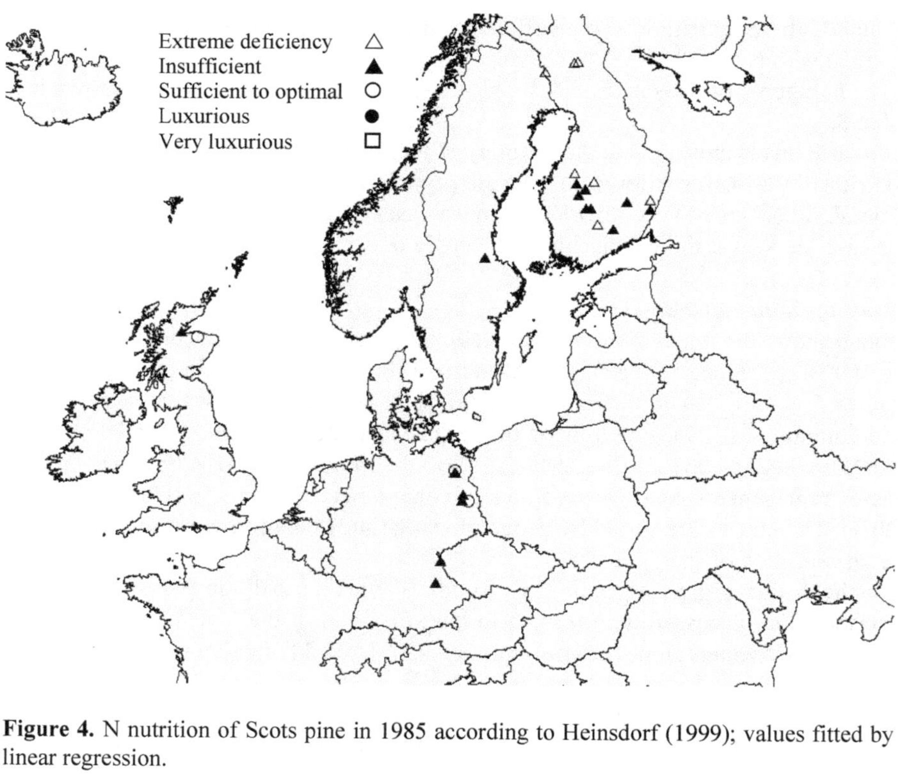

Figure 4. N nutrition of Scots pine in 1985 according to Heinsdorf (1999); values fitted by linear regression.

Table 5. N nutritional status according to Heinsdorf (1999) for Scots pine stands in Central Europe (CE), UK, and Northern Europe (NE). Numbers of sites in each N supply class.

		N nutrition class according to Heinsdorf (1999)			
		1	2	3	Total
REGION	Central Europe		7	2	9
	UK		1	2	3
	Northern Europe	7	9		16
Total		7	17	4	28

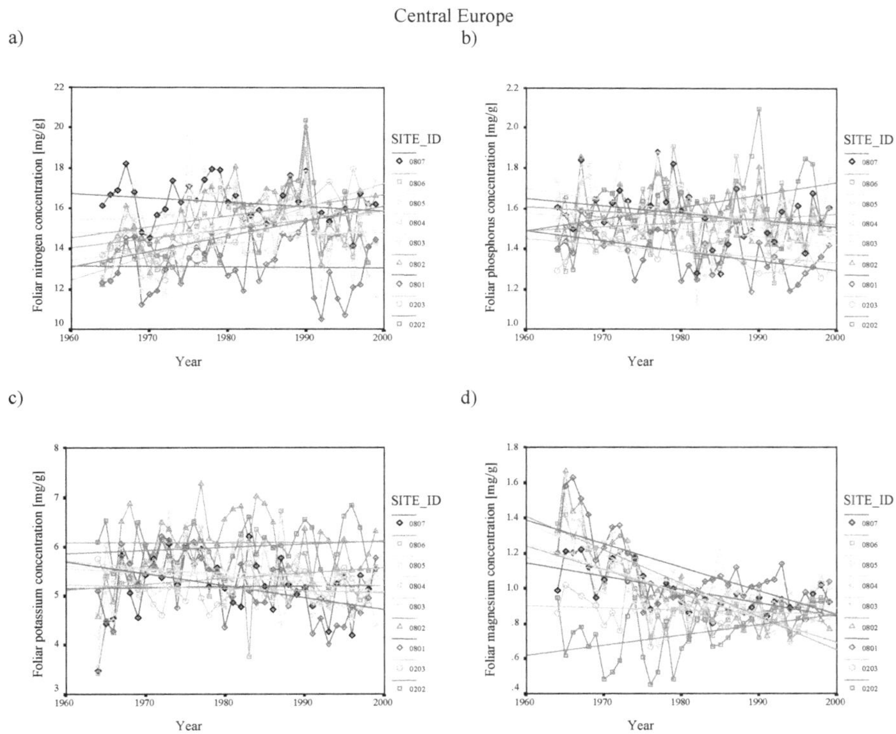

Figure 5. Time series of foliar N (a), P (b), K (c), and Mg (d) concentrations [mg/g] for the Scots pine stands included in the HDI in Central Europe. Presented are annual data and linear trends for each investigated site.

potential causes (changed supply, concentration or dilution effects) for observed changes of foliar nutrient concentrations. Dilution effects occur, when the supply of a formerly limiting element has improved, and the supply of other nutrients is not limiting, but remains constant. As a result, needles or leaves grow bigger, and the

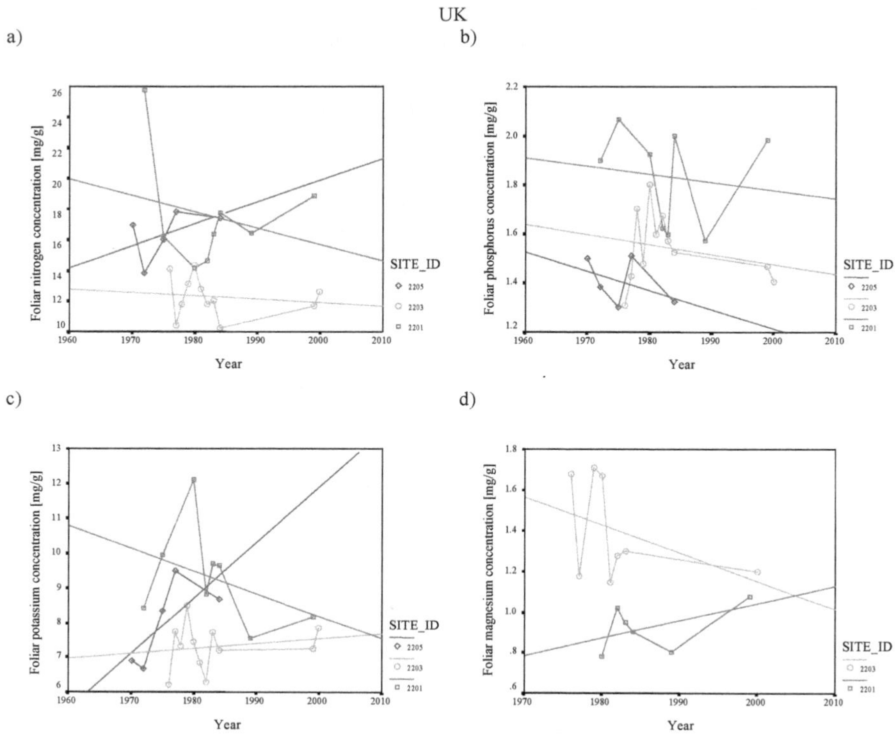

Figure 6. Time series of foliar N (a), P (b), K (c), and Mg (d) concentrations [mg/g] for the Scots pine stands included in the HDI in the UK. Presented are annual data and linear trends for each investigated site.

concentration of other nutrients in the foliage decreases. This dilution effect has often been observed in N fertilisation experiments, as foliage growth is usually strongly affected by an improved N supply.

The 100-needle weight may also be used as indicator for changes of element pools [kg ha^{-1}] in the foliage of forest trees and/or forest stands. However, an increased mass per needle may be associated with a decreased number of needles on a tree. Therefore, the 100-needle weight is only a weak indicator of changes in total nutrient element pools in trees or stands.

There was a significant increase in the 100-needle weight for seven of nine cases of sites in CE and one case in the UK. In contrast, the temporal trends of the 100-needle-weight in NE are more heterogeneous (Table 6).

In five of nine stands in CE, foliar N concentrations (N in Table 6) showed significant increases; no significant change was observed in the remaining four stands (Table 6 and Figure 5a). In the UK foliar N concentrations decreased in two stands and increased in one stand (Table 6 and Figure 6a). In NE, foliar N concentrations increased in six stands, decreased in nine stands, and in one stand remained unchanged (Table 6 and Figure 7a).

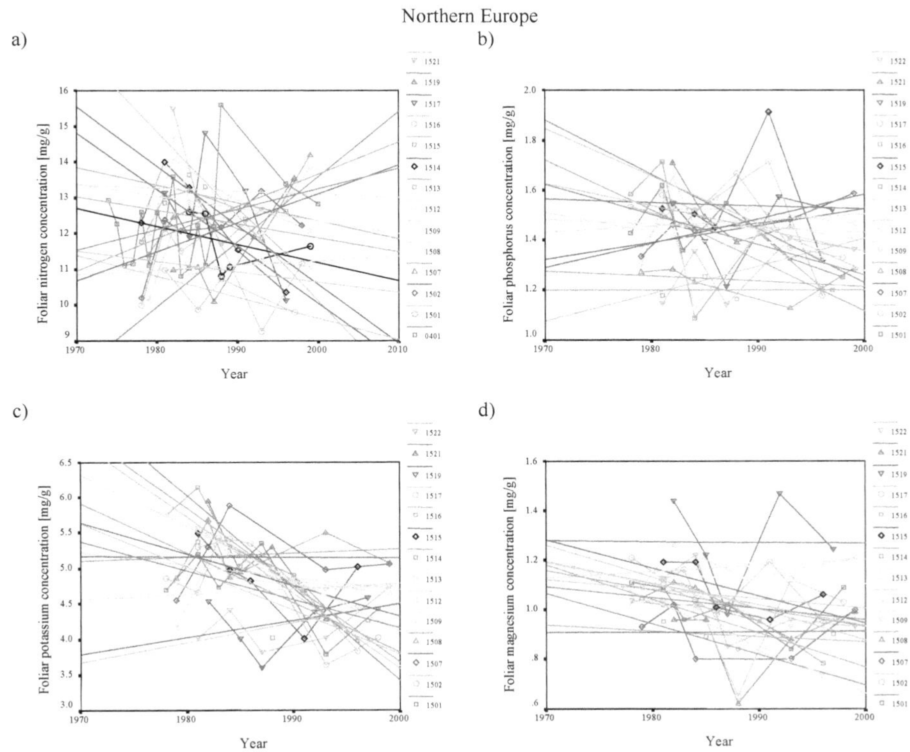

Figure 7. Time series of foliar N (a), P (b), K (c), and Mg (d) concentrations [mg/g] for the Scots pine stands included in the HDI in Northern Europe. Presented are annual data and linear trends for each investigated site.

In CE, the foliar N content (N_100 in Table 6) of most investigated Scots pine stands showed clear systematic changes (Table 6). In seven stands it increased significantly. The situation for the foliar N content of Scots pine stands in the UK and NE was heterogeneous.

The foliar P concentration (P in Table 6) in the Scots pine stands in CE and the UK did not show any strong changes during the investigation period (Figures 5b, 6b and Table 6). For NE the situation was more heterogeneous (Figure 7b and Table 6).

Foliar P contents (P_100 in Table 6) increased significantly in seven out of nine Scots pine stands in CE mainly due to the increase of needle-weight. There were no obvious patterns of change for foliar P content for the UK and NE.

Most Scots pine stands showed negative trends of foliar Mg levels (Mg in Table 6) (Figures 5d, 6d, 7d; Table 6). This decline is especially pronounced in CE, where in seven out of nine cases the decreases of foliar Mg concentrations were significant. In one case a significant positive trend was observed. In NE, significant negative trends were identified in three stands.

A general tendency to decrease was also detected for the foliar K levels (Figures 5c, 6c, 7c; Table 6) and foliar Ca levels (Table 6). The decreases in foliar K were

Table 6. Status and trends in nutrition and level of atmospheric N deposition of the investigated Scots pine stands. Table is sorted by latitude of the sites. Reg.: Region. CE: Central Europe. NE: Northern Europe. N status: N status in 1985 according to Heinsdorf (1999). G: availability of height growth data. N Dep: N deposition; data from measurement on site or EMEP modelling. w100: 100 needle weight. N, P, Mg, K, Ca: trend of foliar level according to Heinsdorf (1999). N_100, P_100 Mg_100, K_100, Ca_100: trend of nutrient content in 100 needles. Significance level of linear regression p<0.05, cell colour dark grey; p<0.1, cell colour light grey.

Site name	ID	G	Obs.	from	to	Reg.	N Status	N Dep	w100	N	N_100	P	P_100	Mg	Mg_100	K	K_100	Ca	Ca_100	N : P	N : Mg	N : K	N : Ca
Pustert	203	yes	23	1964	98	CE	■■	■■■	▲▲	▲▲▲	▲▲▲	↘↘	▲▲	↘↘	▲▲	↘	▲▲	▽	↗	▲▲	▲▲	▲▲	▲▲▲
Pfaffenwinkel	202	yes	35	1964	98	CE	■	■■■	▲▲	▲▲	▲▲▲	↗↗↗	▲▲	▲	▲▲▲	↗↗	▲▲	▽▽	↗↗↗	↗↗	↘↘↘	↗↗↗	▲▲
Briescht	802	yes	35	1964	99	CE	■■	■■■	▲▲	▲	▲▲	↘	▲	▽▽	▽	⇔	▲▲	↘↘	▲▲	▲	▲▲▲	↗↗↗	▲
Löpten	801	yes	36	1964	99	CE	■	■■■	▲▲▲	⇔	▲▲▲	↘↘	▲▲	▽	⇔	▽	▲▲	▽	▲	↗↗↗	▲▲	↗↗↗	▲
Triebsch	803	yes	35	1964	99	CE	■	■■■	▲▲	↗	▲▲	↘↘	▲▲	▽▽	▽	↘↘↘	▲▲	↘↘↘	▲	↗↗↗	▲▲▲	↗↗↗	▲
Gandenitz II	805	yes	36	1964	99	CE	■	■■■	↘↘↘	⇔	↘↘↘	↘↘	▽	↘↘↘	▽	↘↘	↘↘↘	↘	↘	↗↗	▲	↗↗	↗
Gandenitz I	806	yes	36	1964	99	CE	■	■■■	▲	▲	▲▲	↗	▲	▽	▽	↗↗↗	▲▲	⇔	▲▲	↗↗	▲▲	⇔	↗↗↗
Alt Placht	804	yes	36	1964	99	CE	■	■■■	▲▲	▲	▲▲	↘↘	▲▲	▽	↘↘	↘↘	▲▲	⇔	▲▲▲	▲	▲▲▲	↗↗↗	↗↗
Aalkasten	807	yes	36	1964	99	CE	■■	■■	⇔	↘↘	↘	↘↘	↘↘	↘↘↘	▽	⇔	↗↗↗	▲	▲	↗	▲	↘↘	▽
Rosedale	2205		5	1970	84	UK	■■	■■	▲▲▲	▲▲	▲▲▲	↘↘↘	▲▲			▲▲▲	▲▲▲			▲▲		▽▽	
Rosarie	2201		8	1972	99	UK	■■	■	▲	▽	⇔	↘	↗↗↗	▲	▲▲▲	▽	↘			↘↘↘	↗↗↗	⇔	
Black Isle	2203		8	1976	00	UK	■	■	▽▽	▽	▽▽	↘↘	▽▽	▽	▽▽▽	↗↗	▽	▲▲	↗	⇔	▲▲	↘↘↘	▽▽
Jädraås	401		17	1974	00	NE	■	■■		▲▲													
Vierumäki	1521		4	1982	93	NE	■	■	▲▲▲	▽▽▽	▲▲▲	▽	▲▲▲	▽▽	▲▲▲	▽▽▽	▲▲▲	▽▽	▲▲▲	↘↘↘	↘↘	▲▲	↗↗↗
Vesijako 26	1502		6	1978	98	NE	■	■	▽	▲▲▲	⇔	↘↘↘	▽	↘↘↘	▽	▽▽	▽▽	▽	▽▽	▲▲	▲▲	▲▲▲	▲▲
Ruokolahti 64	1512		4	1979	99	NE	■	■	▲▲▲	▽	▲▲▲	↗	▲▲▲	↗	▲▲▲	↗↗↗	▲▲▲	▽▽	▲▲▲	↘↘↘	↘↘↘	↘↘↘	▲
Punkaharju	1501	yes	6	1978	98	NE	□	□	▲▲	▽▽	▲▲	↘↘↘	▲	↘↘	▲▲▲	▽▽	▲	↘↘	▲▲▲	▲	↗↗	▲	⇔
Juva 68	1513		5	1978	96	NE	■	□	▽▽▽	▽	▽▽▽	↘↘↘	▽▽▽	▽▽	▽▽▽	▽▽	▽▽▽	⇔	▽▽▽	↗↗	▲▲▲	▲▲	↘↘↘
Kuorevesi 52	1507		5	1979	99	NE	■	■	▲▲▲	▲▲▲	▲▲▲	↗↗↗	▲▲▲	⇔	▲▲▲	⇔	▲▲▲	▲	▲▲▲	⇔	▲	▲	↗
Vilppula 53	1508		5	1979	99	NE	■	□	↗↗↗	↘↘	↗↗↗	↘↘	↗↗↗	↘↘↘	↘↘↘	↗↗	↗↗↗	▽	↘↘↘	↗	▲	↘↘	↗↗↗
Ähtäri 98	1515		5	1981	96	NE	■	□	▽	▽▽	▽▽	⇔	▽	▽	▽▽	▽▽	▽▽	↘	▽	↘	▲	▲	↗↗↗
Pylkönmäki 73	1514		5	1978	96	NE	□	□	↗↗↗	▽▽▽	▽▽	▽	▽	▽	▽▽	▽▽▽	▽▽	▽▽	▽▽	⇔	↗↗	▲	↗
Kannonkoski	1509		5	1979	99	NE	□	□	▲▲	▲	▲▲	↘	▲▲	↘↘↘	▲	▽	▲▲	▽▽	↗	↗↗	▲▲	↗↗↗	▲▲
Alajärvi 108	1517		5	1981	96	NE	■	□	▽	▽▽▽	▽▽	▽	▽▽	↘↘	▽▽	▽▽▽	▽▽▽	▽	▽▽	⇔	▽	▲	▽
Halsua 106	1516		4	1981	97	NE	□	□	▽	▽▽▽	▽▽	⇔	▽	↘↘	▽▽	▽▽▽	▽▽▽	▽▽▽	▽▽▽	▽	↘↘↘	▲	▲
Kolari 135	1519	yes	5	1982	97	NE	□	□	▽▽▽	▲▲▲	▽▽	↗↗↗	▽▽▽	⇔	▽▽▽	▲▲	▽▽▽	▲▲▲	▽▽	▲▲	▲▲	▲	▽
Sodankylä 184	1523	yes	5	1984	99	NE	□	□	▽▽	▽▽	▽▽	↘↘	▽▽	↘	▽▽	▲▲▲	↘↘	▽▽▽	▽▽▽	↘	▽	▽▽	▲
Sodankylä 161	1522		5	1981	99	NE	□	□	↘	▲▲▲	↗↗↗	▲	▲	▽	▽▽	▲▲	↗↗↗	▽▽	▽	↘↘↘	▲▲	↘↘	▲▲

rate of change per 10 years: %	pos.	neg.	%	pos.	neg.
<0.625	⇔	⇔			
0.625–<1.25	↗	↘	5–<10	▲	▽
1.25–<2.5	↗↗	↘↘	10–<20	▲▲	▽▽
2.5–<5	↗↗↗	↘↘↘	≥20	▲▲▲	▽▽▽

significance of change:
p = 0.10
p = 0.05

N status:	
extreme deficiency	□
insufficient	■
sufficient. to good	■■
luxurious supply	■■■

Annual N deposition [kg ha^{-1}]	
0-5	□
5-10	■
10-20	■■
≥20	■■■

generally more pronounced in NE than in CE. However, it is noted that particularly for K positive trends were also identified in some stands in NE.

In many cases, the negative trends of foliar base cation *concentrations* turn into positive trends, if the foliar base cation *contents* are considered (Table 6). This is true especially for K and Ca in CE, but a similar change is also frequent in NE. Here, however, shifts from negative to positive trends and vice versa were observed.

In CE a general tendency of increasing ratios of foliar N concentrations over the foliar concentrations of other nutrient elements like P, K, Mg, and Ca was present (Table 6). This temporal development is particularly pronounced for the N:Mg ratio, where the positive trend is significant in eight of nine cases. It is also noted that the ratios of foliar N concentrations over other elements also tended to increase, but there were not as many significant changes. The ratios generally showed smaller increases in NE than in CE, and also more sites with decreasing ratios were identified in NE than in CE. The three studied stands in the UK showed no clear trends for the ratios of N over P, K, Mg or Ca.

With respect to a potential effect of changes in the nutritional status of the sample stands on their growth changes, the following conclusions can be drawn:

According to their N supply and to the N deposition that they receive, the investigated stands can be attributed to two different regions:

1. stands in CE are characterised by relatively high N deposition rates (10–>20 kg $ha^{-1}yr^{-1}$), but in most cases also by insufficient supply of N; and
2. stands in NE and the UK are characterised by low N deposition rates (<10 kg $ha^{-1}yr^{-1}$) and in most cases by pronounced N deficiency – an exception is the Rosedale site in the UK, where a high (10–20 kg $ha^{-1}yr^{-1}$) N deposition rate goes along with a sufficient-optimal N nutritional status.

The nutritional trends for the two groups mentioned above can be summarised as follows:

1. *Central Europe*
   Due to a combination of poor initial site conditions (often degraded by human activity) and high atmospheric N deposition during recent decades, the N nutrition of the Scots pine stands in CE in most cases has improved considerably. Interestingly, also the foliar levels of other nutrients than N (e.g. P, K, Mg and Ca) have increased in many cases, but usually to a smaller extent than the N supply. This results in a widespread increase of the ratios of foliar N over other macronutrients. The observed trends indicate that the changes in the nutritional status of Scots pine stands in CE are mainly driven by an interaction of elevated atmospheric N input and natural recovery of the sites from former devastative land-use practices such as litter raking. Consequently, nutritional imbalances will be likely to occur in many of these stands under current conditions in the long term. Sites that are supposed to exhibit growth acceleration during the past decades due to N nutrition improvement are: site 203 Pustert, site 202 Pfaffenwinkel, site 802 Briescht, site 806 Gandenitz I, site 804 Alt Placht, site 803 Triebsch, and site 801 Löpten.
2. *Northern Europe and the UK*
   The picture of nutritional trends for the investigated Scots pine stands in NE and the UK is much more heterogeneous. A shift in nutrition probably

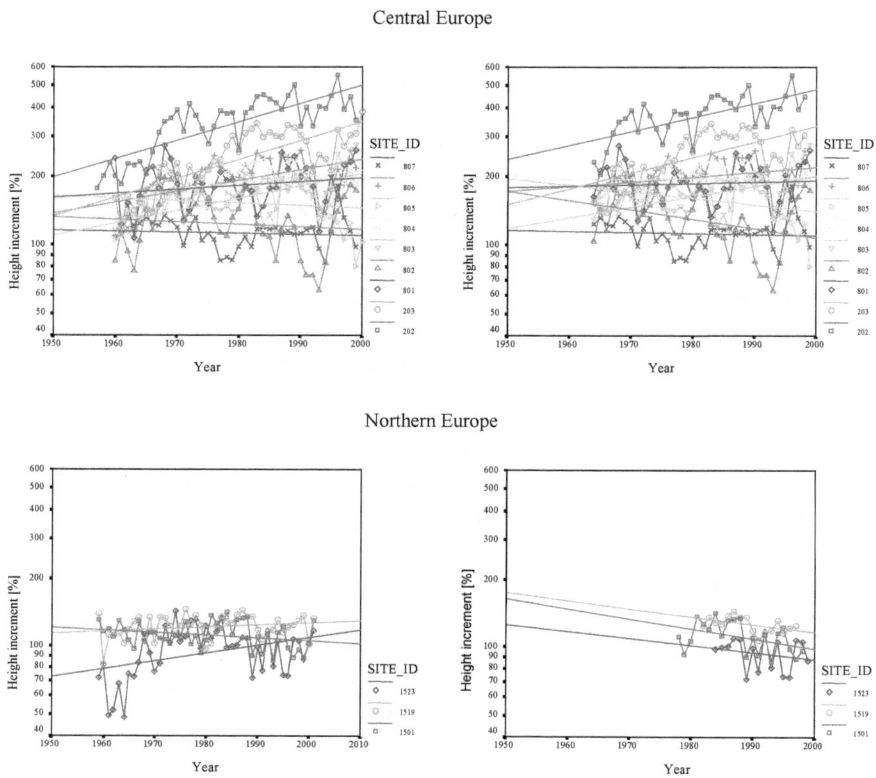

Figure 8. Time series of referenced height increments for the Scots pine stands included in the HDI. Presented are annual data and linear trends for each investigated site. The graphs on the left show annual data and linear trends for each investigated site for the entire growth data set. The graphs on the right show data and trends for the period of foliar monitoring.

induced by elevated atmospheric N deposition and by site recovery may have occurred at some sites, but evidently N is far less important as a potential driving factor for long-term growth changes of Scots pine stands in NE and the UK than in CE. Because of the limited number of observations in most regions, judgements about the situation at specific sites are uncertain.

4.1.4.3 Temporal variation and trends of referenced height increment

The time series of referenced height increment for the Scots pine stands are shown in Figure 8. A value of 100% means that the annual height increment expected according to the reference corresponds exactly to the observed height increment of the studied trees. Dominant height site index curves from regional yield tables were used as a reference (see Chapter 2.2). Referenced growth data are available from

Table 7. Trends of referenced height increments for the Scots pine stands included in the HDI. Table is sorted by latitude of the sites: CE: Central Europe; NE: Northern Europe. Trends larger than ± 0.625%/10 years ↗ or ↘; ± 1.25%/10 years ↗↗ or ↘↘; ± 2.5%/10 years ↗↗↗ or ↘↘↘; ± 5%/10 years ▲ or ▽; ± 10%/10 years ▲▲ or ▽▽; ± 20%/10 years ▲▲▲ or ▽▽▽. Significance level of linear regression $p<0.05$, cell colour dark grey; $p<0.1$, cell colour light grey.

Site ID	Site name	Nutrition obs.	from	to	Region	1951–2000	Observation period
203	Pustert	23	1964	1998	CE	▲▲▲	▲▲
202	Pfaffenwinkel	35	1964	1998	CE	▲▲▲	▲▲
802	Briescht	35	1964	1999	CE	↘↘	▽
801	Löpten	36	1964	1999	CE	↗↗↗	↗↗
803	Triebsch	35	1964	1999	CE	▲▲	▲▲
805	Gandenitz II	36	1964	1999	CE	↘↘	↘↘↘
806	Gandenitz I	36	1964	1999	CE	▲▲	▲
804	Alt Placht	36	1964	1999	CE	▲	↗
807	Aalkasten	36	1964	1999	CE	↔	↘
1501	Punkaharju 25	6	1978	1998	NE	↘↘	▽
1519	Kolari 135	5	1982	1997	NE	↗↗	▽
1523	Sodankylä 184	5	1984	1999	NE	▲	▽

1960 onwards (Figure 8a), but monitoring of the foliar nutrient levels usually started later (Figure 8b). For the CE sites the period of foliar nutrient monitoring started in 1964 (Table 7), and therefore the period for the entire growth data set and the period of foliar nutrient monitoring overlap to a large extent, and the trends in referenced height increment are similar for both periods. For the NE sites the period of foliar nutrient monitoring started much later, and the trends in referenced height increment differ markedly for the two periods (Figures 8c and 8d).

Many of the Scots pine stands in CE showed referenced height increments significantly above the level of expected growth. Six of nine stands in this region experienced height growth acceleration during the second part of the 20th century, which in five cases was significant (Table 7). Growth increases were most pronounced at site 202 Pfaffenwinkel and site 203 Pustert in southern Germany.

In contrast to CE, the referenced height increment trends of Scots pine stands in NE for the period of foliar nutrient monitoring were all negative; however, for the entire growth set (1960–2000) a significant acceleration of height growth was identified in one case. A major problem for the evaluation of Scots pine stands in NE was the limited number of observations and the short length of the period in which stand nutrition was monitored. This fact illustrates the difficulty within the HDI approach to derive sound conclusions for different European regions with incomplete and deviating databases.

4.1.4.4 Relationships among the temporal trends of nutrition, climate and tree growth

In Table 8 trends of referenced height growth and of selected growth factors are summarised and compared for those sample stands, for which height increment development has been assessed.

Table 8. Comparison of trends of the potential main factors for tree growth acceleration of Scots pine stands for which height growth measurements are available. Growth and climate trends are given for both periods, 1950–2000, and for the observation period of foliage nutrients. Abbreviation of climate variables: t: average air temperature, p: sum of precipitation. Periods for t and p are: yr: year; 59: May to September; 46: April to June. Trends larger than ± 0.625%/10 years ↗ or ↘; ± 1.25%/10 years ↗↗ or ↘↘; ± 2.5%/10 years ↗↗↗ or ↘↘↘; ± 5%/10 years ▲ or ▽; ± 10%/10 years ▲▲ or ▽▽; ± 20%/10 years ▲▲▲ or ▽▽▽. Significance level of linear regression $p<0.05$, cell colour dark grey; $p<0.1$, cell colour light grey.

		GROWTH		NUTRITION				CLIMATE											
Site name	Site ID	1950–2000	Obs. Per.	N	N_100	w100	CO_2	t yr	t yr obs	t 59	t 59 obs	t 46	t 46 obs	p yr	p yr obs	p 59	p 59 obs	p 46	p 46 obs
Pustert	203	▲▲▲	▲▲	▲▲▲	▲▲▲	▲▲	↗↗↗	↗↗↗	▲	↗↗	↗↗↗	↗↗	↗↗↗	↘	↘	↘↘	↘	↘↘↘	▽
Pfaffenwinkel	202	▲▲▲	▲▲	▲▲	▲▲▲	▲▲	↗↗↗	↗↗↗	▲	↗↗	↗↗	↗↗	↗↗↗	↘↘	↘↘↘	↘↘↘	↘↘↘	↘↘↘	▽
Briescht	802	↘↘	▽	▲	▲▲	▲▲	↗↗↗	↗↗↗	↗↗↗	↗	↗↗	↗↗	↗↗	↗	⇔	↘	↘	↗↗↗	↘
Löpten	801	↗↗↗	↗↗	⇔	▲▲▲	▲▲▲	↗↗↗	↗↗	↗↗	↗	↗	↗	↗↗	↗	⇔	↘	↘↘	↗↗	↘↘↘
Triebsch	803	▲▲	▲▲	↗	▲▲	▲▲	↗↗↗	↗↗↗	↗↗↗	↗↗	↗↗	↗↗	↗↗↗	↗	⇔	↘	⇔	↗↗↗	⇔
Gandenitz II	805	↘↘	↘↘↘	⇔	↘↘↘	↘↘↘	↗↗↗	↗↗	↗↗↗	↗	↗	↗	↗↗	↗	⇔	⇔	⇔	⇔	↘↘
Gandenitz I	806	▲▲	▲	▲	▲▲	▲	↗↗↗	↗↗	↗↗↗	↗	↗	↗	↗↗	↗	⇔	⇔	⇔	⇔	↘↘
Alt Placht	804	▲	↗	▲	▲▲	▲▲	↗↗↗	↗↗	↗↗↗	↗	↗	↗	↗↗	↗	⇔	⇔	⇔	⇔	↘↘
Aalkasten	807	⇔	↘	↘↘	↘	⇔	↗↗↗	↗↗	↗↗↗	↗	↗	↗↗	↗↗	↗↗	↗↗	⇔	↗↗↗	↗	↘↘
Punkaharju 25	1501	↘↘	▽	▽▽	▲▲	▲▲	↗↗↗	▲	▲▲▲	⇔	⇔	↗↗	↘↘	▲	▲	↗↗↗	▲	▲	↘↘
Kolari 135	1519	↗↗	▽	▲▲▲	▽▽	▽▽▽	↗↗↗	▲▲	▲▲▲	↗	↗	↗↗↗	▽▽	↗↗	↗↗	⇔	↗	↗↗	↗↗
Sodankylä 184	1523	▲	▽	▽▽	▽▽	▽▽	↗↗↗	▲	▲▲▲	↘	↘↘	↘	▽▽▽	↘↘	↘↘↘	↘↘↘	↘↘↘	↗	▲▲

Central Europe

The **N nutrition** of the Scots pine stands in CE in many cases has improved during the past four decades. Due to a coincidence of poor initial site conditions, which in many cases had been caused or aggravated by exploitative past land-use practices, and high atmospheric N deposition during the recent decades, a positive effect of the improved supply with N on growth should be expected. The positive trends of air temperature lead to the conclusion that either a positive effect on growth due to warming in the growing season (especially in spring) or a decline due to aggravation of drought may have occurred. There is no indication that changes in annual precipitation might be responsible for observed tree growth increases, since rainfall slightly decreased during the growing season. Consequently, two factors were identified (N nutrition and temperature) that may have potentially improved conditions for the growth of Scots pine in CE. Due to the fact that the time series of precipitation suggest a slight decrease of water availability during the growing season at least at some sites, a general growth acceleration cannot be expected. Indeed, the temporal trends of referenced height growth show that tree growth was not accelerated at all sites, but in two cases (site 802 Briescht and site 805 Gandenitz II) declined during the observation period. Hitherto the interaction of factors within the process of growth regulation could not be recognised. This will be discussed in the section on the results of the regression models for Scots pine.

Northern Europe

The pattern of growth trends for Scots pine stands in NE is much more heterogeneous than in CE. Only one of the three Finnish stands for which height growth was analysed, experienced a significant growth acceleration during the observation period. The only plausible factor that could explain this acceleration is the increased annual air temperature at this specific site. A general statement about the changes of growth and their regulating factors is not possible for NE on the basis of the present data.

4.1.4.5 Regression model

A. Model addressing ***general changes of Scots pine height growth*** *(all sites included)*

The main results of the models are:

1. The precipitation at the respective sites during the past years is – according to a backward variable selection procedure – the most significant influence for referenced height growth (Final Model 1). The hypothesis that nitrogen nutrition explains some of the variance of referenced height growth was tested within Model 2 with N_1_5 as the sole variable. Model 3, with Prec_1_5 and N_1_5 as predictor variables, demonstrates the dominant role of precipitation within Model Complex A.
2. The referenced height growth of the studied Scots pine stands was not significantly influenced by inter-annual or by longer-term (1 to 5 years) temperature fluctuation.
3. The model residual for a particular year is (auto-)correlated with the residual in the previous year.
4. The variability of referenced height growth is different at different sites.

Table 9. Estimates of the significant variables for the model on changes of referenced height growth of Scots pine stands in Europe (Model A1; results after variable selection). Prec_1_5 is the 5-year backward running average (linear degressive) of the precipitation sum for May to September. N_1_5 is the 5-year backward running average (linear degressive) of the foliar N concentration of current-year needles. For further explanation of the variable abbreviations see section 4.1.1.

Model		Parameter	Estimate	p-value	n
Final model	Model A1	Intercept	103.37	0.0037	434
		Prec_1_5	0.25	0.0003	
	Model A2	Intercept	92.14	0.1540	268
		N_1_5	7.32	0.0364	
	Model A3	Intercept	83.19	0.1940	261
		Prec_1_5	0.18	0.0362	
		N_1_5	4.10	0.2585	

Table 10. Estimates of the significant variable of Model B on changes of referenced height growth of Scots pine stands in Europe. N_1_5 is the 5-year backward running average (linear degressive) of the foliar N concentration of current-year needles. For explanation of the variable abbreviations see the section 4.1.1.

Model		Parameter	Estimate	p-value	n
Final model	Model B1	Intercept	90.48	0.2896	171
		N_1_5	9.92	0.0410	

*B. Model addressing **acceleration** of Scots pine height growth (only those sites where increased referenced height growth was observed are included)*

As stated in the section on Model Formulation (section 4.1.1), the more general formulation (Model A) addresses causes of tree growth *changes*. This approach was supplemented with a separate data analysis for those sites that show an increased tree growth to examine the causes of tree growth *acceleration* (Model B).

The main results of this model (Table 10) are:

1. The N supply of the investigated stands as expressed by N_1_5 (the foliar N concentrations during the past five years) is the most significant factor affecting referenced height growth.
2. Precipitation presented by Prec (current-year value) and Prec_1_5 (preceding years) was rejected as a significant influencing factor for referenced height growth within the backward selection procedure.
3. The referenced height growth of the investigated Scots pine stands was not significantly influenced by inter-annual or by longer-term (1 to 5 years) temperature fluctuation.

4. The residual for one particular year is (auto-)correlated with the residual in the previous year – i.e. the assumption of independence is violated.
5. The variability of referenced height growth is different at different sites.

Whereas the results for models A and B differ in point 1 and 2 (relevance of different growth-driving factors for stands with and without growth acceleration), they agree with regard to (the lack of) temperature effects, auto-correlation among residuals, and site diversity.

With the help of the following examples, the plausibility of model results is assessed, based on the general trends found for the different variables used in the HDI. The Model A result that precipitation influences tree growth is obvious. The example of the site Löpten (Figure 9) shows that there is a close relation between precipitation and the referenced height growth of that stand; a finding which supports the result of Model A. However, as emphasised by the results of model B, precipitation is not an adequate variable to explain tree growth acceleration (see Chapter 4.1.1.2). An example for a site with a positive trend of referenced height growth is site 203 Pustert in southern Germany. Figure 10 shows a pattern that was observed at most Scots pine stands in CE: an acceleration of tree growth is accompanied by an improvement of N nutrition (Chapter 4.1.3.4.1). At Pustert, the improvement of N nutrition is not affected by any longer-lasting or repeated severe drought. On the other hand, N nutrition and tree growth at site 202 Pfaffenwinkel in southern Germany, was significantly reduced by repeated droughts in the 1990s (Figure 11). Figures 9 and 10 support our hypothesis that within the interaction between the strongly fluctuating precipitation and the general improvement of N nutrition, the extent of acceleration of referenced height growth is controlled by water availability.

The interaction between the growth factors water and N nutrition is most obvious for these examples (site 801 Löpten, site 203 Pustert, and site 202 Pfaffenwinkel); however, even for these examples the general relationship could not be confirmed statistically. The reason for the lack of a statistical proof is probably that the data set was too small for the examination of interactions. Additionally, it must be stated that the described patterns and interactions among the factors N nutrition, water supply, and tree growth were not always as clear as in the examples shown above.

At first glance it is surprising that we did not detect any influence of temperature on tree growth, although many study sites experienced a temperature increase during recent decades (Tables 1 and 2). The reason for the absence of a significant temperature effect on tree growth at the study sites is probably the ambivalence of the factor temperature with regard to growth. On the one hand, an increased temperature may result in a prolonged growing season and reduced temperature limitation on tree growth and humus mineralisation. On the other hand, it causes increased plant respiration, increased evapotranspiration stress and drier soil conditions. Depending on specific site conditions (temperature and precipitation levels as well as soil water-storing capacity), increased air temperature may thus result in insufficient soil moisture for optimal humus mineralisation and/or water uptake by the trees, and thus in reduced tree growth. The complex site-specific interaction of growth-promoting and growth-reducing effects of increased air temperature and the limitation of the data set are probably the major reasons for the lack of any clear temperature effect on tree growth at the studied sites.

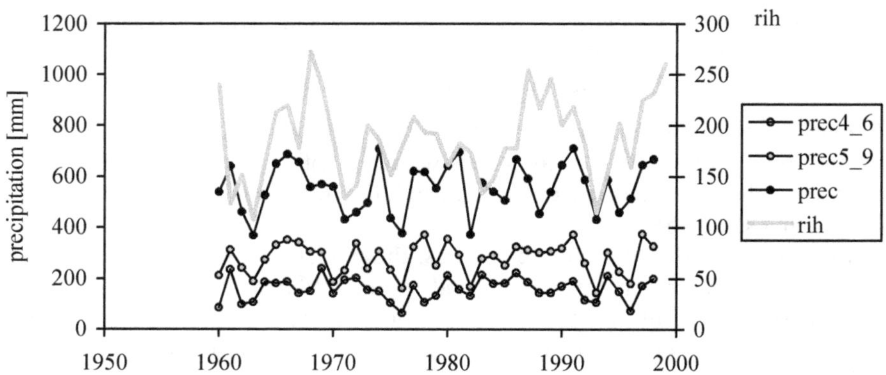

Figure 9. Time series of annual precipitation (prec), precipitation during the growing season (May-September) (prec5_9) and during spring (April-June) (prec4_6) and referenced height increment (rih) at site 801 Löpten.

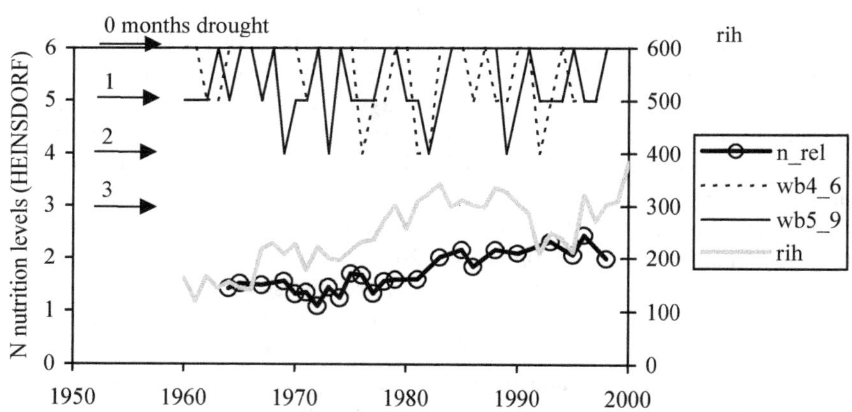

Figure 10. Time series of drought (number of months with negative water balance) during the growing season (May–September) (wb5_9) and during spring (April–June) (wb4_6). Levels of N nutrition according to Heinsdorf (1999) (n_rel) and referenced height increment (rih) at site 203 Pustert. Heinsdorf Levels: 1: extreme deficiency; 2: insufficient supply; 3: sufficient to optimal supply; 4: luxurious supply; >4: very luxurious supply.

4.1.5 Norway spruce (*Picea abies*)

4.1.5.1 General remarks

Some 23 Norway spruce sites fulfilled the requirements for an analysis within the HDI approach (Chapter 2.1). The sites are located in Austria (3), Czech Republic (1), Denmark (4), Finland (9), France (2), Germany (3) and Sweden (1). Data

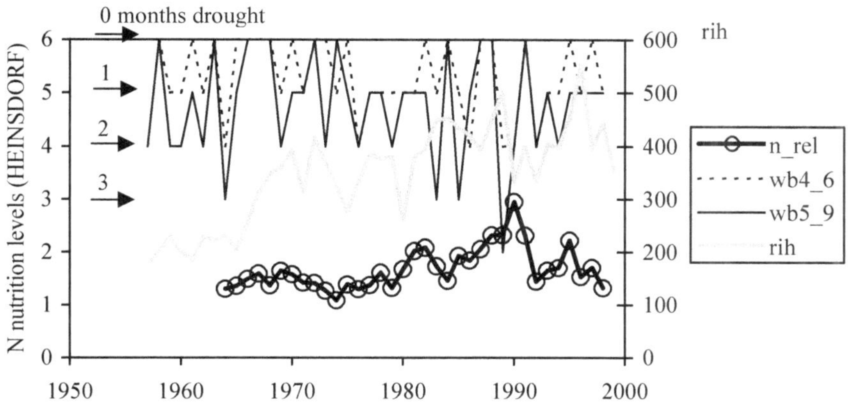

Figure 11. Time series of drought (number of months with negative water balance) during the growing season (May–September) (wb5_9) and in spring (April–June) (wb4_6). Levels of N nutrition according to Heinsdorf (1999) (n_rel) and referenced height increment (rih) at site 202 Pfaffenwinkel. Heinsdorf Levels: 1: extreme deficiency; 2: insufficient supply; 3: sufficient to optimal supply; 4: luxurious supply; >4: very luxurious supply.

regarding changes of tree growth were available for three sites in Finland (site 1510 Kannonkoski 58, site 1524 Kemijärvi 194, site 1525 Sodankylä 197), two sites in Germany (site 201 Höglwald, site 301 Pfalzgrafenweiler), and two sites in Austria (site 601 Helfenberg, site 602 Karlstift).

4.1.5.2 Temporal variation and trends of stand nutrition

Nutritional status of the sample stands

Foliar N values indicate that the investigated stands of Norway spruce in 1985 were generally well supplied with nitrogen: the N supply was sufficient to luxurious in 21 out of 23 stands (Heinsdorf, 1999). The N supply was insufficient in only two stands (Figure 12). The numbers of stands in each N supply class in CE and NE are shown in Table 11.

The foliar P levels for all stands are at least sufficient or better in all stands, and in all but three stands are luxurious or very luxurious. The supply of base cations (Ca, Mg, K) is at least sufficient in all but two stands (Figure 12). Ca supply was insufficient in one stand (site 601 Helfenberg), and K supply was insufficient in one stand (site 2033 Dalgas).

Trends of stand nutrition

There were only 11 stands (two in CE and nine in NE) where data is available describing temporal changes of the 100-needle weight of Norway spruce (Table 12). When comparing tendencies in the time series it has to be taken into account that the length of the data series differs considerably between sites, and that these

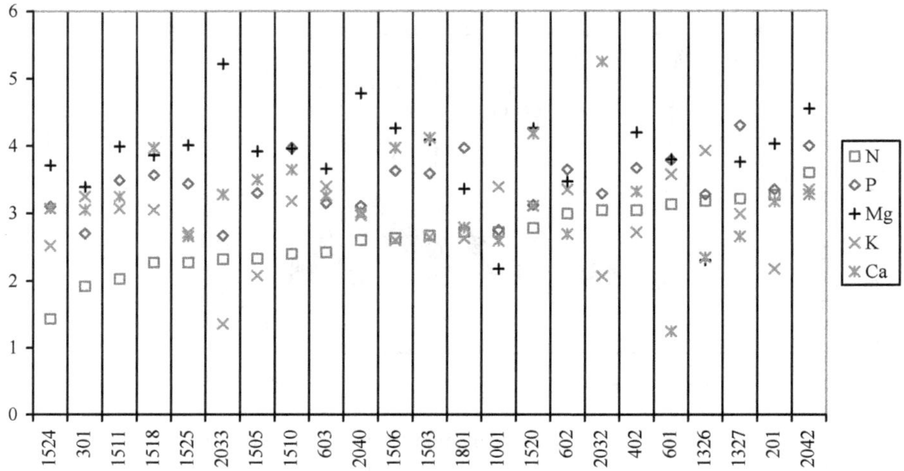

Figure 12. Nutritional status of 23 European Norway spruce stands (reference year 1985) according to Heinsdorf (1999). Heinsdorf Levels: 1: extreme deficiency; 2: insufficient supply; 3: sufficient to optimal supply; 4: luxurious supply; >4: very luxurious supply.

Table 11. N nutrition of Norway spruce stands according to Heinsdorf (1999) in Central Europe (CE) and Northern Europe (NE). Numbers represent numbers of investigated stands in each nutrition class.

		N nutrition class according to HEINSDORF (1999)			
		2	3	4	Total
REGION	Central Europe	1	4	4	8
	Northern Europe	1	10	3	15
Total		2	14	7	23

comparisons are based on a small number of observations. The two sites in CE with available data on the 100-needle weight indicate positive trends, which in one case is statistically significant. In NE, decreases (not significant) of the 100-needle weights predominate (six of nine cases). On the other three sites in NE the 100-needle weights increased (in one case significantly).

Although no significant negative trends were identified for the nitrogen nutrition (foliar N concentration) of Norway spruce stands in CE, the development is characterised rather by a general tendency of decrease than of increase (Table 12, Figure 14). The only Norway spruce stand in CE exhibiting a positive trend is at site 1001 Solling (Germany), where the N concentrations increased significantly between 1968 and 1997. In NE, the trends of foliar N concentration were negative for 11 of the

Figure 13. N nutrition of Norway spruce stands according to Heinsdorf (1999); average nutrition in 1985 values fitted by linear regression.

14 sites (Table 12, Figure 15). There were increases (but not significant) in foliar N concentration for two Danish sites, and a slight increase at one Swedish site.

Due to the limited number of available 100-needle weight data, the foliar N contents (N_100, the product of 100-needle weight and N concentration) could only be calculated for a few sites. The results for site 201 Höglwald and site 301 Pfalzgrafenweiler in CE (Table 12) indicate that the foliar N contents of both stands increased during recent decades and thus developed in a similar way to Scots pine (see section 4.1.4.2). A different pattern was observed for Norway spruce stands in Finland, where the foliar N contents decreased markedly or remained at the same level.

The temporal development of phosphorus nutrition for the studied Norway spruce stands in CE closely resembles the variation of the nitrogen supply (Table 12, Figure 14). Compared to nitrogen, the observed negative trends for phosphorus were more pronounced and often significant. Only in one case (site 1001 Solling), did the foliar P concentration increase. In contrast, the Norway spruce stands in NE show a more heterogeneous pattern.

For both Norway spruce stands in CE (Höglwald and Pfalzgrafenweiler), for which 100-needle weight data were available (Table 12), the observed increases of this parameter turn the generally negative trend of foliar P concentrations into strong positive trends of foliar P content (P_100, the product of foliar P

Central Europe

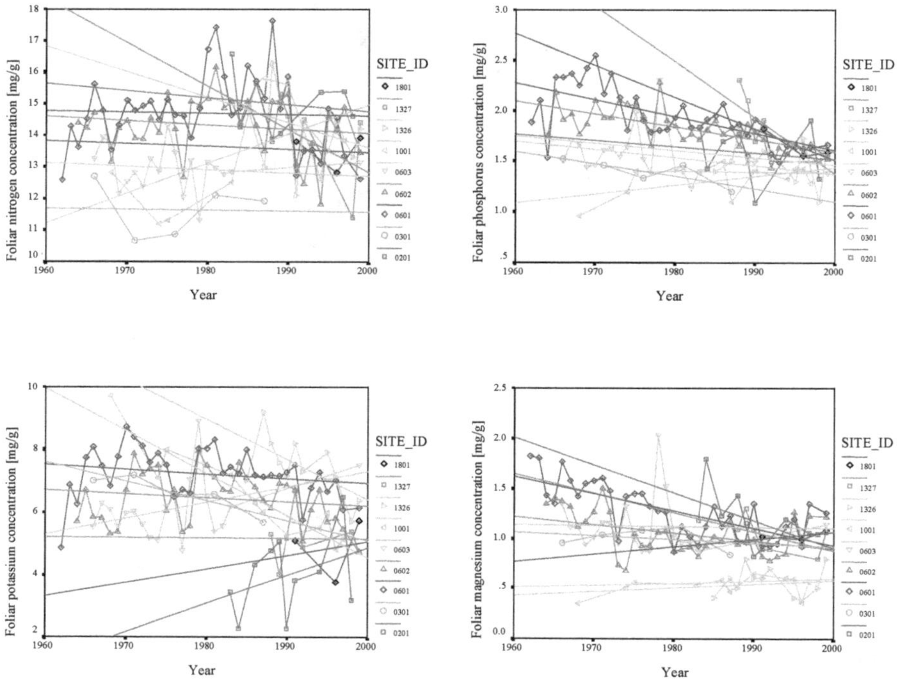

Figure 14. Time series of foliar N, P, K, and Mg concentrations [mg/g] for the Norway spruce stands included in the HDI in Central Europe. Presented are annual data and linear trends for each investigated site.

concentration and needle weight). This is a clear indication that the observed decreases of foliar P concentrations were caused by dilution effects and that in most cases P was not limiting. Also for the Norway spruce stands in Finland (Table 12, Figure 4) with available 100-needle weights, in three of nine cases a decline of foliar P concentrations was associated with increasing foliar P contents. Nevertheless, negative (but not significantly so) trends of foliar P contents predominate in Finland (five of nine cases).

The investigated Norway spruce stands in CE and NE showed a heterogeneous pattern with regard to foliar Mg concentrations (Table 12, Figure 14, Figure 15). The status and the temporal variation of Ca, Mg, and K nutrition of Norway spruce stands in Europe seem to be predominantly governed by local factors.

For both Norway spruce stands in CE, for which 100-needle weight data were available, a comparison of the trends of foliar Mg and K contents with the respective trends of foliar Mg and K concentrations indicates the presence of dilution effects. In contrast to CE, both dilution and concentration effects were observed for the Norway spruce stands in Finland (Table 12).

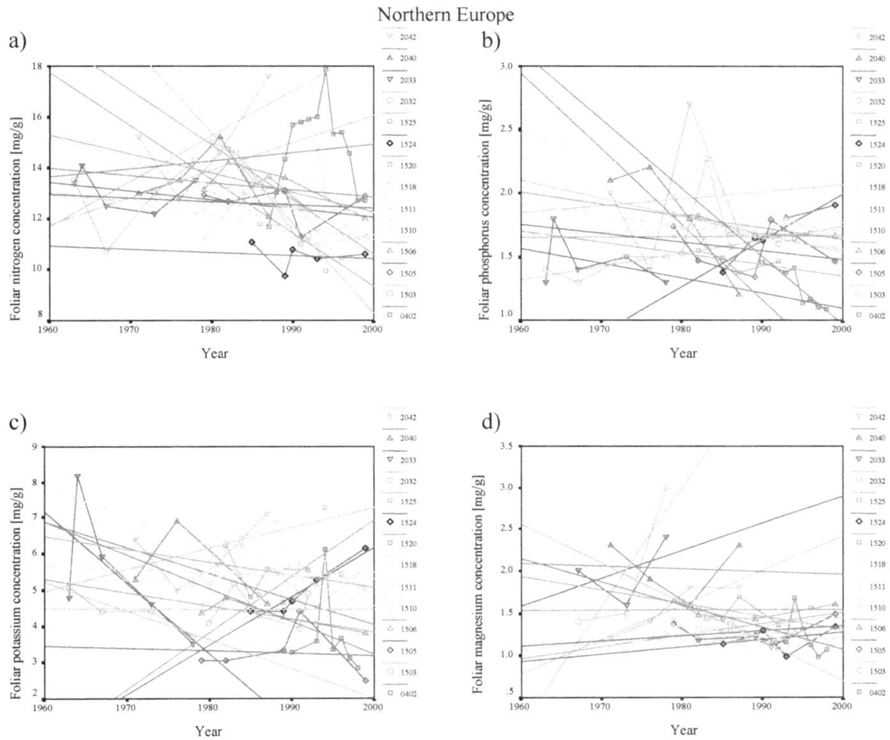

Figure 15. Time series of foliar N (a), P (b), K (c), and Mg (d) concentrations [mg/g] for the Norway spruce stands included in the HDI in Northern Europe. Presented are annual data and linear trends for each investigated site.

As a general pattern, a trend of an increasing N:P ratio in the foliage was identified for almost all investigated Norway spruce stands in CE: the N:P ratio increased in eight out of nine cases (three cases significantly so), and decreased slightly in only one case. There were also more stands with an increasing trend for N:Mg and N:Ca ratios. However, it should be noted that Norway spruce stands with pronounced negative trends were also observed in CE. For the N:K ratio the number of stands with increasing or decreasing trend was balanced.

With regard to the N:P, N:K, and N:Ca ratios in the foliage of Norway spruce in NE the temporal trends are different (Table 12): Whereas increases in these ratios prevail in the Danish and Swedish stands, declines prevail in Finland. Declines in the N:Mg ratio prevail at all sites in NE.

With respect to a potential contribution of changes in the nutritional status of the investigated Norway spruce stands to observed growth changes, the following conclusions can be drawn:

- Norway spruce stands in CE are usually characterised by relatively high N deposition rates and they are well supplied with N, whereas

Table 12. Status and trends of nutrition and level of atmospheric N deposition of the investigated Norway spruce stands. Table is sorted by latitude of the sites. N Dep: N deposition, data from measurement on site or EMEP modelling; w100: 100 needle weight, N, P, Mg; K, Ca: trend of nutrition level according to Heinsdorf (1999); N_100, P_100 Mg_100, K_100, Ca_100 = Trend of nutrient content in 100 needles; N status 1985 according to Heinsdorf (1999).

Site name	Site ID	G	Obs.	from	to	Region	N Status	N Dep	w100	N	N_100	P	P_100	Mg	Mg_100	K	K_100	Ca	Ca_100	N : P	N : Mg	N : K	N : Ca
Dobrowa	603		19	1966	1999	CE	■■	■■		↘↘		↘↘↘		↔		↗↗↗		↗↗		▲	↗	▽	▽
Bonhomme	1326		11	1985	1999	CE	■■■	■■		▽		↘↘↘		↗↗↗		▽		↘↘		↗↗	▽	▲▲	↘
Gemaingoutte	1327		7	1988	1999	CE	■■	■■		▽▽		▽▽▽		▽		↔		▽		▲▲▲	▲	▽▽	▲
Höglwald	201	yes	8	1983	1998	CE	■■■	■■■	▲▲	↘↘↘	▲▲	↘↘↘	▲▲▲	▽▽	▲▲	▲▲▲	▲▲▲	↘	▲▲	↗↗↗	▲▲	▽▽▽	↗↗
Pfalzgrafenweiler	301	yes	5	1966	1987	CE	■	■■■	▲	↔	↗↗↗	▽	▲	↘↘	▲▲▲	▽	▲	↘↘↘	▽	▲▲	▲	▲▲	▲
Helfenberg	601	yes	37	1962	1999	CE	■■■	■■■		↔		▽		▽		↘		↘↘		▲	▲▲	↗↗	↗↗↗
Karlstift	602	yes	35	1962	1999	CE	■■	■■■		↘↘		▽		↘↘↘		↘		▽		▲	▲	↗	▲▲
Rajec	1801		3	1991	1999	CE	■■	■■■		↘↘		▽▽		↗↗↗		▲		▲▲▲		▲▲▲	▽	▽	▽▽▽
Solling-Fichte	1001		11	1968	1997	CE	■■	■■■		▲▲		▲		▲		▽		▽		↘	↔	▲▲▲	▲▲▲
Sdromme	2040		4	1971	1987	NE	■■	■■		↘↘↘		▽▽▽		↘		▽		▽▽		▲▲▲	↗↗	▲	▲▲▲
Djbplantage	2042		4	1971	1987	NE	■■■	■■		▲▲▲		↗↗		▲▲		▲		▽		▲▲	▽▽	↗↗	▲▲▲
Dalgas	2032		4	1963	1978	NE	■■■	■■		▲▲		▲		▲▲▲		▽▽		▲▲▲		↗	▽▽▽	▲▲▲	▽▽▽
Dalgas	2033		5	1963	1978	NE	■■	■■		▽		▽		▲		▽▽▽		↘↘↘		↗↗↗	▽	▲▲▲	▲▲
Skogaby	402		13	1987	1999	NE	■■■	□		↗↗↗		▽▽▽		▽		▽▽		▽▽▽		▲▲▲	↔	↔	▲▲▲
Heinola 113	1518		8	1978	1998	NE	■■	■	↘↘↘	↘↘↘	↘↘↘	↗	▽	↗↗↗	▲	↗	↘↘↘	↘↘	▽	↘↘	▽	↘↘	↗↗↗
Sahalahti 138	1520		4	1982	1992	NE	■■	■	▲▲	▽▽▽	↘↘↘	↘↘↘	▲	↘↘↘	↗	↘↘↘	▲	↘↘↘	↔	▽	▽	▽	↘↘↘
Moinniemi 49	1506		4	1979	1999	NE	■■	■	▲	▽▽	↔	↘↘↘	↗	↔	▲	▽	↘↘↘	▲	▲▲▲	↔	▽	↗↗↗	▽▽
Patasalo 31	1503		5	1980	1992	NE	■■	□	▲▲▲	▽▽▽	↘↘↘	▽	▲▲	▽▽	▽▽	↔	▲▲▲	↔	▽	▽▽	▲	▽▽▽	▽▽▽
Sulkava 34	1505		5	1979	1999	NE	■■	□	▽	↘↘	▽	↘↘↘	▽▽	↗↗	↘↘	↘↘	▽▽	↗↗	↘↘↘	↗↗↗	↘↘↘	▲	▽
Kannonkoski 60	1511		5	1979	1999	NE	■	□	▽	▽	▽	↘↘↘	▽▽	↗↗↗	↗	↘	▽	▲	▲▲	↗↗	▽▽	↘↘	▽▽
Kannonkoski 58	1510	yes	7	1979	1999	NE	■■	□	▽▽	▽▽	▽▽	▽	▽▽▽	↘↘↘	▽▽	▽	▽▽▽	▲	↘↘↘	↗↗↗	↔	▲	▽▽
Kemijärvi 194	1524	yes	5	1985	1999	NE	■	□	▽▽	↘↘↘	▽▽	▲▲	▲▲	↗↗↗	↘↘↘	▲▲▲	▲▲	▽▽	▽▽▽	▽▽▽	▽	▽▽▽	▲▲▲
Sodankylä 197	1525		4	1986	1996	NE	■	□	▽▽▽	▽▽▽	▽▽▽	↔	▽▽▽	↗↗↗	▽▽▽	▲▲▲	▽	▲▲▲	↗↗	▽▽	▽▽▽	▽▽▽	▽▽▽

rate of change per 10 years:	%	pos.	neg.	%	pos.	neg.
	<0.625	↔	↔			
	0.625–<1.25	↗	↘	5–<10	▲	▽
	1.25–<2.5	↗↗	↘↘	10–<20	▲▲	▽▽
	2.5–<5	↗↗↗	↘↘↘	≥20	▲▲▲	▽▽▽

significance of change
p = 0.10
p = 0.05

N status	
extreme deficiency	□
insufficient supply	■
sufficient to good	■■
luxurious supply	■■■

Annual N deposition [kg ha^{-1}]	
0–<5	□
5–<10	■
10–<20	■■
≥20	■■■

- Norway spruce stands in Finland generally receive low N deposition, but are nevertheless in most cases adequately supplied with N. The stands in Denmark and Sweden resemble the Central European stands.

The nutritional trends can be summarised as follows:

Central European and Danish stands
At first glance, the dominant negative trend of foliar nitrogen concentrations in Norway spruce stands in CE is surprising, since the high atmospheric N deposition in these stands should result in an improved N supply. According to the observed increases of foliar N contents this prevailing trend of decreasing foliar N concentrations is most likely due to dilution. This is an indication that N was not limiting growth at these sites. An even more pronounced dilution effect is indicated by increasing foliar N:P, N:Mg, N:K and N:Ca in many cases. Presumably, increased deposition loadings and recovery from former devastative land use changed the nutritional status of most Norway spruce stands in CE. However, this does not necessarily lead to growth acceleration in the investigation period, since the N supply of the studied stands probably was sufficient and did not significantly limit growth already at the start of the foliar N assessments.

Finnish stands
The heterogeneous pattern of nutrient trends in Norway spruce stands in Finland, the usually adequate N-status, and the absence of a clear trend of improving nitrogen nutrition make it unlikely that changes in nitrogen nutrition are responsible for growth changes of these stands.

4.1.5.3 Temporal variation and trends of referenced height growth

Figure 16 and Table 13 show the time series of referenced height increment of Norway spruce in NE and CE. The earliest referenced growth data are available from 1960 (Figure 16a) but the foliar analysis usually started only in the 1960s (CE) or 1980s (NE) (Figure 16b). As with Scots pine, there are also differences between the trends of both periods for Norway spruce, but they are relatively small. At one Finnish site (Sodankylä 1525) there was a positive trend for the whole data set, and a negative trend for the period of foliar nutrient monitoring.

In CE, three of four investigated Norway spruce stands grew significantly faster than was expected according to yield tables (Table 13). During the second half of the 20th century, the most pronounced increase and a high general level of referenced height increment was detected for Pfalzgrafenweiler (site 301) in southern Germany. The two Austrian sites (Helfenberg (site 601) and Karlstift (site 602)) show also considerable acceleration of height growth, but at different levels. The overall trend for height increment of the Höglwald stand (site 201) is slightly negative.

The height increment of the three Norway spruce stands in Finland varies approximately at the expected level for the period 1960–1985. After that period, height growth seems to increase in one stand (Sodankylä 197), whereas it decreases at the other two stands (Kannonkoski 58 and Kemijärvi 194). If one considers only the period of foliar nutrient monitoring, all stands in Finland exhibit a decline of referenced height increment.

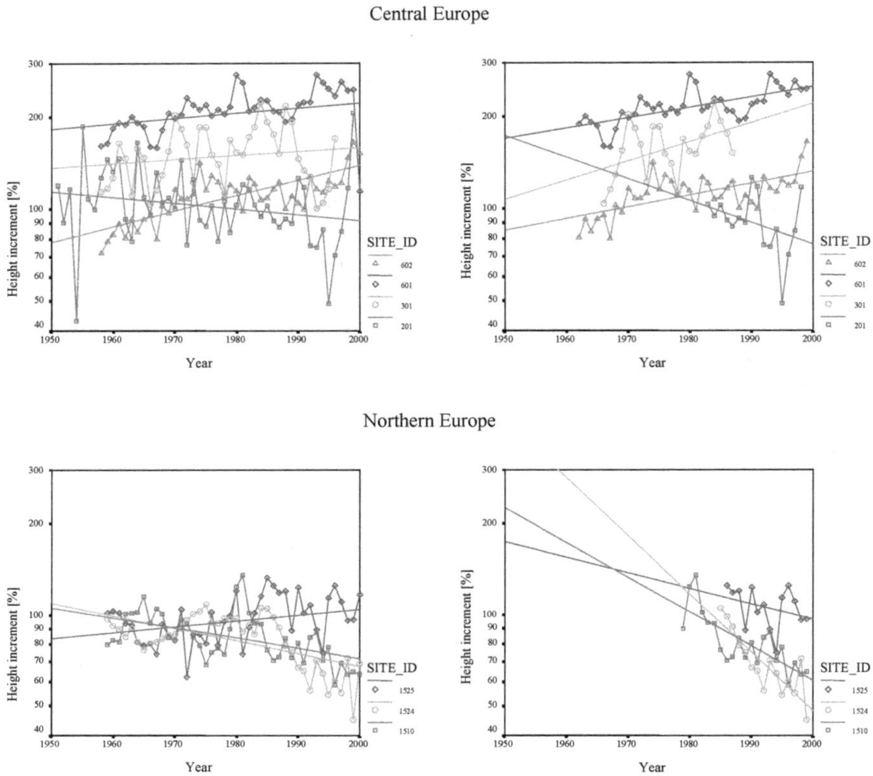

Figure 16. Time series of referenced height increments for the Norway spruce stands included in the HDI. Presented are annual data and linear trends for each investigated site. The graphs on the left show annual data and linear trends for each investigated site for the entire growth data set. The graphs on the right show data and trends for the period of foliar monitoring.

Table 13. Trends of referenced height increments for the Norway spruce stands included in the HDI. Table is sorted by latitude of the sites: CE: Central Europe, NE: Northern Europe. Trends larger than ± 0.625%/10 years ↗ or ↘; ± 1.25%/10 years ↗↗ or ↘↘; ± 2.5%/10 years ↗↗↗ or ↘↘↘; ± 5%/10 years ▲ or ▽; ± 10%/10 years ▲▲ or ▽▽; ± 20%/10 years ▲▲▲ or ▽▽▽.

Region	Site ID	Site name	Nutrition obs.	from	to	1951–2000	Observation period
CE	201	Höglwald	8	1983	1998	↘↘↘	▽▽
CE	301	Pfalzgrafenweiler	5	1966	1987	▲▲	▲▲▲
CE	601	Helfenberg	37	1962	1999	▲	▲
CE	602	Karlstift	35	1962	1999	▲▲	▲▲
NE	1510	Kannonkoski 58	7	1979	1999	▽	▽▽▽
NE	1524	Kemijärvi 194	5	1985	1999	▽	▽▽▽
NE	1525	Sodankylä 197	4	1986	1996	▲	▽▽

4.1.5.4 Relationships among the temporal trends of nutrition, climate and tree growth

As already stated above, the observed positive temperature trends may result either in a positive effect on tree growth due to a prolonged and warmer growing season, or in a negative effect due to an enhanced likelihood and duration of periods with insufficient water supply of the stand (Figure 1).

Of the four sites in CE where growth analysis was conducted, annual precipitation decreased at two sites (Höglwald and Karlstift), or increased slightly (Pfalzengrafenweiler) or remained unchanged (Helfenberg). Spring precipitation increased only at Pfalzgrafenweiler (see Table 3).

None of the stands where annual height growth had been measured (Figure 16 and Table 13) showed positive trends of N nutrition as indicated by increased foliar N concentrations (Table 12). However, at least for the stands Pfalzgrafenweiler and Höglwald, where information about 100-needle weight is also available, it was shown that 100-needle weights and foliar N pools increased. The absence of an additional increase of foliar N concentrations for these stands is due to dilution effects. It is likely that a similar increase in foliar N contents has also taken place at the other Norway spruce stands in CE. However, we were not able to prove this effect due to the lack of 100-needle weight data. The site-specific conditions and changes of growth factors result in different growth reactions. Growth acceleration was observed in the case of improved or at least constant climate factors (1), and unchanged or decreasing growth in case of deterioration of other factors (2). In the case of Pfalzgrafenweiler, an increased supply of N in combination with a slight increase in air temperature and precipitation is probably the reason for the observed growth acceleration. In contrast, at the Höglwald stand, height growth fluctuated strongly, but did not exhibit a systematic positive trend due to an increased frequency and duration of droughts in the 1990s. As there are no needle mass data available for Helfenberg and Karlstift, a rigorous assessment of the causes of the long-term growth developments is not possible for either of the stands. The observed growth acceleration at Karlstift may have been caused solely by temperature increases, but an improved N supply due to increased atmospheric N input and recovery from former devastative land use may additionally have supported the positive growth trend.

4.1.5.5 Regression model

For Norway spruce, no significant results could be found neither with the single site models nor with the Mixed Model procedure.

Even though the absence of any statistically significant influencing factor for the observed long-term changes of Norway spruce growth would not have been expected from an ecological point of view, this result is plausible due to the fact that also the graphical assessment of time series did not indicate any general control factor or any interaction of factors. Additionally, with exception of the two Austrian sites (Karlstift and Helfenberg) the investigated Norway spruce stands generally had shorter observation periods and a smaller number of observations than the studied Scots pine stands. For that reason the chances of identifying influencing factors for the observed growth changes were already considerably smaller for Norway spruce than for Scots pine.

4.1.6 Beech (*Fagus sylvatica*)

4.1.6.1 General remarks

For beech, data from five stands were available for analysis within the HDI. Four of these stands are situated in eastern France, and one (Solling) is located in Germany. Unfortunately, for none of these stands data about temporal changes of referenced height growth had been made available within the period of our evaluation. Due to the lack of available referenced height growth data, there was no analysis of temporal variation of referenced height growth, or the effects of temporal variation of nutrition and climate on tree growth.

4.1.6.2 Temporal variation and trends of nutrition

Nutritional status of the sample stands

Nitrogen nutrition in 1985 was at least sufficient in all five stands of beech (Figure 17). Insufficient supply was identified for Ca (two cases), Mg (three cases), and P (three cases). Consequently, it must be hypothesised for these stands that further improvement of N nutrition will not necessarily lead to increased growth, but may cause nutrient imbalances between N and base cations or P.

Trends of stand nutrition

In four of five stands (Table 14), the N nutritional status of beech improved considerably during the monitoring period. P nutrition decreased at all sites, and at most sites the supply of base cations also decreased. These trends are reflected in the trends of the ratios of N over other nutrients, which in all cases are positive. The conclusion with regard to a possible contribution of nutritional changes to any growth changes of the investigated beech stands is that an additional input of N is likely to cause nutrient imbalances which may retard tree growth rather than accelerate it.

4.1.7 General discussion of historical developments

4.1.7.1 Methodology

In the Historical Development Investigation (HDI) of Recognition, retrospectively assessed referenced height growth data of dominant and predominant trees in Scots pine, Norway spruce, and beech stands dispersed over a large part of Europe were correlated with long-term data series of foliar nutrient levels and measured or modelled climate data in order to determine whether recent changes of height

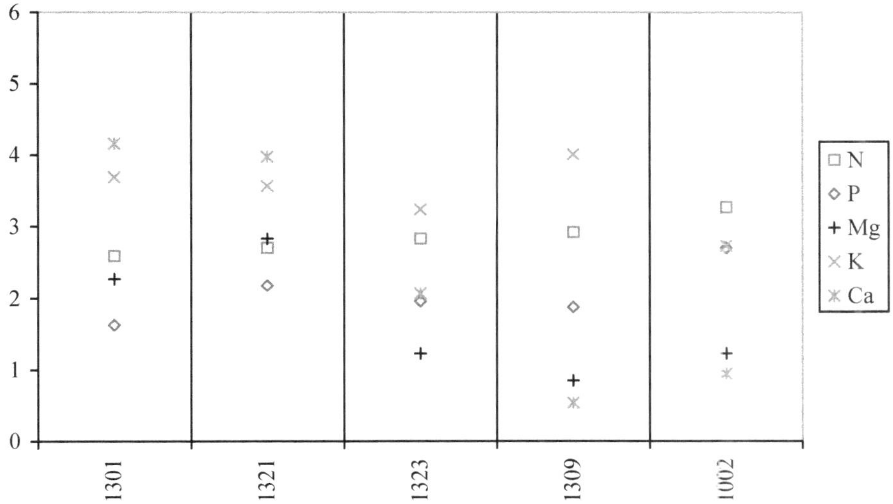

Figure 17. Status of nutrition of beech stands according to Heinsdorf (1999). Fitted values for 1985; Level: (1) extreme deficiency, (2) insufficient supply, (3) sufficient to optimal supply, (4) luxurious supply, (>4) very luxurious supply.

growth can be attributed to recent changes in the nutritional status of these tree species and changes of temperature and precipitation at the respective sites on a European scale. A similar investigation was conducted for southern Germany by Heitz et al. (2001), who had within the small frame of the project the possibility to complete at least some of the missing data. On the European-wide scale of Recognition, this was possible only for a very limited number of variables. This is especially true for the variable referenced height growth, which could be assessed only for a part of the stands included in the HDI (stem analysis).

Results from investigations like HDI, which are based on heterogeneous sources of data, and where stand nutrition has been assessed by different research groups with different methods and for different time intervals, must be interpreted with caution. The main problems concerning the representativity and comparability of the results, and potential sources for misinterpretations are:

- Most forest stands studied in the HDI approach comprise control plots of long-term (old) fertilisation experiments. These experiments had often been installed at particularly poor sites and for tree species which were usually cultivated in such situations. This meant that of the stands included in HDI, most were Scots pine stands, there were fewer Norway spruce stands, and only a few beech stands.
- The data sources are often unbalanced – i.e. they comprise different numbers of observations, different times of commencement, and a different length of the periods of foliar nutrient monitoring, and/or different observation intervals.

Table 14. Status and trends of nutrition and level of atmospheric N deposition of the investigated beech stands. Table is sorted by latitude of the sites. G: Height growth data. N Dep: N deposition, data from measurement on site or EMEP modelling; N status □ very poor, ■ poor, ■■ medium, ■■■ rich. N, P, Mg, K, Ca: trend of nutrition level according to Heinsdorf (1999). Trends stronger than ± 0.625%/10 years ↗; ± 1.25%/10 years ↗↗; ± 2.5%/10 years ↗↗↗; ± 5%/10 years ▲; ± 10%/10 years ▲▲; ± 20%/10 years ▲▲▲; significance level of linear regression: p<5% cell colour dark grey, p<10% light grey.

Site name	Site ID	G	Obs.	from	to	Region	N Status	N Dep	N	P	Mg	K	Ca	N : P	N : Mg	N : K	N : Ca
Epinal (88)	1309		4	1969	1997	CE	■■	■■	▲▲▲	▽▽	▽▽	↗↗↗	⇔	▲▲	▲▲▲	↗↗↗	▲
Solling-Buche	1002		4	1969	1997	CE	■■	■■	↗↗↗	▽▽	▽▽▽	↘↘	▽▽▽	▲▲	▲▲▲	↗↗↗	▲▲▲
Ste Hélène (88)	1323		4	1969	1997	CE	■■■	■■	▲▲▲	▽▽	▽▽▽	↘↘	▽▽	▲▲	▲▲▲	▲▲	▲▲▲
Souilly (55)	1321		4	1969	1997	CE	■■	■■	▲	▽▽	▽▽	▽	⇔	▲▲	▲▲▲	▲	↗↗↗
Auberive (52)	1301		12	1968	1998	CE	■■■		↘↘↘	▽▽	▽▽	▽▽	▽▽	▲	▲	▲▲▲	▲

rate of change	%	pos.	neg.	%	pos.	neg.	significance of change	N status		Annual N deposition [kg ha^{-1}]	
per 10 years:	<0.625	⇔	⇔				p = 0.10	extreme deficiency	□	0–<5	□
	0.625–<1.25	↗	↘	5–<10	▲	▽	p = 0.05	insufficient supply	■	5–<10	■
	1.25–<2.5	↗↗	↘↘	10–<20	▲▲	▽▽		sufficient to good	■■	10–<20	■■
	2.5–<5	↗↗↗	↘↘↘	≥20	▲▲▲	▽▽▽		luxurious supply	■■■	≥20	■■■

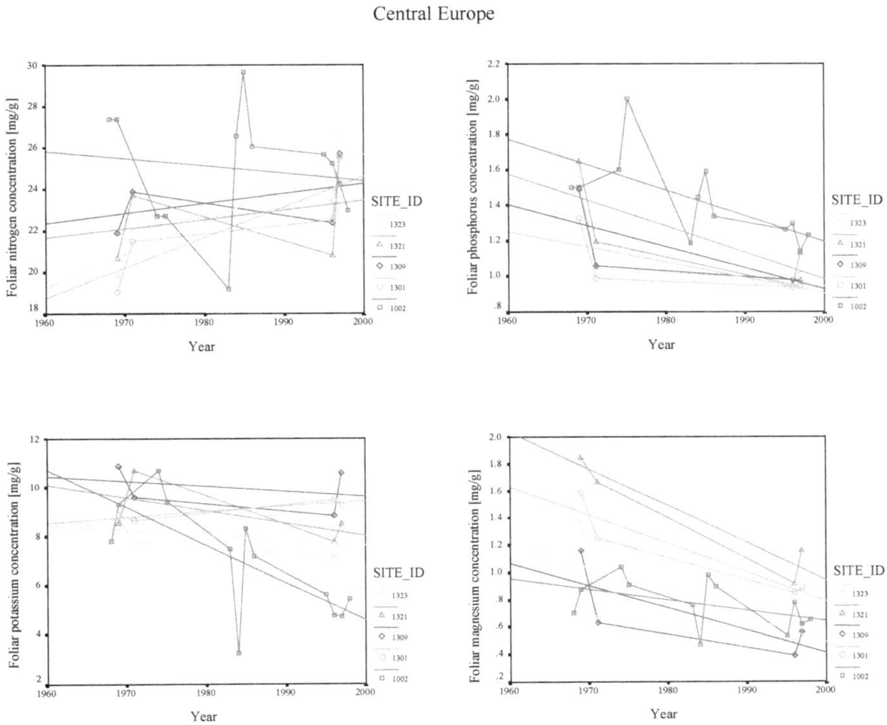

Figure 18. Time series of foliar N, P, K, and Mg concentrations [mg/g] for the beech stands included in the HDI in Central Europe. Presented are annual data and linear trends for each investigated site.

- For many studied stands, data regarding the unit dry mass of needles or leaves (100-needle weights) were not available; therefore the discrimination of changes in the supply of a nutrient from dilution or concentration effects often was not possible.
- For most sites, measured climate data were not available and had to be replaced with data from nearby meteorological stations or modelled data.
- The high temporal resolution of climate data had to be adapted to the lower resolution of growth and foliage nutrition data, which was only one observation per year.
- Changes in height increment are identified as deviations between observed height increment and expected height increment. For the analyses presented here the retrospectively derived height increment of the sample trees has been compared with the expected height increment according to site index curves from yield tables. The appropriate site index has been determined by the age-height relationship given in the reference year 1960. The yield tables used have been selected following the suggestions made by growth and yield experts familiar with the local growth conditions. Since it was not possible within the project to

verify the specific suitability of the used site index curves an uncertainty remains whether observed changes in height increment are really due to changes in growth or due to the comparison with inappropriate site index curves.

The minimum criteria for a stand to be included in the HDI were at least four observations over a period of at least 15 years. The reason for that rather weak qualification threshold was the restricted number of long and intensively observed stands. The fact that all these heterogeneous data should finally be compared with each other in a synoptic evaluation had the consequence to restrict some analyses on the smallest common denominator.

Another critical issue of the HDI approach was the fact that climate data measured on-site were in most cases not available, and data obtained from the closest meteorological station or modelled climate data had to be used as surrogates. Additionally, these data had to be aggregated to an annual time scale. These circumstances lead to the decision to use monthly averages of climate data (mean air temperature, precipitation sum) as basic data for our statistical evaluations. However, that way of data aggregation has the drawback that extreme short-term weather events were not recognised in our approach of historic climate evaluation; for example, frost events with a duration of only one or a few days, which can cause damage to trees and can therefore affect tree growth. This disadvantage, however, is not critical, as the main task of HDI is not to investigate inter-annual growth fluctuations, which are supposed to be considerably influenced by this kind of short-term events, but rather to assess the influencing factors for systematic long-term changes of tree growth.

The limitations of the data also affect the possibilities for the data analysis. It must be stated that only a few sites completely met the specific requirements of the applied procedures of statistical data analysis with regard to the number of observations. Generally, the available data did not allow a (desirable) separation of short-term fluctuations from long-term trends by means of spectral analysis and low/high pass filtration – as was applied by Mäkinen et al. (2002). In many cases, even simple auto-regressive regression models could not be applied in a useful way due to the limited availability of data. This was particularly true for Norway spruce and beech. Therefore, a combination of explorative and comparative data analysis using trend analysis and regression analysis was the most important tool used within the HDI.

4.1.7.2 Results

Due to the considerable heterogeneity of the temperature, precipitation and nutrition trends within the observation periods for the various studied stands, a uniform and close relationship between growth trends on the one hand and climate and/or stand nutrition trends on the other for all tree species all over Europe was not to be expected. This is especially true for Norway spruce and beech, for which the available data were very limited. Thus, for the eight Norway spruce stands, for which referenced height growth data were available, no significant factor influencing referenced height increment could be identified. However, concerning the four stands which showed a growth acceleration, nutrition with N, P, K and Mg was obviously improved at one stand (Pfalzgrafenweiler in CE), even though the *concentration* of foliar nutrient elements stagnated or decreased, since the 100-needle weight and thus the nutrient *contents* increased for that stand. By contrast, the growth increase of the Norway spruce stand Sodankylä 197 (NE) has been rather due to improved climate conditions than to an improved nutritional status. In most cases, however, the data were not adequate (few observations, high heterogeneity, 100-needle-weights not available) to make judgements about nutrition improvement as a factor for growth acceleration of Norway spruce. However, the growth-enhancing influence of atmospheric N deposition is supposed to have been minor for the Norway spruce stands as compared to the Scots pine stands, since all Norway spruce stands were generally characterised by a better *a priori* N nutritional status. Therefore, their sensitivity to further improvement of N nutrition is less pronounced. This agrees with results of Anders et al. (2002), who identified site specific-growth reactions to nitrogen deposition for different Scots pine forests in northeastern Germany with a growth acceleration of medium or well supplied stands only taking place if atmospheric N inputs exceed 20 kg $ha^{-1} yr^{-1}$.

Referenced height growth data for beech were not available within the duration of Recognition.

In summary, the available information about Norway spruce and especially beech was insufficient to make regionalised judgements concerning relationships among recent changes of the growth and those of growth-driving factors within the HDI approach of Recognition. However, regionalised statements were possible for Scots pine, the tree species which had been studied most intensively over longer periods in fertilisation experiments. According to the results of the HDI approach, most Scots pine stands in CE showed significant growth acceleration. There is much evidence that this increase was predominantly caused by improved nitrogen nutrition. The latter is probably due to: (1) a recovery from former devastative land-use practices (such as litter raking); and (2) elevated atmospheric N deposition over large parts of that region.

The results of a combined approach, using both graphical evaluation techniques and procedures of regression analysis, suggest that precipitation appears to be a key factor for water availability and thus for inter-annual and short-term (1- to 5-year) fluctuations of tree and stand growth, but not for the long-term growth trends. In our study the extent of any nutrient-driven growth acceleration was effectively controlled by water availability. In a regional study conducted in various Scots pine stands in Bavaria, Heitz et al. (2001) identified a similar interaction of the growth factors nutrient and water supply regarding referenced volume growth, but they could not detect any significant influences on height growth. This finding suggests that biomass growth (which resembles rather volume growth than height growth) is the most representative parameter of tree growth. Therefore, the ideal concept of future HDI-like research approaches would be to relate changes of climate and stand nutrition to changes in biomass increment of the respective stands. However, due to a lack of information about past forest management (e.g. thinning) at the HDI sites, which strongly affects standing biomass, referenced height growth was used as indicator for tree growth in our study, since it is robust against biases due to management effects.

The increasing atmospheric concentration of CO_2 has also probably contributed to the observed growth acceleration of most investigated Scots pine stands in CE.

However, we assume that the increased supply of CO_2 was only of secondary relevance for tree growth. This judgement is based on two observations:

(1) the spatial pattern of referenced tree growth changes reveals that there is no consistent growth acceleration over Europe as a whole. Under the assumption that no other factor is limiting, such a widespread consistent growth acceleration would be expected if the increased supply of atmospheric CO_2 was the major driving factor. However, growth increases were mainly restricted to those stands where N nutrition improved.
(2) During the 1990s, when the rate of CO_2 increase was especially high, there was a stagnation or even a decrease of growth rates rather than a growth increase for most studied stands.

However, these findings do not provide a proof for a subordinate role of CO_2, since the effect of this nutrient could also be confounded by other driving factors – such as water availability – in a way that any direct and positive effect of an increased supply of atmospheric CO_2 may have been masked by growth-decreasing side effects of additional CO_2 like increased air temperatures associated with decreased water availability to the forests. The limited number of investigated sites within the HDI approach does not allow investigation of the influence of an increased atmospheric CO_2 concentration on tree growth by spatial differential diagnostics. The PSA approach (see Chapter 4.2) provides the possibility to investigate that issue.

Within the HDI analysis, we could not detect any influence of the air temperature on referenced height growth of the studied Scots pine and Norway spruce stands. We explain this fact by the ambivalence of the growth factor air temperature. On the one hand, increased air temperature may be associated with a prolonged growing season. On the other hand, it leads to an increase of evapotranspiration and to drier soil conditions, thus reducing water availability for forest trees and stands as well as for soil micro-organisms, and thus inducing a growth decrease. This might be the reason for the negative correlation between (high) summer temperatures and tree growth found by Kahle (1994) for Norway spruce stands in southern Germany, and by Anders et al. (2002) for Scots pine stands in eastern Germany. Also Heitz et al. (2001), who used the number of frost-free days as indicator for temperature changes, could not find any positive effect of an increased air temperature for the volume growth of Scots pine stands in Bavaria, but rather an indication for a negative influence of increased drought stress. However, the absence of a significant correlation between tree growth and air temperature does not mean that the length of the growing season has no impact on tree growth. Because of the lack of appropriate data, this issue could not be directly examined within the HDI approach.

At the investigated sites in NE and the UK, an increased nitrogen supply was evidently far less important as a driving factor for long-term growth changes of Scots pine stands than in CE. Nevertheless, nitrogen nutrition obviously plays a prominent role at all sites where N deposition is high, especially in Denmark and southern Sweden. A differentiated judgement for that region is difficult due to: (1) a limited number of observations; (2) a comparably short period of foliar nutrient monitoring; and (3) a considerable deviation of referenced height growth trends observed for the period of foliar nutrient monitoring from the overall trend during the entire second part of 20th century for most sites.

4.1.7.3 Outlook

At present the results of the HDI approach cannot provide a comprehensive and regionalised overview about the relationships between recent changes of forest growth and recent changes of potential growth-regulating factors such as nutrient supply, water supply and climate. However, they still constitute a considerable increase of knowledge concerning these interrelations in different parts of Europe. Moreover, the HDI concept itself and the presented techniques of data evaluation may provide a standard for future research utilising a considerably larger number of available data on tree growth and site conditions. In the long term, the Level II monitoring programme of the European Commission and the UN/ECE (de Vries et al., 2002; http://www.icp-forests.org) will provide an ideal database to investigate that issue.

References

Anders, S., Beck, W., Bolte, A., Hofmann, G., Jenssen, M., Krakau, U. and Müller, J. 2002. Ökologie und Vegetation der Wälder Nordostdeutschlands. Verlag Dr. Kessel, Oberwinter, 283 p.

de Vries, W., Reinds, G.J., Dobben, H. van, Zwart, D. de, Aamlid, D., Neville, P., Posch, M., Auée, J., Voogd, J.C.H. and Vel, E.M. 2002. Intensive Monitoring of Forest Ecosystems in Europe – Technical Report 2002. Geneva, UN/ECE, EC, Forest Intensive Monitoring Coordinating Institute (FIMCI).

Dorsch, B. 2002. Statistische Analyse des Wachstums von Wäldern in verschiedenen Regionen Deutschlands. Diploma thesis at the Department of Statistics, Ludwig-Maximilans-University Munich. 95 p.

Heinsdorf, D. 1999. Düngung von Forstkulturen auf Lausitzer Kippen, LAUBAG Eberswalde 54 p.

Heitz; R., Rehfuess, K.-E., Küsters, E. and Pretzsch, H. 2001. Forschungsvorhaben "Wachstumstrends von Kiefern und Fichte in Bayern in Abhängigkeit von Ernährungszustand, Standort und Depositionsbelastung" Abschnitt II: Standortkundlicher Teil, Final Report, Technical University Munich. Unpublished report.

Kahle, H.P. 1994. Modellierung der Zusammenhänge zwischen der Variation von klimatischen Elementen des Wasserhaushalts und dem Radialzuwachs von Fichten (Picea abies (L.) Karst.) aus Hochlagen des Südschwarzwalds: Eine standortbezogene dendroökologische Untersuchung unter besonderer Berücksichtigung von Trockenstreß. Diss. Freiburg, 184 S.

Mäkinen, H., Nöjd, P., Kahle, H.-P., Neumann, U., Tveite, B., Mielekäinen, K., Röhle, H. and Spiecker, H. 2002. Radial growth variation of Norway spruce (Picea abies (L.) Karst.) across latitudinal and altitudinal gradients in central and corthern Europe. Forest Ecology and Management 171: 243–259.

Schönwiese, C.D., Rapp, J., Fuchs, T. and Denhard, M. 1993. Klimatrend-Atlas Europe 1891–1990. Berichte des Zentrums für Umweltforschung 20. 228 p.

Thornthwaite, C.W. and Mather, J.R. 1955. The water balance. Climatology – Drexel Institute of Technology, Lab. of Climate 8(1). 104 p.

4.2 Temporal trends and spatial patterns of height growth changes in relation to changes in air temperature and precipitation, and in relation to levels of foliar nitrogen and nitrogen deposition

Hans-Peter Kahle[1], Heinrich Spiecker[1], Rüdiger Unseld[1], Pedro-José Pérez-Martínez[1], Jörg Prietzel[2], Karl-Heinz Mellert[2], Ralf Straussberger[2] and Karl-Eugen Rehfuess[2]

[1]*Institute for Forest Growth, University Freiburg, Germany*
[2]*Chair of Soil Science, Technical University Munich, Germany*

4.2.1 Introduction

The correlative approach within the Recognition project comprises the Historical Development Investigation (HDI), and the Present State Analysis (PSA) (Chapter 2.1 and 2.2). Results of the HDI are presented in Chapter 4.1. This section presents results of the PSA: temporal trends and spatial patterns of height growth changes in relation to changes in air temperature and precipitation, and in relation to levels of foliar nitrogen and nitrogen deposition on the European scale. The results concern growth data for Norway spruce, Scots pine and beech that were collected on sites across Europe for the purpose of this research study. The underlying research hypotheses and concepts are described in Chapters 2.1 and 2.2.

Age-related height growth is used as an indicator of forest site productivity. Therefore, differences in the average tree height at a certain base age (i.e. differences in site index) of trees with different germination dates growing on a similar site are regarded as indicative for changes in site productivity (Chapter 2.2). At each sampling plot two age classes of trees (cohorts) were sampled and analysed. This allows comparison of the average height of trees at the same age, but at different points in time.

4.2.2 Data and analysis methods

The retrospective growth data analysed within the PSA refer to the European level and were compiled during the lifetime of the project. The sampling strategy and quasi-experimental treatment design applied in this observational study, as well as the measurement and analysis methods, are described and discussed in detail in Chapter 2.1 and 2.2 of this volume.

Sampling sites for the retrospective growth analysis were selected adjacent to the plots of the EC-UN/ECE Forest Intensive Monitoring Programme in Europe, Level II programme (EC-UN/ECE, 1998; Lorenz, 1995; de Vries et al., 2000; 2001; 2002). For the Norway spruce, Scots pine and beech dominated Level II plots baseline plot data and data of selected core and optional assessments carried out until 1998 have been provided to the project in 2000 by the Forest Intensive Monitoring Co-ordinating Institute (FIMCI) and the responsible National Focal Centres (NFCs).

The data set 'CRU 0.50 lat./long. gridded monthly climate data' was supplied by the Climate Impacts LINK Project on behalf of the Climatic Research Unit of the University of East Anglia, UK. Thornthwaite's formulae (Thornthwaite and Mather, 1955 and 1957) was applied to the CRU 0.50 data to calculate time series of potential evapotranspiration and seasonal climatic water balance for all sampling sites. Homogenized gridded data on nitrogen deposition rates over Europe are obtained from the Co-operative Programme for Monitoring and Evaluation of the Long-Range Transmission of Air Pollutants in Europe (EMEP, http://emep.int).

According to the main research hypotheses three putative causal factors for forest growth changes were considered as treatment factors: (i) change in seasonal air temperature, (ii) change in seasonal precipitation, (iii) nitrogen nutrition status. The distributions of the levels of these factors as represented by the level II plots were subdivided into percentile classes and intersections of these different factor level classes were used as primary criteria for the selection of sample plots. Climate change data were calculated from the CRU 0.50 data: absolute deviations for temperature (in K) and relative deviations for precipitation totals (%). The spatial distributions of long-term changes in mean seasonal air temperature (May to September) and mean seasonal precipitation totals between 1961–1990 and 1901–1930 over Europe are mapped in Figure 1. We applied a uniform definition of growing season all over Europe, namely the time period May to September, for the site classification with respect to seasonal climate. The periods 1901–1930 and 1961–1990 were chosen according to the "climatic normal periods" of the World Meteorological Organization.

Despite the fact that both maps in Figure 1 share the same support, the spatial distribution of changes in seasonal air temperature appears more clustered than the one for precipitation. Meso-scale structure is the dominant feature of the precipitation map, whereas macro-scale patterns are the prominent characteristic of the temperature map. In both cases, substantial spatial differences in the extent of change occur over Europe: the range of change in temperature is –1.3K to + 1.8K, and for precipitation ± 13%. However, there is not a unique and prominent geographical gradient of change for Europe for either air temperature or seasonal precipitation.

Foliar nitrogen nutrition was considered as the third classification variable for the selection of sample sites. Data of the level II network on foliar nitrogen concentration [N] from the nutrition assessments 1993–1998 have been aggregated to time-averaged stand-wise mean values. Maps of the foliar [N]-levels of the investigated species (one-year-old needles of Scots pine and Norway spruce; and leaves of beech) are presented in Figures 2 and 3.

In total 62 sites adjacent to the level II plots have been sampled for the purpose of this study. Figure 4 shows the sites at which sample trees have been taken.

The target variable for the tree sampling and the response variable in the regression analysis is *change in tree height growth*. In order to identify changes in

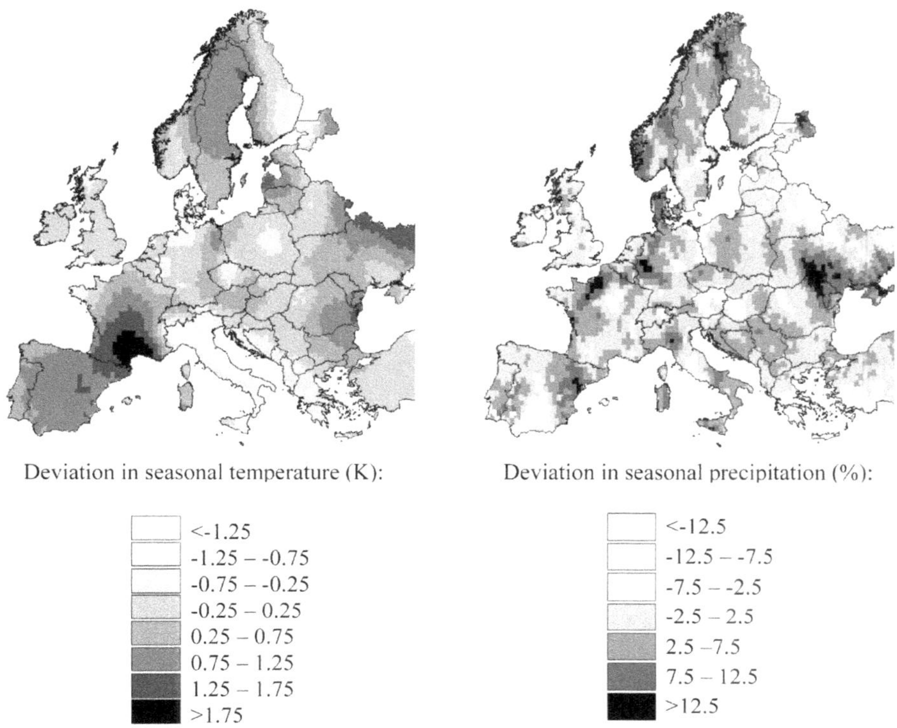

Figure 1. Deviation of mean seasonal (May–September) air temperature (in K) and mean seasonal precipitation (as a percentage) between 1961–1990 and 1901–1930 (data source: the CRU 0.50 lat./long. gridded monthly climate data has been supplied by the Climate Impacts LINK Project on behalf of the Climatic Research Unit, University of East Anglia).

tree height growth a cohort comparison approach has been applied: height of trees at the same age but at different points in time is compared (for details see Chapter 2.2). Therefore, at each sampling site a sample of younger and a sample of older trees have been taken. In order to be qualified for the further analysis a sample had to fulfil the following requirements: (i) for the age-overlap period the tree cohorts had to be similar with respect to age composition and growth history; and (ii) the young trees had to be at least 40 years old. However, in some cases it was not possible to verify these conditions before the felling of the trees in the field. Plots which did not fulfil these requirements were excluded from the analyses.

The temporal coverage of the sample tree cohorts is described in Table 1. The mean calendar year of the innermost tree ring in breast height for the cohort of old and of young trees is for spruce 1906 and 1945, for pine 1922 and 1954, and for beech 1903 and 1941, respectively. There is considerable variation among the different plots within the strata.

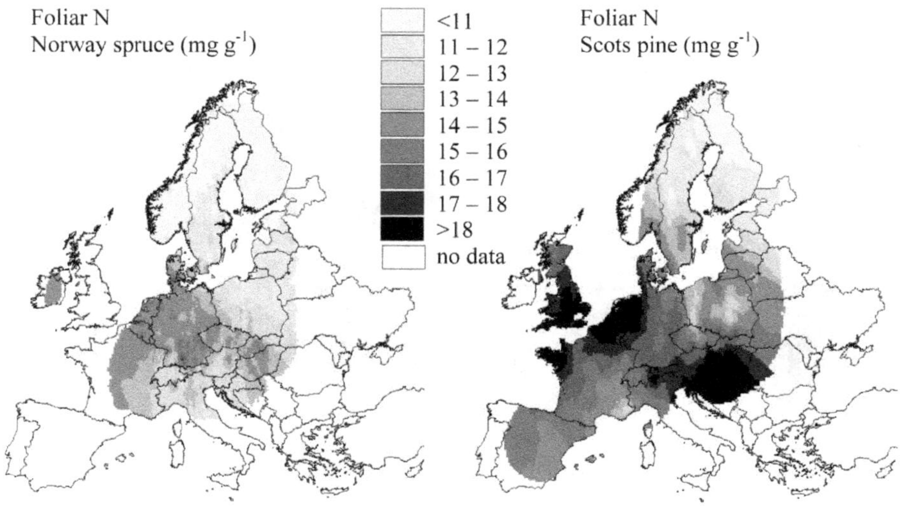

Figure 2. Mean foliar nitrogen concentration (mg g^{-1}) of one-year-old needles of Scots pine and Norway spruce from assessments within the EC-UN/ECE Forest Intensive Monitoring Programme in Europe (Level II) during 1993–1998 (data from n=192 Scots pine and n=192 Norway spruce dominated Level II plots have been spatially interpolated by applying ordinary Kriging. Delineation of predictions is limited to areas where sample coverage and Kriging variance allow reliable prediction (Data provided by FIMCI, 2000).

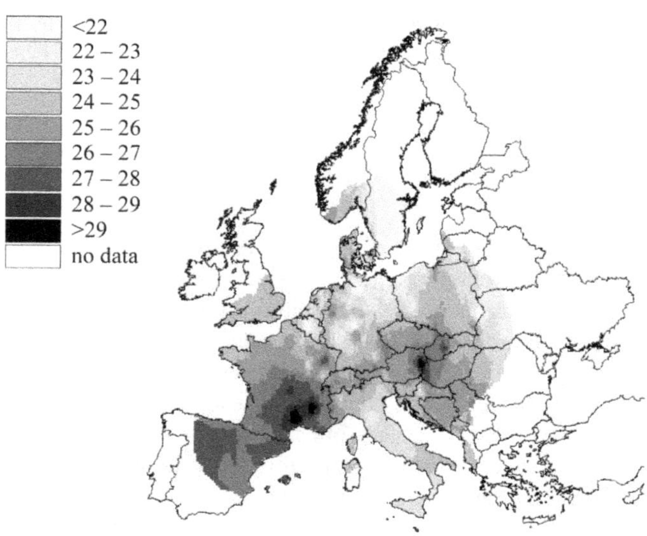

Figure 3. Mean foliar nitrogen concentration (mg g^{-1}) of leaves of beech from assessments within the EC-UN/ECE Forest Intensive Monitoring Programme in Europe (Level II) during 1993–1998. Data from n=97 Level II plots have been spatially interpolated by applying ordinary Kriging. Delineation of predictions is limited to areas where sample coverage and Kriging variance allow reliable prediction (Data provided by FIMCI, 2000).

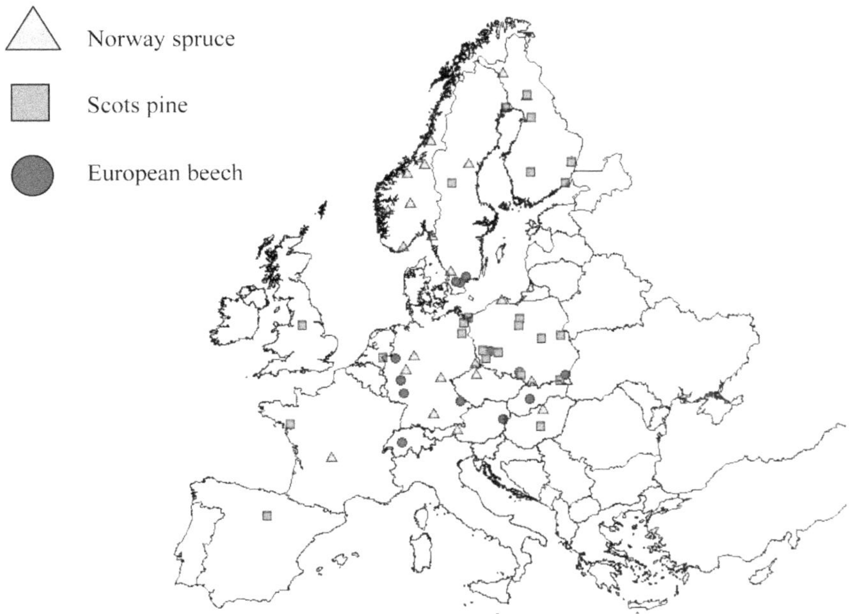

Figure 4. Sample sites of the growth assessment for the Present State Analysis. Sites were selected adjacent to the plots of the EC-UN/ECE Forest Intensive Monitoring Programme in Europe (Level II). At each indicated site sample trees have been selected and felled for stem analyses. Since not all data sets fulfil the requirements for the cohort-comparison, the total number of sites sampled differs from the number of sites included in the analysis in this report (number in brackets): Norway spruce n=24 (17); Scots pine n=24 (19); beech n=14 (9).

Table 1. Temporal coverage of sample tree cohorts.

Species	Time span: cohort of old trees			Time span: cohort of young trees		
	Calendar year of innermost tree ring in breast height					
	Min	Max	Mean (±SD)	Min	Max	Mean (±SD)
PCAB[1]	1837	1943	1906 ± 27.9	1870	1978	1945 ± 25.5
PISY[1]	1876	1958	1922 ± 19.8	1917	1974	1954 ± 15.7
FASY[1]	1853	1934	1903 ± 25.4	1900	1960	1941 ± 17.7

[1] PCAB: Norway spruce, PISY: Scots pine, FASY: Beech.

To trace possible changes in the competitive status of the spruce and pine sample trees in the past and to assess the similarity of the sample tree cohorts with respect to their growth history, the development of the height-to-diameter-ratios (h/d-ratio: tree height divided by tree diameter in breast height inside bark) was analysed. Figure 5 presents box-plot time series of the plot-wise means of the h/d-ratios

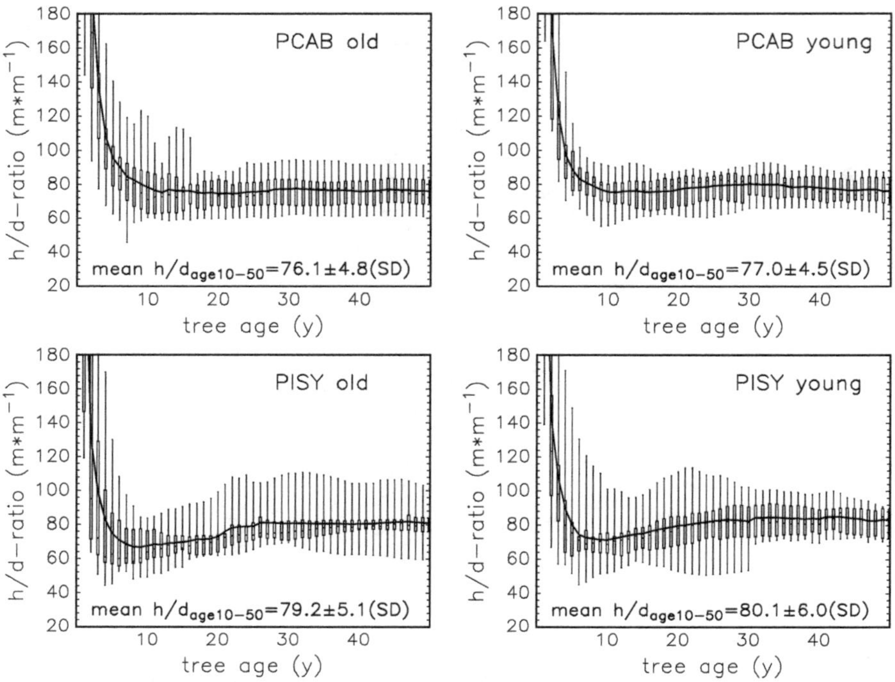

Figure 5. Box-plot time series of plot-wise mean height-to-diameter-ratios (*h/d-ratio* in m·m^{-1}) versus tree age (age in breast height $t_{1.3}$) for the two age cohorts (old: left, young: right) of the Norway spruce (PCAB, top) and Scots pine sample trees (PISY, bottom). Mean and standard deviation (SD) refer to the age period 10–50 years. Only series longer than 15 years are included (PCAB: n_{max} = 17; PISY: n_{max} = 18). The top and bottom of the boxes mark the empirical quartile ranges, boxes are divided by the median, the absolute maximum and minimum limits are taken from the actual limits of the data. Solid lines represent the arithmetic mean. There is no difference in the mean h/d-ratio in any year over the age period 10–50 years ($h/d_{age10-50}$) between the old and young age cohorts for both species ($P(|t_{\alpha(2),(v1+v2)}| \geq t) > 0.05$).

versus age by species and age cohorts. All curves follow a pattern which is typical for dominant trees under intraspecific competition in even-aged stands: at young age the h/d-ratio decreases until diameter growth slows down due to increasing age and increasing competition during stand closure. After an initial decline the development of open grown trees would be characterized by a slowly but consistently decreasing trend in the h/d-ratio (Abetz, 1976; Hasenauer et al., 1994). However, the h/d-ratios of the spruce and pine sample trees stabilize on a ± constant (spruce PCAB) or slowly increasing (pine PISY) level, a development which is typical for dominant trees under even-aged stand dynamics. There is no difference in the mean h/d-ratio between the old and young age cohorts for both species in any year over the age period 10-50 years. Hence, the sample tree strata are similar with respect to the level of the h/d-ratio as well as its change in the course of time.

Change in height growth is defined here as the percentage deviation of the mean annual height increment (*ih*) in year *i* of the cohort of younger trees from that of the older tree cohort at the same mean age, referred to the value of the older tree cohort, formally called $\Delta ih_i(\%)$. The time-subscript *i* is skipped in the following for reasons of simplicity. According to this definition change in height growth in fact refers to change in annual height increment. Since *ih* is a measure of growth velocity (v), Δih measures the acceleration (a) of growth ($a = dv/dt$). Since we were also analysing changes in height increment with respect to time, another measure has to be introduced: the rate of change of acceleration with respect to time, i.e. the time derivative of the acceleration vector a (Chapter 2.2).

Retrospective height increment data are derived from stem analysis: current annual height increment for the conifers, and periodic height increment for beech (Chapter 2.2). Since we were interested in long-term changes in height growth, the time series of the measurement data were smoothed by applying a rather stiff cubic spline approximation (50%-frequency cut-off: 30 years) on an individual tree basis prior to the calculation of height growth changes. By this procedure the amount of short-term variation in the increment time series is substantially dampened and the medium- and long-term signals are amplified and preserved for the subsequent analyses. The degree of smoothing was determined as a compromise between sufficient flexibility to preserve medium-term variation and acceptable stiffness to eliminate high-frequency noise from the searched for growth trend signal.

4.2.3 Temporal trends in height growth changes

4.2.3.1 Temporal trends in height growth changes of the different tree species

Figure 6 shows time series of the mean height growth changes for the Norway spruce, Scots pine and beech samples. It has to be considered that some minor shifts in the aggregated data series result from unbalanced data, e.g. as a result of differences in the length of the age-overlap between the tree cohorts, the number of observations in the first years of the time series is not constant over time.

During the period 1960–2000 height increment of young Norway spruce, Scots pine and beech sample trees is significantly and consistently larger than that of the older sample trees at the same site. Over the 41-year period the median change in annual height increment ($\Delta ih_{1960-2000}$) is 22.8 ± 20.3 (SD) % for Norway spruce, 24.8 ± 36.3 (SD) % for Scots pine, and 25.1 ± 38.3 (SD) % for beech. The means over the whole period are for all three species significantly larger than zero ($p < 0.05$) (Figure 7).

The absolute value ranges of $\Delta ih_{1960-2000}$ differ between species: the difference between maximum and minimum values is largest for Scots pine, reflecting considerable heterogeneity in this sample. However, the relative coefficient of variation ($CV(\%) = (SD/\bar{x})/\sqrt{n} * 100$) of $\Delta ih_{1960-2000}$ is largest for beech ($CV_{1960-2000}(\%)$=50.9%), followed by pine ($CV_{1960-2000}(\%)$=33.6%) and spruce ($CV_{1960-2000}(\%)$=21.6%), indicating that the precision of the estimated growth trend signal is highest for spruce and lowest for beech.

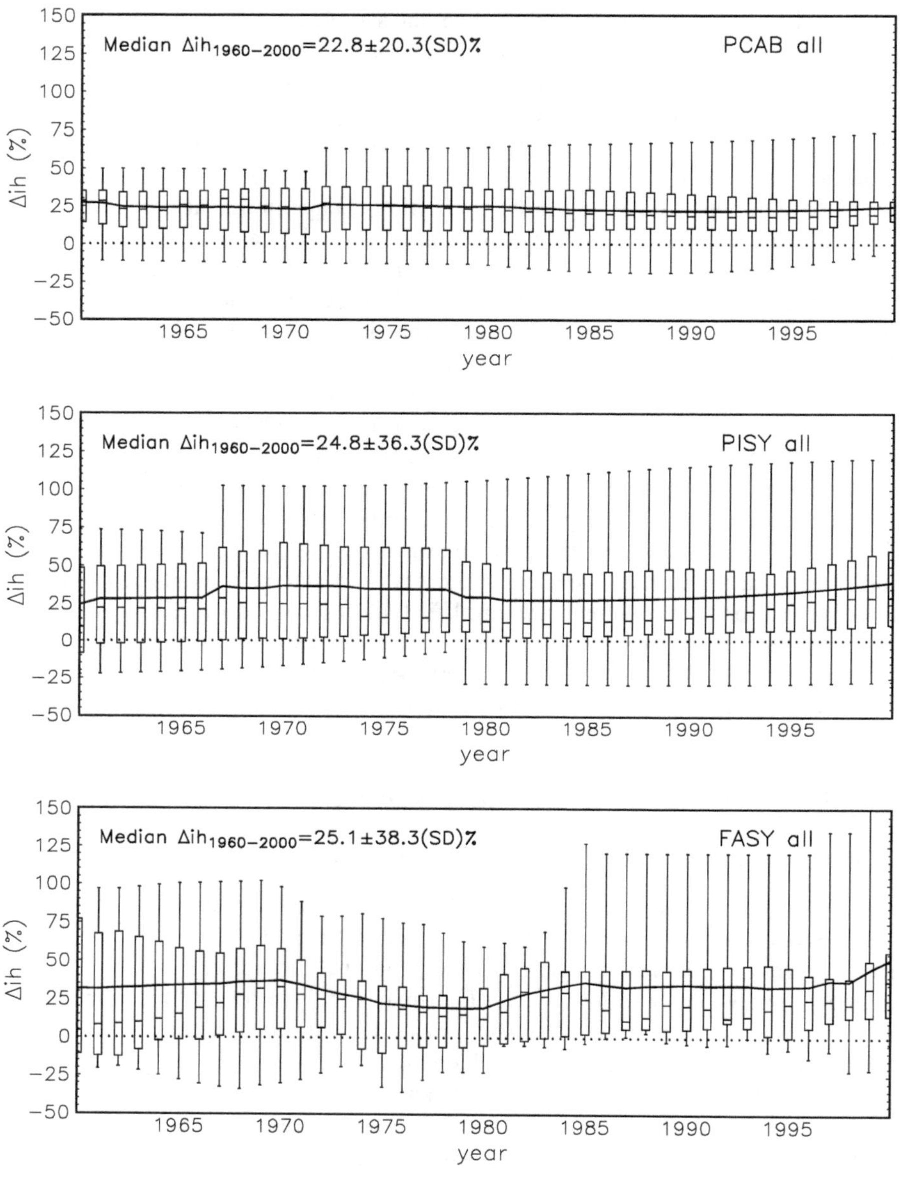

Figure 6. Time series of plot-wise mean change in height growth (*Δih%*: annual height increment of young trees compared to old trees at the same age). Norway spruce (PCAB): n_{max}=17; Scots pine (PISY): n_{max}=19; beech (FASY): n_{max}=9. Mean and standard deviation (SD) refer to the period 1960–2000. Only series longer than 10 years are included. According to the overall test the regression models for all three species do not coincide (Table 3). See Figure 5 for further explanation of the box-plot diagram.

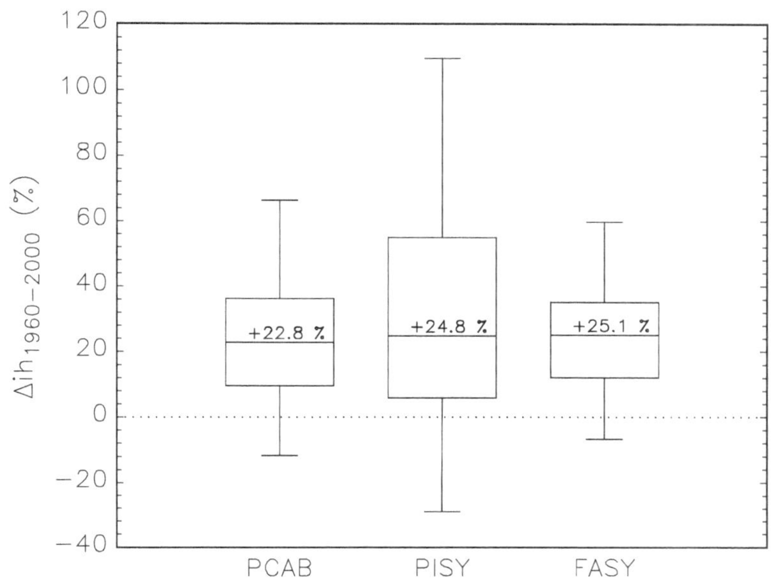

Figure 7. Plot-wise mean height growth acceleration for the period 1960–2000 of Norway spruce (PCAB), Scots pine (PISY), and beech (FASY) (further explanation see Figure 5).

The level of change in annual height growth is relatively stable over the whole period. However, a small yet significant increase in height growth acceleration is inferred for Norway spruce ($\Delta ih_{1960-2000}$: +0.26% y^{-1}) and beech ($\Delta ih_{1960-2000}$: +0.22% y^{-1}) (Table 2). The slopes of both regression lines are statistically not different ($P(|t_{\alpha(2),(v1+v2)}| \geq t = \frac{b_1 - b_2}{s_{b1-b2}}) > 0.05$; Zar, 1974). Though, according to the overall test of coincidence of the individual regression models of Δih versus time the null hypothesis of equality of the linear regressions among all three species is rejected (Table 3) (Zar, 1974).

The beech and pine samples, and to a minor degree also the spruce sample, show a slight increase in height growth acceleration during the last decade. However, this feature should not be overstressed since changes in slope at the beginning and/or end of the curves could be over-accentuated as well as be masked due to end-fit problems of the spline interpolation.

4.2.3.2 Modelled effect of changes in height growth on total wood volume production potential

Figure 8 shows the increase in total wood volume production potential (TVP, $m^3 \cdot ha^{-1}$) calculated from the observed increases in stand top height (h_o, m) for the three tree species according to selected yield tables commonly applied in central Europe: (1) Norway spruce (PCAB), Wiedemann, developed 1936/1942 (published 1949); (2) Scots pine (PISY), Wiedemann, developed 1943 (published 1948); and (3)

Table 2. Figures on the sample tree strata and regression analyses of $\Delta ih(\%)$ versus time t.

	Obs[2]	ΔTp_{5-9}[3]	$\%Pr_{5-9}$[4]	$[N_0]$[5]	Intercept[6] b_0	Slope[6] b_1
	n	1/10 K (±SD)	% (±SD)	mg g^{-1} (±SD)		
Overall:						
PCAB[1]	17	0.53 ± 2.55	100.6 ± 3.69	13.7 ± 2.17	27.875**	0.2551**
PISY[1]	19	0.12 ± 2.08	100.0 ± 4.08	14.2 ± 3.76	17.651**	*n.s.*
FASY[1]	9	0.49 ± 1.67	99.9 ± 1.78	24.5 ± 3.80	15.078**	0.2199*
Stratification according to change in seasonal temperature (ΔTp_{5-9}):						
PCAB Tp high	10	2.14 ± 1.75[a]	100.9 ± 2.67	13.2 ± 2.16	21.735**	–0.1555**
PCAB Tp low	7	–1.99 ± 1.10[a]	100.2 ± 5.14	14.6 ± 2.00	31.155**	–0.5174**
PISY Tp high	9	1.39 ± 1.61[b]	100.2 ± 3.54	15.1 ± 4.61	41.949**	–1.2072**
PISY Tp low	10	–1.26 ± 1.64[b]	99.8 ± 4.79	13.3 ± 2.41	*n.s.*	0.9762**
FASY Tp high	5	2.09 ± 0.89[c]	100.6 ± 2.32	26.5 ± 3.57	41.896**	–0.8803**
FASY Tp low	4	–0.98 ± 0.54[c]	99.2 ± 1.26	22.1 ± 3.40	*n.s.*	0.7527**
Stratification according to change in seasonal precipitation ($\%Pr_{5-9}$):						
PCAB Pr high	11	0.73 ± 2.75	102.8 ± 1.95[a]	13.1 ± 2.11	21.875**	–0.0635**
PCAB Pr low	6	0.13 ± 2.29	96.3 ± 2.02[a]	14.9 ± 1.92	24.345**	–0.2727**
PISY Pr high	11	0.45 ± 1.62	102.9 ± 3.08[b]	13.1 ± 3.48	4.588**	0.3447**
PISY Pr low	8	–0.24 ± 2.54	96.8 ± 2.06[b]	15.5 ± 3.81	17.679**	0.3952*
FASY Pr high	5	1.17 ± 2.01	101.3 ± 1.54[c]	24.3 ± 2.78	9.456**	0.3844*
FASY Pr low	4	–0.06 ± 1.53	98.5 ± 0.89[c]	24.3 ± 5.40	38.115**	–1.1687**
Stratification according to foliar nitrogen concentration ($[N_0]$):						
PCAB N high	8	–0.61 ± 2.38	99.0 ± 4.05	15.7 ± 0.97[a]	8.239**	–0.2913**
PCAB N low	9	1.67 ± 2.29	102.2 ± 2.59	11.8 ± 0.54[a]	31.655**	–0.2805**
PISY N high	9	0.37 ± 1.17	98.5 ± 3.13	17.0 ± 3.14[b]	25.298**	*n.s.*
PISY N low	10	–0.14 ± 2.82	101.6 ± 4.52	11.2 ± 0.86[b]	–2.205*	0.6892**
FASY N high	5	1.37 ± 1.53	99.8 ± 1.11	27.3 ± 1.84[c]	35.429**	*n.s.*
FASY N low	4	–0.61 ± 1.53	100.1 ± 1.11	21.0 ± 1.84[c]	10.644*	*n.s.*

[a,b,c] Superscripts indicate significant differences between adjacent means ($P(|t_{\alpha(2),(v1+v2)}| \geq t)$ bold: <0.001, standard: < 0.05)
[1] PCAB: Norway spruce, PISY: Scots pine, FASY: Beech
[2] Obs: number of sample plots
[3] ΔTp_{5-9}: Difference in mean seasonal air temperature between the age-overlap periods of the sample tree strata (young versus old trees) (data source: CRU)
[4] $\%Pr_{5-9}$: Relative mean seasonal precipitation between the age-overlap periods of the sample tree strata (young versus old trees) (data source: CRU)
[5] $\%[N_0]$: Foliar nitrogen concentration (current years' needles/leaves) (data source: FIMCI, 2000)
[6] Intercept and slope of simple linear regression models of $\Delta ih(\%)$ versus time t ($\Delta ih(\%) = b_0 + b_1 t$; regression through the group-wise medians of $\Delta ih(\%)$). Referring to the following periods, overall: 1960–2000 (n = 41), others: 1970–2000 (n = 31). Significance levels: $P(|t_{\alpha(2),(v1+v2)}| \geq t)$: ** < 0.001, * < 0.05.

beech (FASY), Schober, developed 1967 (published 1975). Assuming that the yield level models – that is the functional relationship between TVP and stand height, regardless of stand age (Assmann, 1961) – of the cited yield tables are valid for the study material, sustained increases in mean annual height increment (*ih,* referred to stand top height h_0) at the magnitude found in this study for the period 1960–2000 (see Figure 7), in the average would correspond to even larger increases in TVP. Compared to the height growth acceleration the magnitude of the increases in TVP is 2.5 times larger for spruce, 1.8 for pine, and 2.9 for beech respectively. For this calculation the median height of the older trees at base age 30 is taken as initial height. The yield table of Wiedemann (1949) for spruce is built upon a single

Table 3. Statistics for the overall test of coincidence of the regression models of $\Delta ih(\%)$ versus time *t*.

Stratum	Regression Residual		Pooled Regression Residual		Total Regression Residual	DF_k [2]	*F*	Prob[3]
	SS_n [1]	DF_n [2]	SS_p [1]	DF_p [2]	SS_t [1]			$F_{á(1),i1,i2}$
PCAB	124.04	39						
PISY	1560.43	39	3797.41	39	327.92	4	26.72	0.0000
FASY	2112.94	39						
PCAB Tp high	13.14	29	144.40	29	32.72	2	22.43	0.0000
PCAB Tp low	131.27	29						
PISY Tp high	2064.25	29	3107.77	29	993.36	2	19.73	0.0000
PISY Tp low	1043.51	29						
FASY Tp high	1469.70	29	6229.79	29	1134.62	2	23.72	0.0000
FASY Tp low	4760.10	29						
PCAB Pr high	15.43	29	196.28	29	32.72	2	24.16	0.0000
PCAB Pr low	180.85	29						
PISY Pr high	256.28	29	1251.33	29	993.36	2	15.48	0.0000
PISY Pr low	1875.01	29						
FASY Pr high	1251.33	29	2557.10	29	1134.62	2	16.13	0.0000
FASY Pr low	1305.76	29						
PCAB N high	399.66	29	416.31	29	32.72	2	26.72	0.0000
PCAB N low	16.65	29						
PISY N high	1492.39	29	1638.15	29	993.36	2	11.42	0.0002
PISY N low	145.79	29						
FASY N high	3875.01	29	6328.10	29	1134.62	2	23.80	0.0000
FASY N low	2453.09	29						

[1] SS_n, SS_p, SS_t: Residual sum of squares for regression, pooled regression, and total regression respectively
[2] DF_n, DF_p, DF_k: Residual degrees of freedom for regression, pooled regression, and total regression respectively
[3] The significance levels for the *F*-test are inflated by autocorrelation, *p*, in the model residuals. According to the Durbin-Watson statistic, *d*, all total-regression model residuals are positively autocorrelated: $d_{overall}$=0.325, d_{PCAB}=0.385, d_{PISY}=0.097, d_{FASY}=0.600. However, the *F*-test results for the overall model and the one for PCAB are significant on the $p = 0.05$ level if for the calculation of the parametric values of DF_p the effective sample size n' ($n' \cong n\frac{1-p_1}{1+p_1}$, Wilks, 1995) is used instead of the uncorrected sample size *n*.

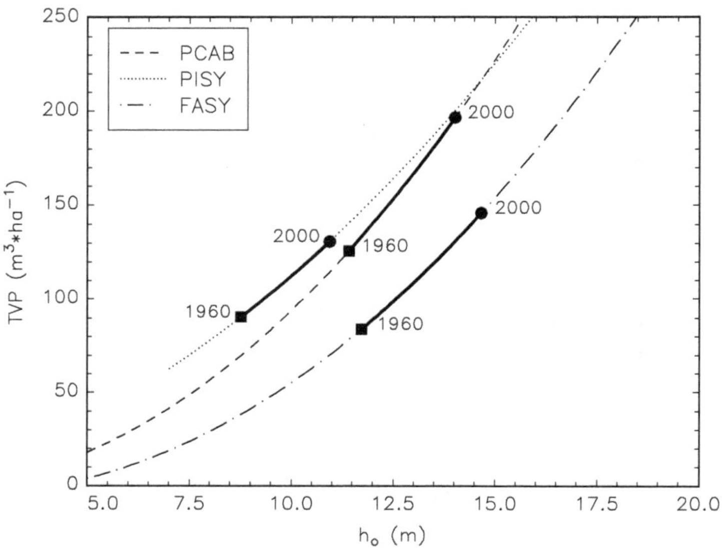

Figure 8. Yield level models (total wood volume production potential (TVP, $m^3 \cdot ha^{-1}$) versus stand top height (h_0, m)) for Norway spruce (PCAB), Scots pine (PISY), and beech (FASY) according to yield tables (thin lines; PCAB: Wiedemann 1936/1942 (published 1949); PISY: Wiedemann 1943 (published 1948); FASY: Schober 1967 (published 1975)). The thick solid lines represent the vectors of change along the yield level curves according to the observed changes in stand height growth between 1960 and 2000. The x-axis values of the solid squares are the median top-heights of 30 year old stands of the particular tree species under the growing conditions in the year 1960, and those of the solid circles in the year 2000, respectively. The predicted change in TVP between 1960 and 2000 is +56.4% for spruce, +44.9% for pine, and +74.0% for beech.

general yield level model, whereas the yield tables for pine and beech (Wiedemann 1948; Schober 1975) consider differences in yield levels with respect to site index and are therefore based on different specific yield levels (Assmann, 1961) or synonymic 'production classes' (Bradley et al., 1966; Hamilton and Christie, 1971). The relation between site index and specific yield level for pine and beech in these yield tables is reciprocal: the higher the site index the lower the specific yield level. For the straightforward model calculations an average yield level, calculated as the mean over the whole range of specific yield levels included in the corresponding yield tables, was used as a reference.

4.2.3.3 Temporal trends in height growth changes with respect to changes in seasonal air temperature and precipitation

Figure 9 and 10 show time series of height growth changes at the sampled sites with respect to changes in seasonal (May to September) air temperature and seasonal precipitation. The data sets are stratified according to whether air temperature or precipitation increased or decreased (see Table 2).

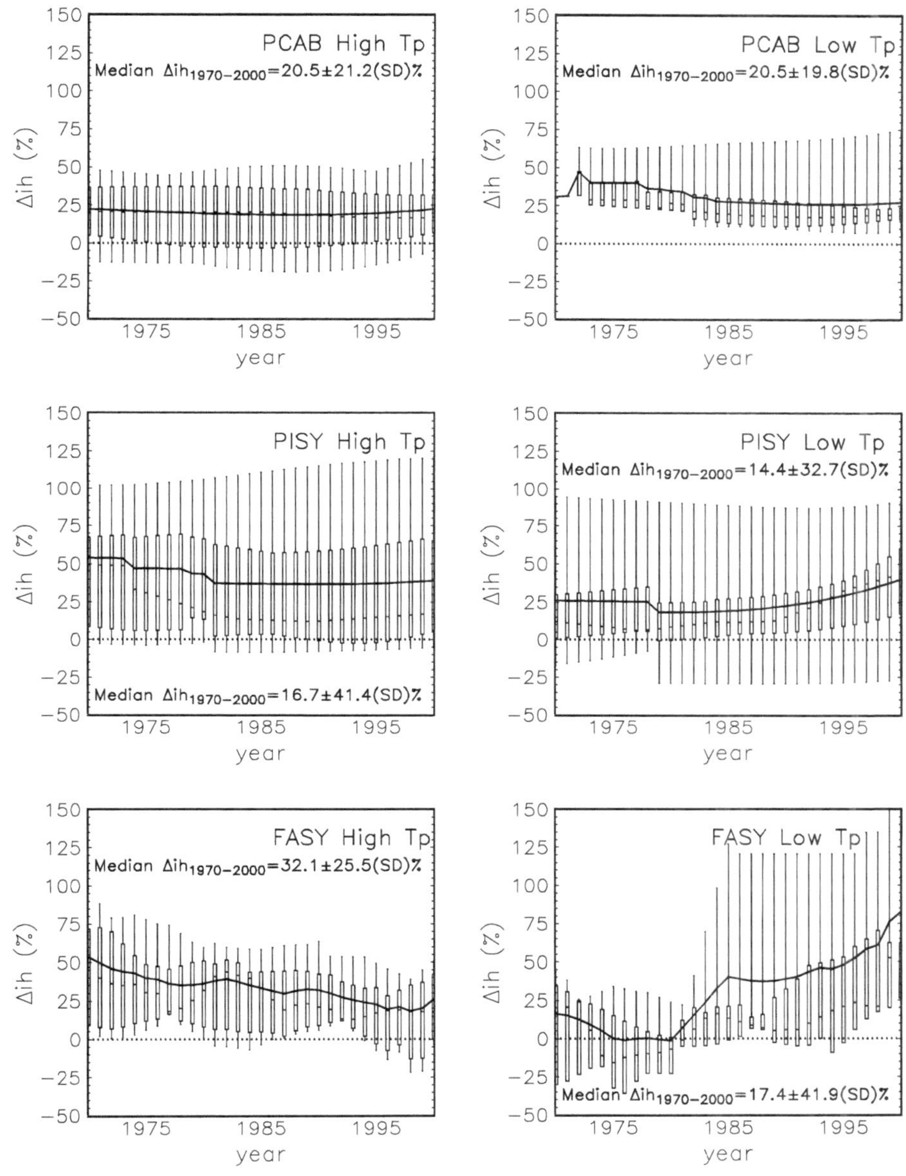

Figure 9. Time series of plot-wise mean change in height growth (*Δih%*) stratified by classes of change in seasonal air temperature (Tp_{5-9}) at the sampling site (high Tp: increase in Tp; low Tp: decrease in Tp). Norway spruce (PCAB) high Tp: n_{max}=10, low Tp: n_{max}=7; Scots pine (PISY) high Tp: n_{max}=9, low Tp: n_{max}=10; beech (FASY) high Tp: n_{max}=5, low Tp: n_{max}=4. The means over the period 1970–2000 are statistically not different between the strata ($P(|t_{\alpha(2),(v1+v2)}| \geq t) > 0.05$), however, the regression models for the two temperature-change strata are different for each of the three species (Table 3).

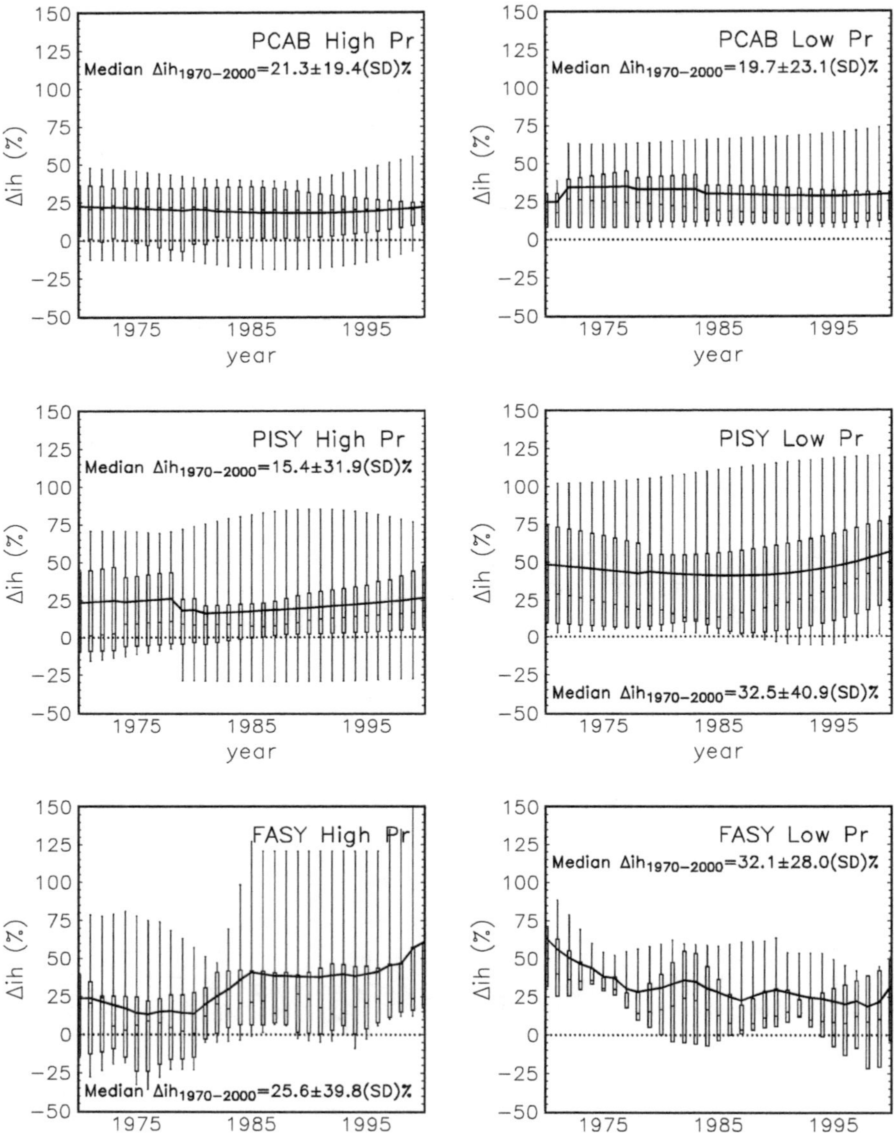

Figure 10. Time series of plot-wise mean change in height growth ($\Delta ih\%$) stratified by classes of change in seasonal precipitation (Pr_{5-9}) at the sampling site (high Pr: increase in Pr; low Pr: decrease in Pr). Norway spruce (PCAB) high Pr: $n_{max} = 11$, low Pr: $n_{max} = 6$; Scots pine (PISY) high Pr: $n_{max} = 11$, low Pr: $n_{max} = 8$; beech (FASY) high Pr: $n_{max} = 6$, low Pr: $n_{max} = 3$. The means over the period 1970–2000 are statistically not different between the strata ($P(|t_{\alpha(2),(v1+v2)}| \geq t) > 0.05$). However, the regression models for the two precipitation-change strata are different for all three species (Table 3).

The treatment variables *change in seasonal air temperature* and *change in seasonal precipitation* were initially fixed as referring to changes between the periods 1961–1990 and the reference period 1901–1930. Accordingly, the levels of changes in climate between these two periods were used as selection criteria in the sampling design. However, only after the trees had been cut and stem analyses had been conducted it was possible to calculate the exact age of the sample trees. Only then it was possible to adjust the climate change variables exactly to the periods of the individual tree-cohort's lifetimes. For example, if the mean age overlap between the cohort of young and old trees is 40 years, the climate change variable has been calculated as the deviation between the climatic conditions during the first 40 years of the lifetime of the younger sample trees and the climatic conditions during the first 40 years of the lifetime of the older sample trees. The resulting *a posteriori* classification as described above differs markedly from the *a priori* classification used for selecting the sample sites.

The mean levels of change in seasonal temperature, $\Delta tp_{5\text{–}9}$, between the temperature-change strata range from –0.13 K to +0.14 K for pine and –0.20 K to +0.21 K for spruce. Mean height growth acceleration over the period 1970-2000 did not differ significantly for any of the three species between the two temperature-change strata (Figure 9). However, the temporal courses of the height growth acceleration as described by linear regressions are statistically different among the strata (Table 3).

Results of the treatment classifications and differences of means tests are summarized in Table 2. Change in air temperature and level of foliar nitrogen concentration [N] are linked: at sites where seasonal air temperature increased, stand foliar nitrogen nutrition tends to be lower in Norway spruce but higher in beech and pine (not significant).

The classification according to the level of change in seasonal precipitation (Figure 10) reveals that there were no significant differences in mean height growth acceleration over the period 1970–2000 for any of the three species. Though, the overall tests indicate that the regression lines for the precipitation-change strata are statistically different (Table 3). The magnitudes of the mean levels of change in precipitation ($\Delta pr_{5\text{–}9}$) range among the strata from –3.7% to +2.8% for spruce, and from –1.5% to +1.3% for beech. A mixing up of the treatment factors is evident for Scots pine: the mean level of foliar nitrogen nutrition seems to be higher at sites where seasonal precipitation decreased, however, the difference in means is not significant (Table 2).

4.2.3.4 Temporal trends in height growth changes with respect to level of foliar nitrogen nutrition

Results of the stratification according to (species specific) classes of mean foliar [N] taken from the assessments during 1993–1998 are presented in Figure 11. Median height growth acceleration ($\Delta ih_{1970\text{–}2000}$) tended to be larger in Norway spruce trees with lower N nutrition at the end of the observation period ($\Delta ih_{1970\text{–}2000}$ = 33.1 ± 18.5 (SD) %) than in trees with higher N nutrition ($\Delta ih_{1970\text{–}2000}$ = 16.5 ± 23.6 (SD) %). However, the difference in means over the period 1970–2000 is not

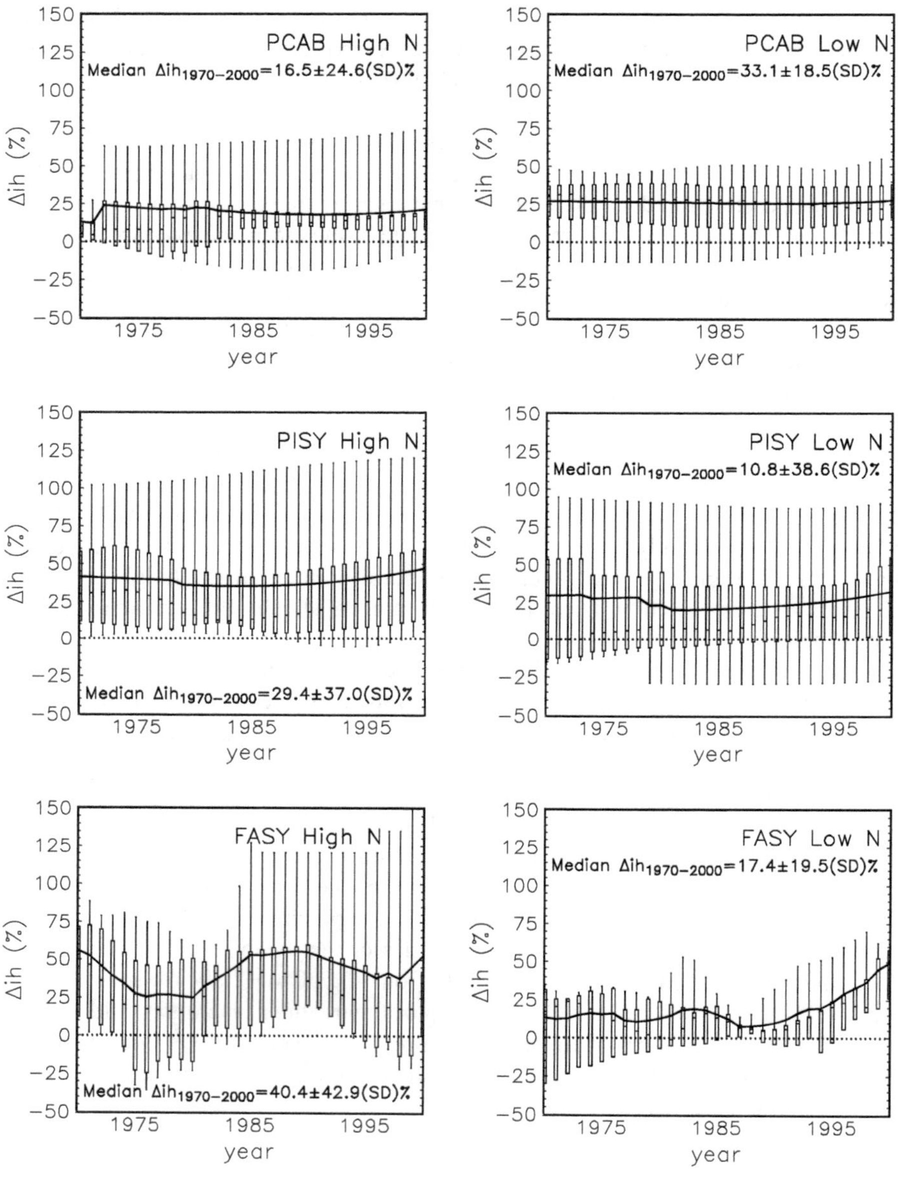

Figure 11. Time series of plot-wise mean change in height growth (Δ*ih%*) stratified by species specific classes of foliar N concentration (high N: high foliar N concentration; low N: low foliar N concentration). Norway spruce (PCAB) high N: n_{max}=8, low N: n_{max}=9; Scots pine (PISY) high N: n_{max}=9, low N: n_{max}=10; beech (FASY) high N: n_{max}=5, low N: n_{max}=4. For beech the difference in means over the period 1970–2000 between the strata is significant ($P(|t_{\alpha(2),(v1+v2)}| \geq t) < 0.05$); for PCAB and PISY not ($P(|t_{\alpha(2),(v1+v2)}| \geq t) > 0.05$). The regression models for the two foliar [N] strata are different for all three species (Table 3).

significant. Results for beech and pine differ from spruce: the median height growth acceleration of beech trees with higher foliar [N] (40.4 ± 42.9 (SD) %) is more than double that of trees with lower foliar [N] (17.4 ± 19.5 (SD) %) (significant). The median height growth acceleration of pine trees with higher foliar [N] is 29.40 ± 37.0 (SD) %, and that of pine trees with lower foliar [N] is 10.8 ± 38.6 (SD) % (not significant).

Detection and identifiability of the effect of foliar nitrogen nutrition level is complicated (1) because of possible dilution effects, e.g. if needle/leaf mass increased proportionally more than the N content, [N] is decreased due to dilution even so total N content might have increased, and (2) because of confounding, particularly with respect to changes in seasonal temperature: sites with high foliar nutrition are linked with seasonal air temperature increase for beech, and decrease for spruce. However, in both cases differences in means are not significant (Table 2).

4.2.4 Mean changes in height growth in relation to levels of putative causal factors

In this section relationships between the plot-wise temporal means of the response variable Δih and the plot-wise mean levels of the treatment factors *change in seasonal air temperature, change in seasonal precipitation, level of foliar N nutrition,* and level of *N deposition* are analysed graphically using bivariate scatter plots, and quantified using regression analysis.

4.2.4.1 Mean changes in height growth in relation to changes in air temperature and precipitation

In Figure 12 and 13 the intercepts, b_0, and slopes, b_1, of simple linear regression models of Δih on time as the predictor variable, fitted individually to each plot-sample, are displayed versus the level of change in mean seasonal temperature and change in seasonal precipitation at the site. The intercepts characterize the starting level of change in height growth at the beginning of the age-overlap period between the sample tree cohorts. The slopes are the coefficients of the regression models, describing the upward (positive) or downward tendency (negative sign) of the Δih-series over time. When assessing the results the inverse inter-dependency of the intercept and the slope coefficients needs to be considered. To parameterize the relationships, simple linear regression models have been calculated for the bivariate plots of intercept and slope (as response variable) versus level of the treatment factor (as predictor variable) (solid lines in Figure 12 and 13). For beech the starting level of height growth acceleration is positively related whereas the slope is negatively related to the change in mean seasonal temperature (p=0.022 and p=0.052 respectively, Figure 12). The other data do not support a significant relationship neither between the starting level nor the slope of height growth changes and the climate change variables (Figure 12 and 13). Also, the pooled time-series cross-sectional analyses with a variable-intercept model (Chapter 2.2) did not reveal significant relations for neither of the two climate change variables.

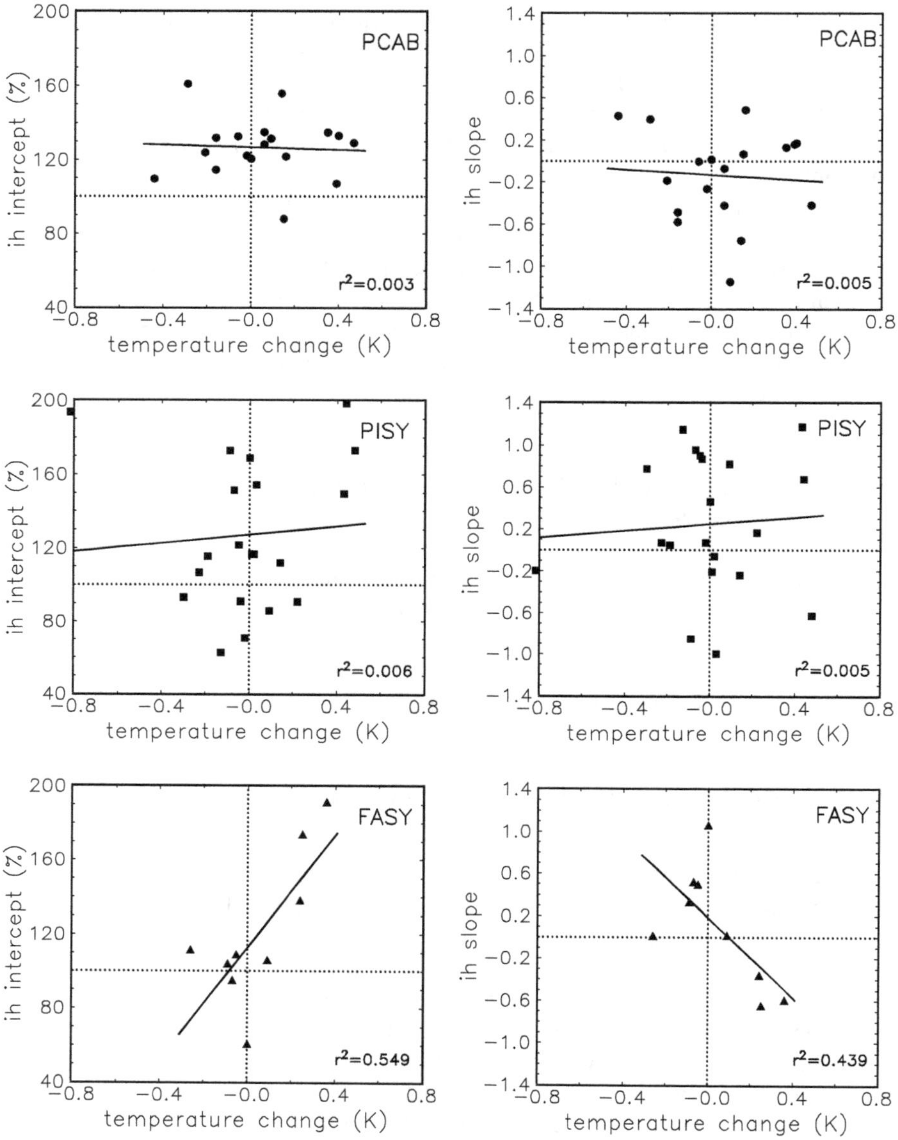

Figure 12. Intercepts (b_0, left) and slopes (b_1, right) of linear regression models of Δih (%) versus time t ($\Delta ih(\%) = b_0 + b_1 t$) for Norway spruce (PCAB), Scots pine (PISY) and beech (FASY) plotted versus change in mean seasonal temperature (Δtp_{5-9} in K) between the periods where the cohort-pairs overlap in age. Lines are fitted by simple linear regression. The model for the intercept of beech is significant ($P(|t_{\alpha(2),v-2}| \geq t) = 0.0224$), all other models are not significant ($P(|t_{\alpha(2),v-2}| \geq t) > 0.05$).

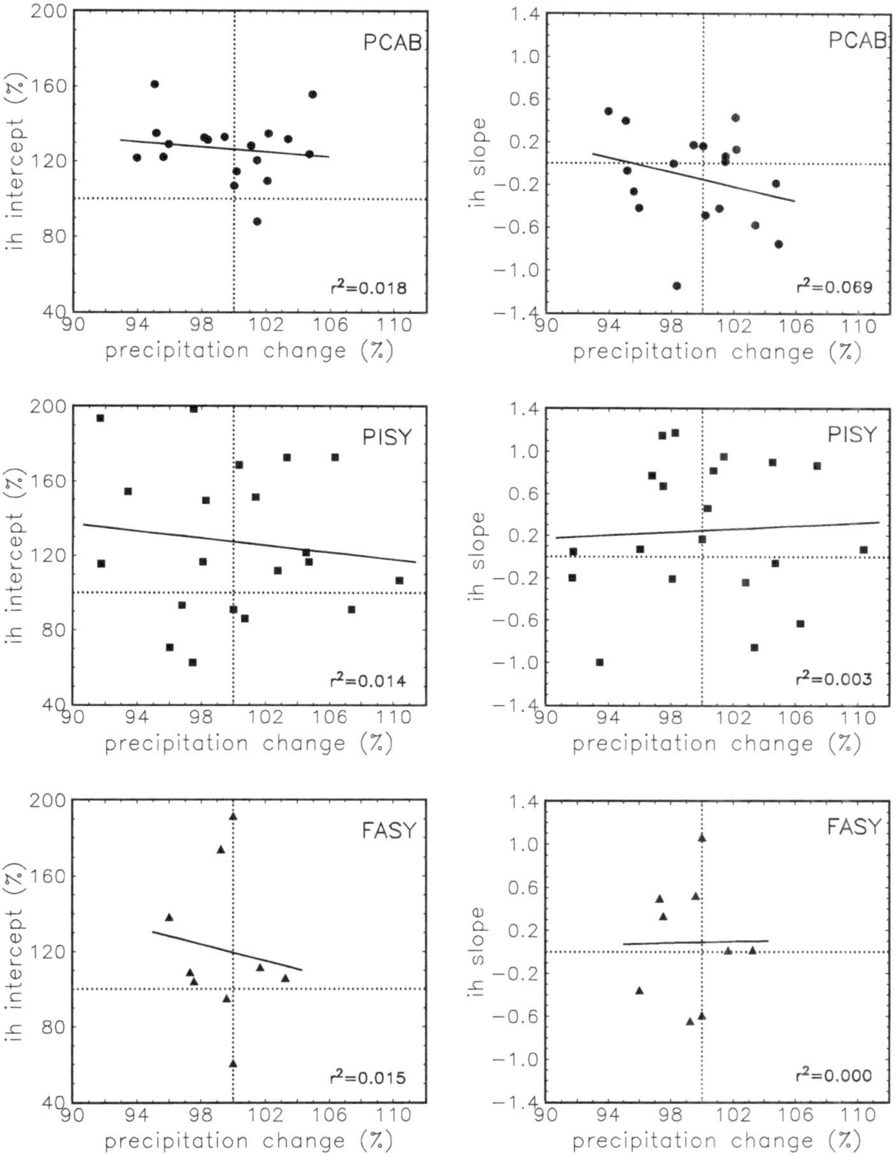

Figure 13. Intercepts (b_0, left) and slopes (b_1, right) of linear regression models of Δih (%) versus time t ($\Delta ih(\%) = b_0 + b_1 t$) for Norway spruce (PCAB), Scots pine (PISY) and beech (FASY) plotted versus change in mean seasonal precipitation (Δpr_{5-9} as a percentage) between the periods where the cohort-pairs overlap in age. Lines are fitted by simple linear regression. None of the models is significant ($P(|t_{\alpha(2),\nu-2}| \geq t) > 0.05$).

4.2.4.2 Mean changes in height growth in relation to changes in climatic water balance

Effects of long-term changes in water supply on changes in tree height growth were also of interest. Since we were lacking suitable long-term data on soil moisture content or soil moisture potential and its temporal variation for all the sampling sites we considered the climatic water balance to be an adequate approximation of the atmospheric water supply for the trees. Therefore Thornthwaite's formulae (Thornthwaite and Mather, 1955 and 1957) were applied to the CRU 0.5 data to calculate time series of the seasonal climatic water balance for each sampling site.

Intercepts and slopes of linear regression models of height growth changes of the Norway spruce, Scots pine and beech sample trees plotted versus the change in seasonal climatic water balance are shown in Figure 14. The starting level of height growth acceleration of Scots pine trees at sites where climatic water balance decreased was significantly higher than at sites where atmospheric water supply was enhanced. The same applies for the slope of Δih of beech: the larger the decrease in climatic water balance the larger the increase in slope.

4.2.4.3 Mean changes in height growth in relation to foliar nitrogen nutrition

In Figure 15 the intercepts and slopes of the Δih-regression models are plotted versus mean foliar [N]. None of the presented regression models (solid lines) is significant, and also the variable intercept model does not indicate significant relations. However, noticeable features in these graphs are the either indifferent (spruce) or increasing tendencies of the intercepts and the downward trends of the slopes with respect to increasing foliar nitrogen nutrition.

4.2.4.4 Mean changes in height growth in relation to nitrogen deposition

Figure 16 shows the spatial distribution of the total N (reduced N and oxidised N) deposition rate over Europe in the year 1999 according to the EMEP deposition model. Total N deposition rates overall Europe in this year ranged between less than 5 kg and more than 30 kg N per hectare and year. The dominant large-scale pattern over Europe is a latitudinal zoning with low deposition rates close in magnitude to the assumed natural background values in the north, maximum deposition rates in the centre and central-western part, and again lower deposition rates in most of the Mediterranean region. From central to eastern Europe deposition rates decrease gradually. We assume that this large-scale spatial pattern would also have been characteristic for the relative N deposition regimes in earlier decades where no such data are available. However, the absolute level of N deposition has changed substantially since then.

The large-scale pattern of N deposition closely resembles the spatial distribution of mean foliar [N] of Scots pine and Norway spruce (Figure 2). Similar relations are described by de Vries et al. (2000). However, the spatial distribution of foliar [N] of beech (Figure 3) does not match as well as for the other species.

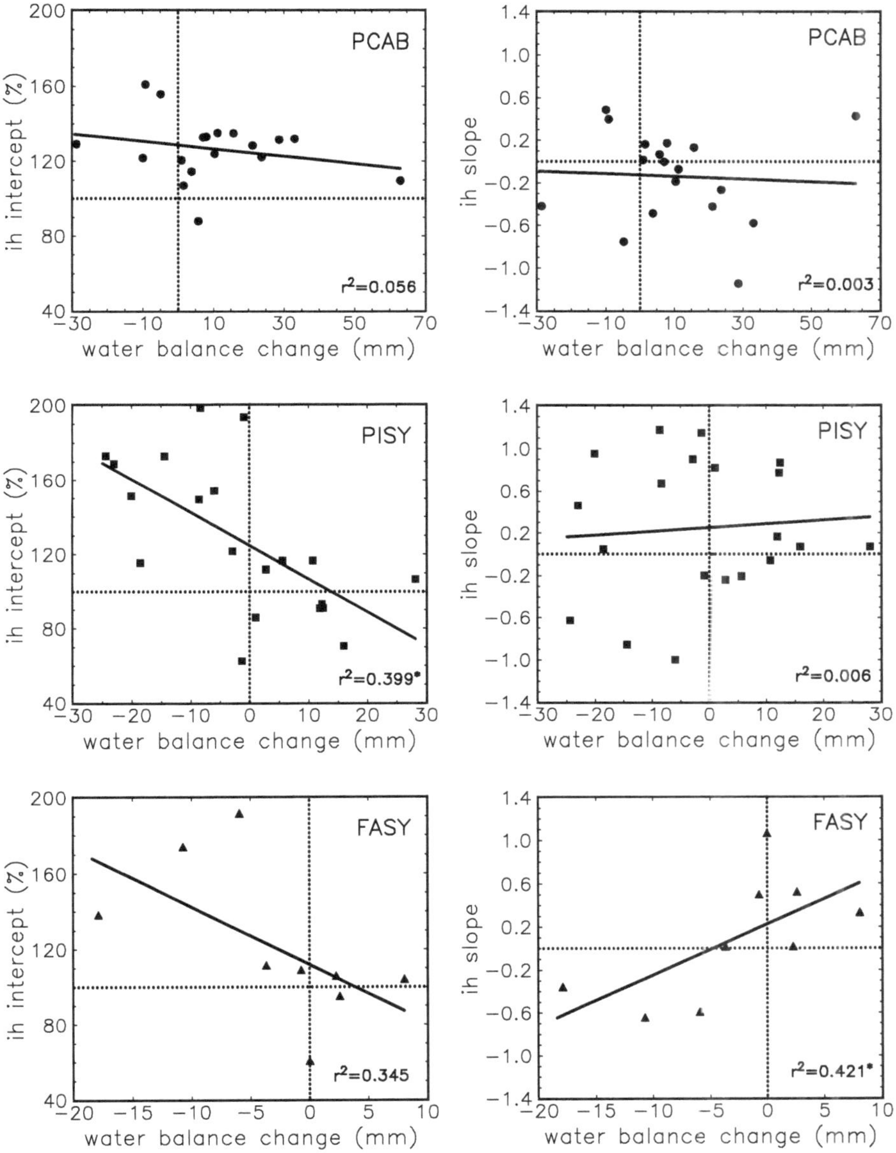

Figure 14. Intercepts (b_0, left) and slopes (b_1, right) of linear regression models of Δih (%) versus time t ($\Delta ih(\%) = b_0 + b_1 t$) for Norway spruce (PCAB), Scots pine (PISY) and beech (FASY) plotted versus change in seasonal climatic water balance (mm) between the periods where the cohort-pairs overlap in age. Lines (solid) are fitted by simple linear regression. The models for the intercept of pine and for the slope of beech are significant ($P(|t_{\alpha(2),19-2}| \geq t)$ =0.0037 and $P(|t_{\alpha(2),9-2}| \geq t)$ =0.0312 respectively). Note the different scalings of the x-axes.

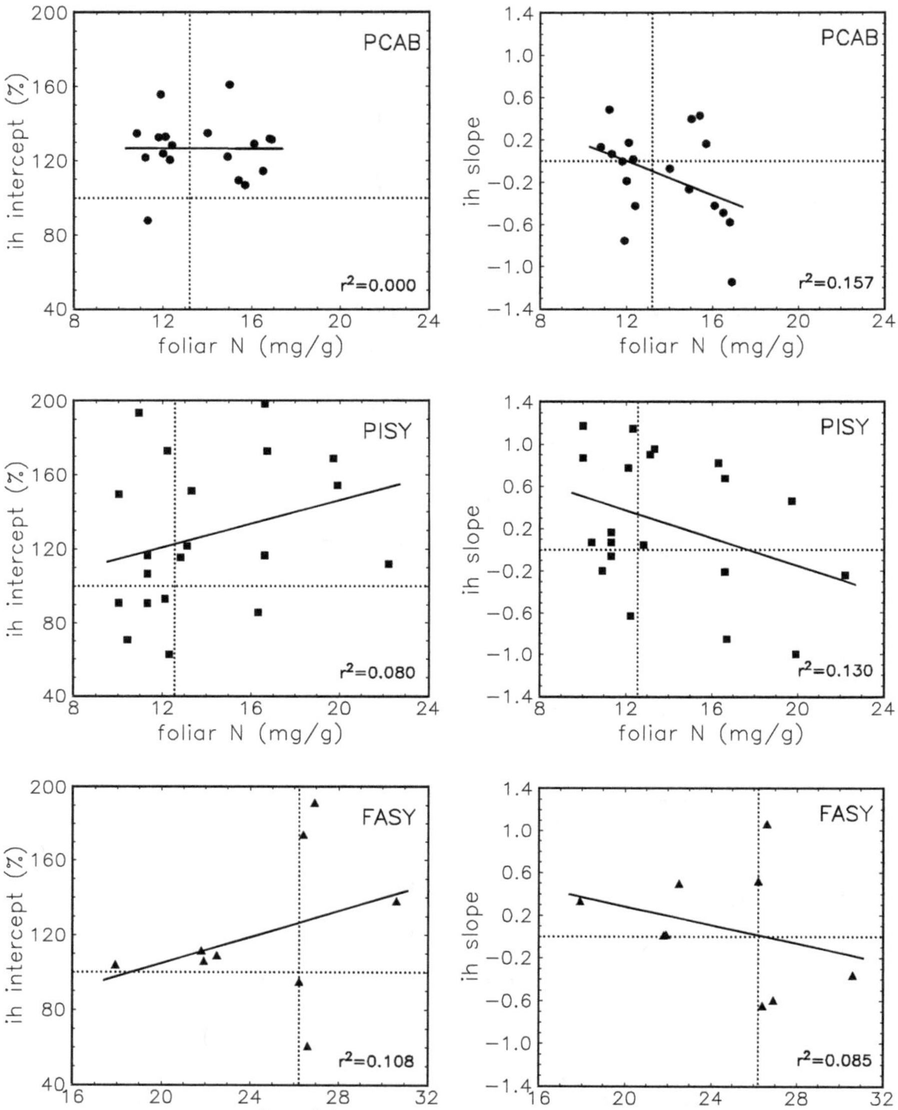

Figure 15. Intercepts (b_0, left) and slopes (b_1, right) of linear regression models of Δih (%) versus time t ($\Delta ih(\%) = b_0 + b_1 t$) for Norway spruce (PCAB), Scots pine (PISY) and beech (FASY) plotted versus mean foliar N concentration of the stand (vertical dotted lines: median foliar N concentration). Lines (solid) are fitted by simple linear regression. None of the models is significant on the 5 %-level ($P(|t_{\alpha(2),v-2}| \geq t) > 0.05$). Note the different scalings of the x-axis.

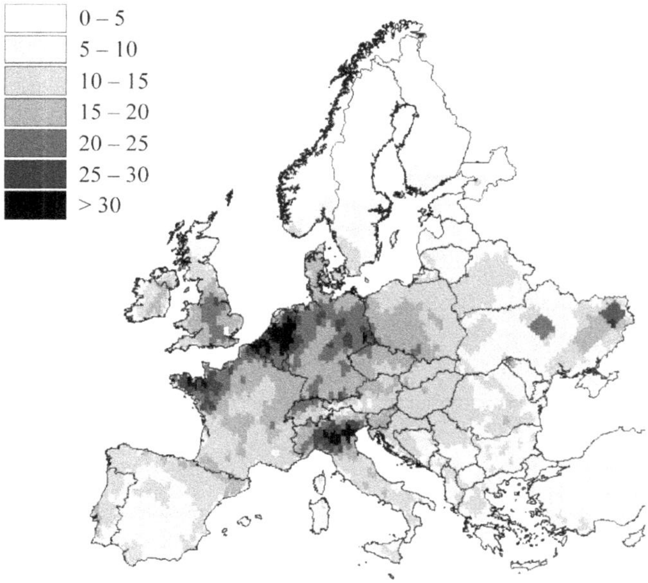

Figure 16. Spatial distribution of the total nitrogen deposition (kg $ha^{-1} yr^{-1}$) (reduced and oxidized nitrogen) in 1999 according to the EMEP deposition model (data source: EMEP, http://emep.int).

Figure 17 shows the bivariate plots of the intercepts and slopes of the Δih-regression models versus total N deposition in the year 1999. The regression model for the slope of Norway spruce is significant (p=0.037): from 1960–2000 the temporal trend of height growth acceleration is negatively related to the N deposition. The same tendency is indicated for the other species.

4.2.4.5 Mean changes in height growth in relation to site index

In Figure 18 intercept and slope of the Δih-regression models are plotted versus height of the older tree cohort at age 30 years (site index h_{30}). The site index of the older trees is regarded as a proxy for the *native* site productivity, i.e. the reference site productivity in the past. Site indices vary largely among the sampled stands and range from h_{30}=5 m to h_{30}=20 m.

For the Scots pine and beech sample, there is a significant inverse relation between the intercept of height growth change and the site index h_{30}: the larger the site index, the lower the level of change in height growth (p=0.005 and p=0.030 respectively). Also for the beech and spruce, height growth acceleration tends to decrease with increasing site index.

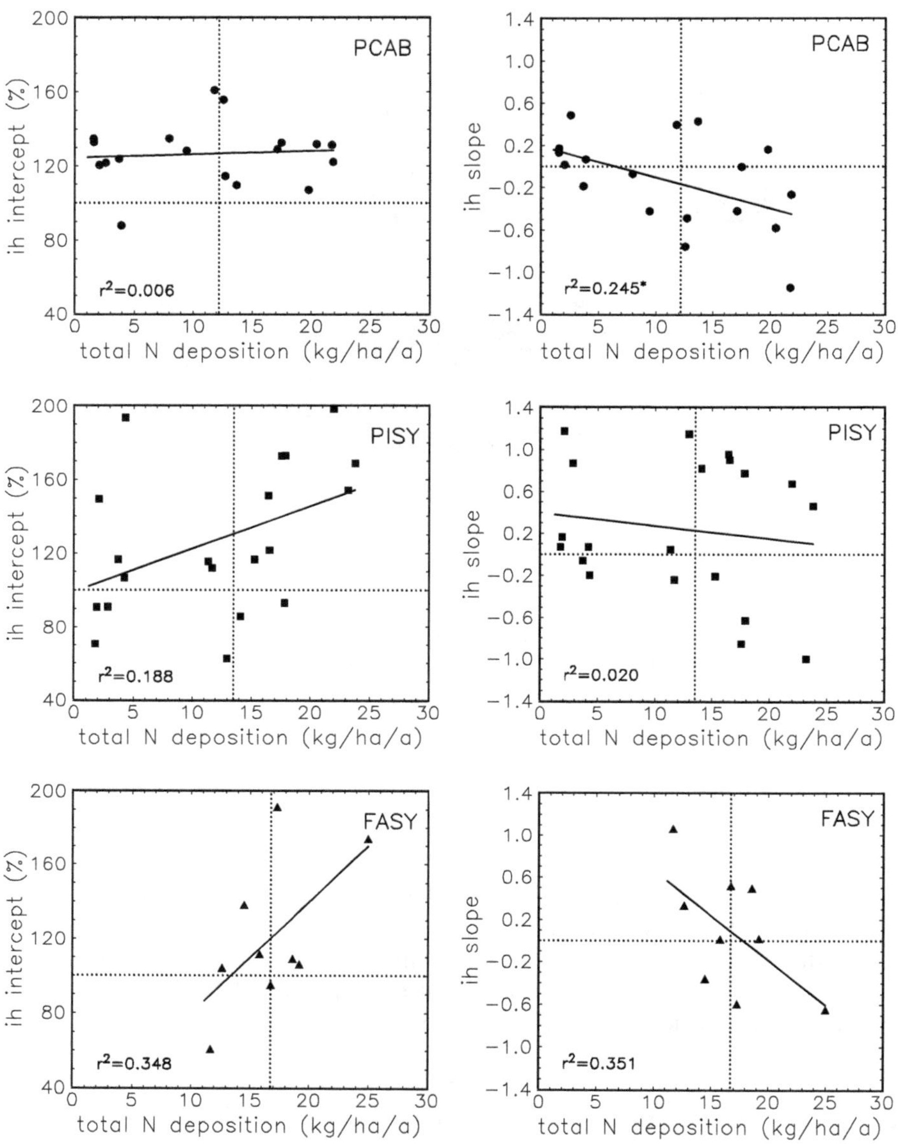

Figure 17. Intercepts (b_0, left) and slopes (b_1, right) of linear regression models of Δih (%) versus time t ($\Delta ih(\%)=b_0+b_1t$) for Norway spruce (PCAB), Scots pine (PISY) and beech (FASY) plotted versus modelled total N deposition rates in the year 1999 (vertical dotted lines: median N deposition rate). Lines (solid) are fitted by simple linear regression. The model for the slope of spruce is significant ($P(|t_{\alpha(2),18-2}| \geq t)=0.0367$).

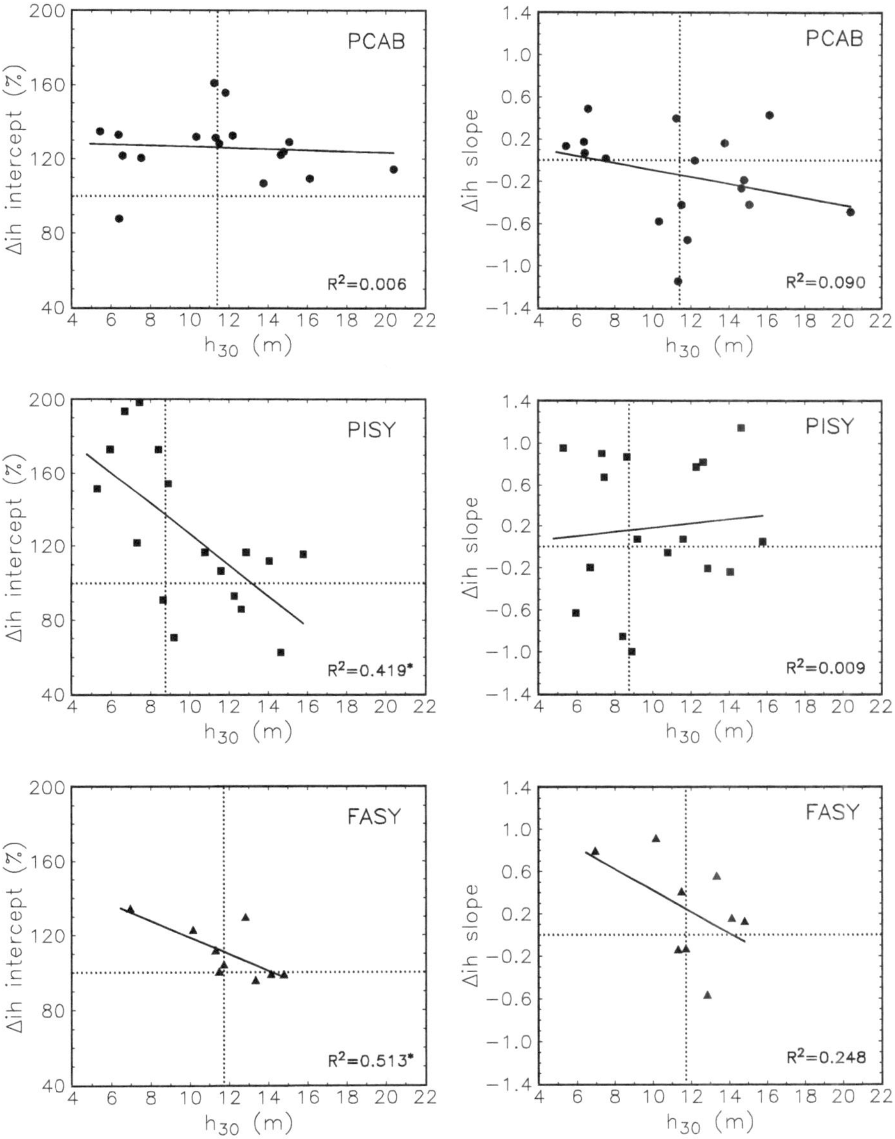

Figure 18. Intercepts (b_0, left) and slopes (b_1, right) of linear regression models of Δih (%) versus time t ($\Delta ih(\%) = b_0 + b_1 t$) for Norway spruce (PCAB), Scots pine (PISY) and beech (FASY) plotted versus site index of older trees (h_{30}: average tree height base age 30) (vertical dotted lines: median site index). Lines (solid) are fitted by simple linear regression. The models for the intercept of pine and beech are significant ($P(|t_{\alpha(2),19-2}| \geq t)=0.0050$ and $P(|t_{\alpha(2),9-2}| \geq t)=0.0300$ respectively).

4.2.5 Spatial patterns of height growth changes

4.2.5.1 Mean changes in height growth in relation to latitude and longitude

Bivariate scatter plots of changes in height growth versus geographical coordinates are presented in Figure 19 and 20. The analysed plots span a wide geographical range from 47.0° to 66.4° North, and –1.8° to 29.3° East. None of the presented regression models explains a significant part of the overall variability in height growth change. However, the following tendencies are indicated: (i) the starting level of change in height growth decreases with increasing latitude for pine and beech; and (ii) during 1960–2000 height growth acceleration tended to increase with increasing latitude for all species. Similar tendencies are indicated in relation to longitude for pine and beech: from West to East the starting levels of change decrease, and height growth acceleration rates increase.

The geographical location is not a growth factor *per se*; rather, it is an expression of the action of a complex of site factors on growth which are linked to latitude (e.g. radiation balance, day-length, level of air temperature) or longitude (e.g. degree of continentality, frequency and intensity of frost events).

4.2.5.2 Spatial pattern of height growth changes of Norway spruce and Scots pine

Based on the assumption that differences in height growth changes between spruce and pine are rather due to differences in site conditions than to differences in specific physiological characteristics, we merged both data sets and parameterized a mixed spatially explicit model. Figure 21 shows a map of the mean annual height increment of Norway spruce and Scots pine in central and northern Europe during the period 1960–2000 in percent of the reference trees. The reference trees were on average 35 years older than the report trees (age difference for spruce: 39 years, for pine: 32 years). For the geostatistical model collocated co-Kriging was applied, in which mean level of foliar [N] in the period 1993–1998 was used as the auxiliary variable.

According to the geostatistical model largest mean increases in height growth rates of Norway spruce and Scots pine occurred in an area extending from central Norway in the north, over eastern Germany, western Poland to the Czech Republic and Slovakia in the south. Another area of growth acceleration is centred on southern Finland. The model suggests a decrease in height growth in northern Finland. However, the isolines of relative height growth rates outlining the area covered by the model tend to be longitudinally oriented, and otherwise do not show a clear latitudinal gradient.

4.2.6 Discussion

In this study we identified and analysed changes in height growth of Norway spruce, Scots pine and beech on sites across Europe. The aim of our approach was not to conduct a representative sample survey on growth trends of the respective

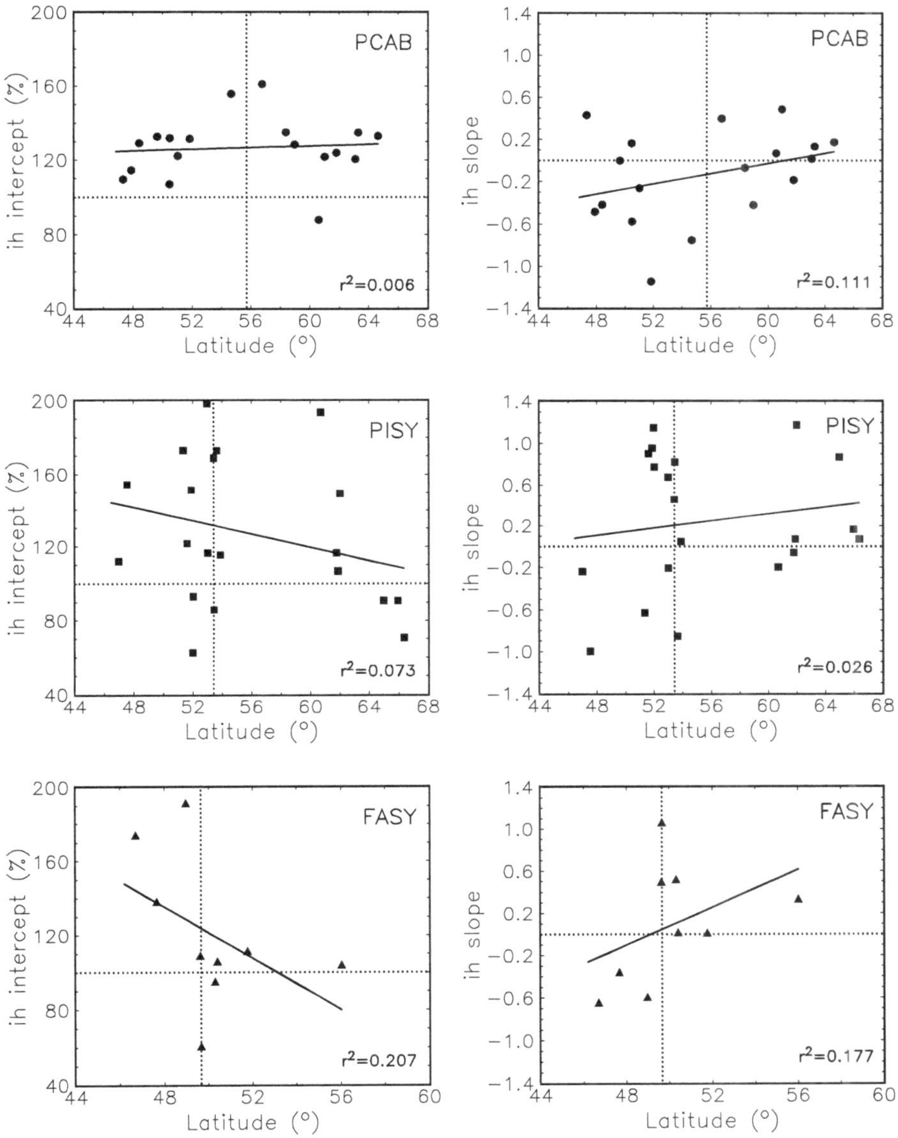

Figure 19. Intercepts (b_0, left) and slopes (b_1, right) of linear regression models of Δih (%) versus time t ($\Delta ih(\%) = b_0 + b_1 t$) for Norway spruce (PCAB), Scots pine (PISY) and beech (FASY) plotted versus latitude (vertical dotted lines: median latitude). Lines (solid) are fitted by simple linear regression. In no case is the slope of the linear regression model significantly different from zero ($P(|t_{\alpha(2),v-2}| \geq t) > 0.05$).

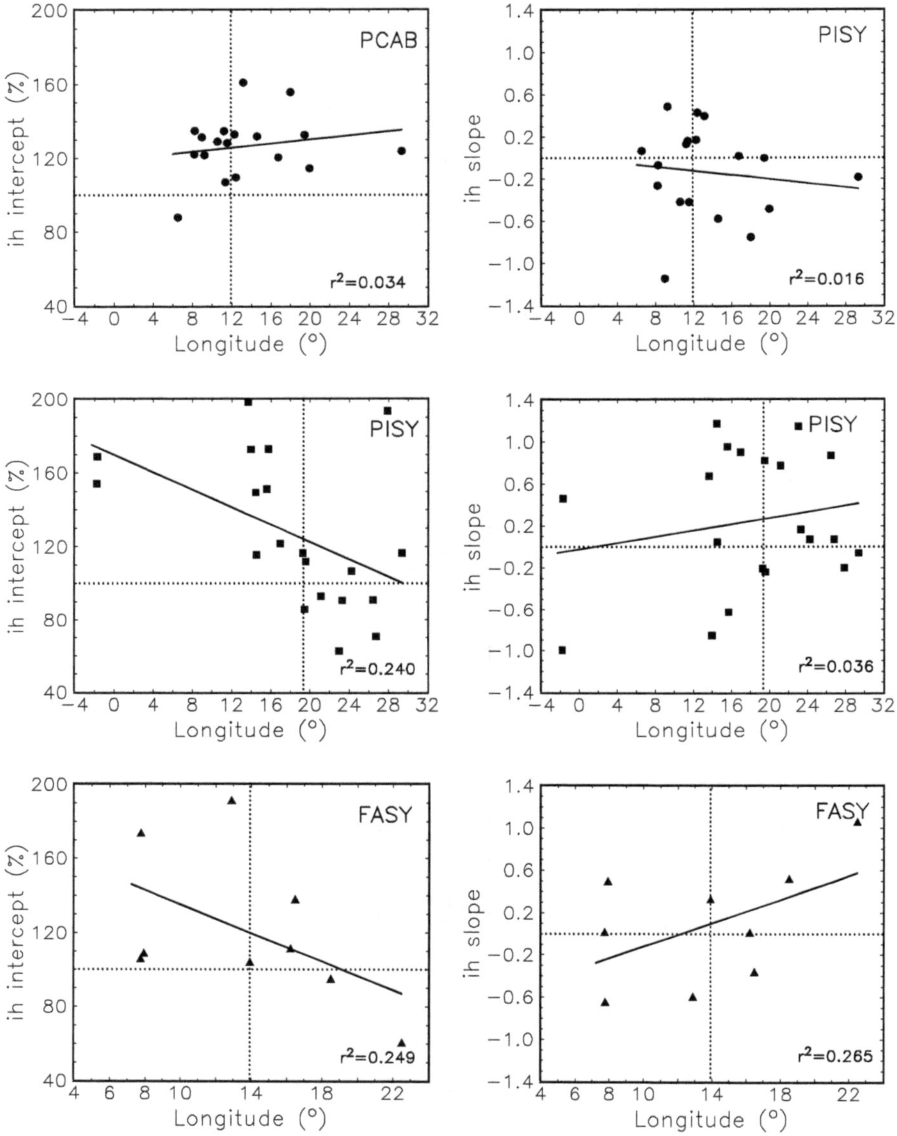

Figure 20. Intercepts (b_0, left) and slopes (b_1, right) of linear regression models of Δih (%) versus time t ($\Delta ih(\%) = b_0 + b_1 t$) for Norway spruce (PCAB), Scots pine (PISY) and beech (FASY) plotted versus longitude (vertical dotted lines: median longitude). Lines (solid) are fitted by simple linear regression. In no case is the slope of the linear regression model significantly different from zero ($P(|t_{\alpha(2),v-2}| \geq t) > 0.05$).

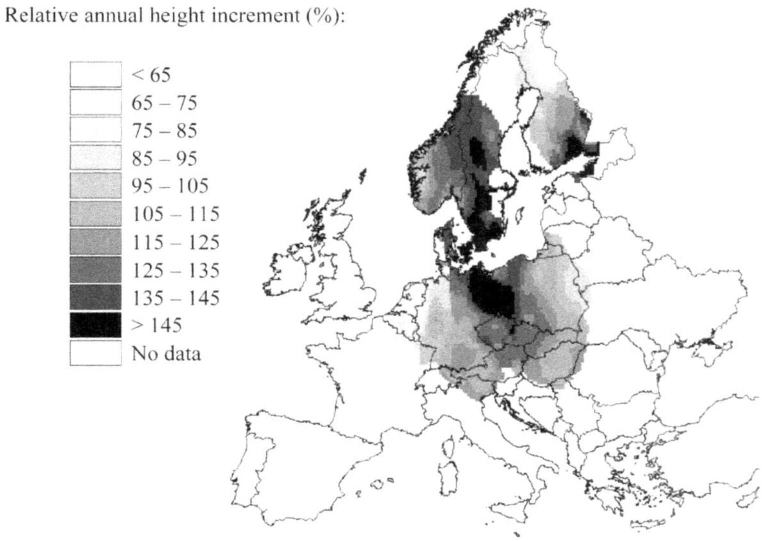

Figure 21. Mean annual height increment during the period 1960–2000 of Norway spruce and Scots pine in percent of the height increment nowadays older trees had when they had the same age. Data are spatially modelled by applying collocated co-Kriging (auxiliary variable: level of foliar [N] 1993–1998). The boundaries of the spatial model are determined by an uncertainty threshold (square root of Kriging variance SD ≤ 13%). Number of plots used for model parameterization: n=30. Two plots have been excluded because of their isolated location (one each in the UK and France). Four plots in the very north of Finland have been excluded in model parameterization, but have been included in model estimation.

tree species in European forests, but to use the heterogeneity in environmental conditions over Europe as a large-scale basis for a quasi-experimental factorial treatment design to investigate possible causes of growth trends (Rehfuess et al., 1999).

4.2.6.1 Temporal trends in height growth changes of the different tree species

We could show that the mean annual height growth rates during the last 40 years for all three species are significantly larger than that of older trees on similar sites at the same age. For the period 1960–2000 the median increase in annual height increment of trees sampled across Europe was +22.8% for Norway spruce, +24.8% for Scots pine, and +25.1% for beech (Figures 6 and 7; Table 2). Assuming adequacy and validity of the applied yield level models – that is the functional relationship between total wood volume production potential (TVP) and stand height – the effect of these changes in height growth on changes in TVP was calculated to be +56.4% for spruce, +44.9% for pine, and +74.0% for beech (Figure 8).

Analysis of the development of the height-to-diameter-ratio (h/d-ratio) of the spruce and pine sample trees revealed no difference either in the level of the h/d-ratio or in its long-term development between the age cohorts (Figure 5). As the h/d-ratio is an indicator of the competitive status and its changes over time, unchanged h/d-ratios provide evidence that there is (i) no difference in the competitive status, i.e. sample trees belong to the dominant crown class, and (ii) no difference in the growth history which is mainly determined by the stand dynamics the trees experienced, between the young and the older sample trees. Since in literature no such relation between the h/d-ratio and the competitive status and its changes has been established for beech, this type of analysis has only been conducted for the coniferous species.

Furthermore, increased height growth (Figure 6 and 7) associated with unchanged h/d-ratios for the spruce and pine sample trees (Figure 5) provide evidence for isometric changes of growth. To achieve isometry the increase in height growth needs to be paralleled by an equal increase in diameter growth. Based on this assumption the observed mean height growth acceleration for the period 1960–2000 results in a proportional mean increase in single tree volume increment at the power of three, which means +85% for spruce, +94% for pine, and +96 % for beech. However, this is only an arithmetic example since it refers to the individual tree level and cannot be directly scaled up to the stand level.

The results achieved here, in general, are well in agreement with the findings compiled and summarized by Spiecker et al. (1996) and more recently reviewed in Spiecker (1999) and Hunter and Schuck (2002). Since methodology, sampling design and data sources as well as regional and temporal coverage often differ among growth trend studies, the magnitudes of reported changes cannot be directly compared. However, in the rest of this section, results of some more recent studies are reviewed to evaluate their magnitudes and ranges.

Based on retrospective data from stem analyses Untheim (1996a) investigated trends in height growth of Norway spruce and beech in two growth regions in south-western Germany. During the period 1950–1995 height growth of dominant Norway spruce increased by 20%, that of beech by 5–10%. The calculated increases in total volume production based on general yield level models were 50–60% for spruce, and about 20% for beech. Pretzsch and Utschig (2000) analysed growth trends of Norway spruce stands in Bavaria (south-eastern Germany) based on data from regional and national forest inventories and from long-term and temporal research plots. In all investigated regions the current and mean annual wood volume increment (MAI) is significantly larger than predicted by yield tables. Especially the presupposed decline in growth rate with increasing age in mature stands is much slower than predicted by yield tables. On many sites in Bavaria mean actual volume growth rate increased by 20–35% as compared to 1930. In another growth trend study recently conducted in Bavaria, Reimeier and Kennel (2001) analysed data from repeated forest inventories in the period 1984–1998. The calculated MAI during this period is referred to the respective values from conventionally used yield tables. According to this study the MAI of stands of average site index, with a 'normal' stand basal area equal to the yield table, was for spruce at the base age of 100 years enhanced by 151%, for pine by 118% and for beech by 125%. The deviation in volume growth rate is larger, the older the stand and the lower the site fertility. For instance, the deviation for spruce on the highest and lowest quality sites is 97% and 200%, respectively. For 50-year-old compared to 100-year-old stands, it is 13–95% and 97–200%, respectively. Since the respective yield tables applied as references in this study have been published in 1963 for spruce (Assmann and Franz), 1948 for pine (Wiedemann), and 1932 for beech (Wiedemann), and the data used for their construction have been collected over several decades, it is not possible to refer the observed increase to a specific period of time.

Cannell et al. (1998) compiled results from several growth studies conducted in the UK during the 1980s and 1990s. These site related studies have shown that the maximum mean annual volume increment of Sitka spruce plantations in the northern UK increased between the 1930s and 1970s by 1.0 $m^3 \cdot ha^{-1} \cdot yr^{-1}$ per decade, and by up to 1.2 $m^3 \cdot ha^{-1} \cdot yr^{-1}$ per decade on better quality sites in Scotland. Given mean values for Sitka spruce over the period of study of 14.0 $m^3 \cdot ha^{-1} \cdot yr^{-1}$ in the northern UK, and 18.9 $m^3 \cdot ha^{-1} \cdot yr^{-1}$ in Scotland, this suggests that between the 1930s and 1970s volume growth rates have increased by about 20–40%.

Kiviste (1999) tested the hypothesis weather site index of forest stands in Estonia changed between the 1950s and the 1990s by using data from repeated inventories. The average increase in site index (stand height at age 50 years) over the approximately 40 years period was 1.6 m or +10% (referred to the mean site index in 1950s) for spruce stands and 2.1 m or +19% for pine stands. Based on extensive sample tree data from the Swedish National Forest Inventory, Elfving and Tegnhammer (1996) found an increase in tree height growth of 0.5–0.8% per year for the period 1953–1992. Acceleration in mean annual height increment at age 50 years over the 40 years period was 34% for Norway spruce, and 24% for Scots pine.

Besides the magnitudes of the changes in growth the temporal development of growth changes during recent decades is of special interest with respect to potential causes. As indicated by the samples of all three species in the presented study the level of the height growth acceleration rate is surprisingly constant over the last 40 years, suggesting that the curvature of the age-height relation remained more or less unchanged, whereas the slope of the curve is steeper and the level of growth is proportionally shifted upwards. Similar changes in height growth have been found by Kenk et al. (1991) for Norway spruce in German forests and Elfving et al. (1996) for Norway spruce and Scots pine in Swedish forests: trees have grown faster, but have still followed selected isomorphic site index curves. Others, such as Pretzsch (1992), Reimeier and Kennel (2001) and Moosmayer et al. (2003), who analysed growth of forests in Germany, identified increasing discrepancies between observed growth and site index curves from yield tables. This is attributed by the authors to both, changes in growth, and to the fact that the yield tables used as references are not – and might have never been – adequate for the analysed material. Schöpfer et al. (1994) found for Norway spruce that the age-height-trajectories of most of the trees growing today follow the same curvature as older trees did in the past, but at an augmented level of growth.

4.2.6.2 Projected effects on total wood volume production potential

Using the observed increase in height growth as the input variable for conventional yield level models, this study shows that the average increase in total wood volume production potential (TVP) between 1960 and 2000 amounts to +56% for spruce,

+45% for pine, and +74% for beech (Figure 7). Because of the non-linear relation between TVP and stand height the relative increase in TVP exceeds the observed relative increase in stand top height on average by 147% for spruce, 81% for pine, and 195% for beech. The following conditions have to be fulfilled that these model results can be regarded as relevant (Assmann, 1959): (i) the sample trees have to be representative for the stand top height, (ii) the underlying yield level models are adequate for the study material, (iii) the yield level models are also valid under conditions of changes in growth. Results from an earlier study indicated that it might not be possible to find individual sample trees which represent within close confidence intervals by their own growth the past volume growth of the whole stand over a long-term period (Spiecker, 1992).

The aim of the sampling was to draw a representative sample of the stand collective of top height trees (see Chapter 2.2). According to the applied yield tables for Norway spruce and Scots pine this is defined as the 200 thickest trees per hectare, for beech the 100 thickest trees per hectare (Ministerium für Ländlichen Raum Baden-Württemberg, 1993). Dominant trees have been selected as sample trees but no breast height diameter measurements have been conducted for the stands to verify if the trees belong to the respective stand collective.

The applied yield tables are based on data from research plots of the former Prussian Forest Research Institute mainly located in central and eastern Germany. It has not been tested if the yield level models are adequate for the samples on the European level. Evidence for differences in yield levels within and between regions can be found in the literature (e.g. Mitscherlich, 1958 and 1959; Assmann, 1959; Assmann and Franz, 1963; Bradley et al., 1966; Hamilton and Christie, 1971; Schmidt, 1973; Kennel, 1973; Bergel, 1985). Mitscherlich (1963) compiled and analysed growth data on Norway spruce from long-term research plots in different European countries stretching from French-Savoyen in the south-west to Romania in the south-east, and from Scotland in the north-west to Finland in the north-east. Remarkable differences in yield levels were found: the total wood volume production at stand height 20 m ranged from 480 $m^3{\cdot}ha^{-1}$ to 950 $m^3{\cdot}ha^{-1}$ with an overall mean of 620 $m^3{\cdot}ha^{-1}$. However, no unique geographical trend in yield level was detectable across the part of Europe covered by the data. In the northern zone a clear trend was found from maximum yield levels in the Atlantic region of the UK (120%, referred to overall mean) to minimum yield levels in continental Finland (83%). The mean for the northern zone was 103%. A similar, but flatter West-East gradient with a lower maximum was identified for the middle zone (zonal mean: 90%). The mean yield level in the southern zone was largest among all three zones (zonal mean: 107%). The yield level of Norway spruce stands at high altitudes in the Alps and the Black Forest (550–1150 m above sea level: 106%) was significantly higher than on lowland sites (<550 m above sea level: 97%).

The approach selected in this study to expand from changes in stand height increment to changes in stand volume production is a rather generic one. Hence, the simplistic assumption of a species-specific, but unique (i.e. independent of site index) general yield level across Europe was adopted.

In order to test the validity of the relation between stand height and total wood volume production and its constancy over time, detailed data from long-term forest research plots need to be examined. Only very few recent studies on this issue can be found in literature. Moosmayer and Ripberger (1990), Weise (1991), and Untheim (1996b) compiled old and more recent data on stand volume growth from long-term research plots in south-western Germany and could not find evidence that the general yield levels for Norway spruce, silver fir, and beech have changed during recent decades.

The figures on the effects of changes in height growth on total wood volume production presented here may overestimate the real magnitude across Europe. The reasons are: (i) typically the yield level decreases with increasing site index, whereas in this study a constant yield level was assumed – this effect could be especially severe for beech where the differences between yield levels are larger than for pine and spruce; and (ii) evidence from literature suggests lower yield levels in the continental and north-east parts of Europe, whereas in this study a species-specific but unique yield level, calibrated for forest stands in Germany, was used across Europe.

4.2.6.3 The role of changes in seasonal air temperature and precipitation

Mean height growth acceleration of spruce and pine did not differ significantly between the two temperature change strata – increase or decrease in seasonal temperature (Figure 9). The same is true for change in seasonal precipitation (Figure 10). For beech the starting level of height growth acceleration was positively related whereas the slope was negatively related to the change in mean seasonal temperature (Figure 12). However, because of confounding with foliar nitrogen nutrition, the effects of changes in temperature and precipitation cannot be unequivocally identified. Also the differences between the treatments are quite small: (i) the range of mean levels of change in seasonal temperature between the strata ranges from –0.20 K to +0.21 K for spruce, –0.13 K to +0.14 K for pine, and –0.10 K to +0.21 K for beech; and (ii) differences in seasonal precipitation range from –3.7% to +2.8 % for spruce, –3.2% to +2.9% for pine, and –1.5% to +1.3% for beech (Table 2).

The reason for the small differences between treatments and for the only weak orthogonality of the data is due to the fact that the *a posteriori* climate change classification differed markedly from the *a priori* classification. The variables *change in seasonal air temperature* and *change in seasonal precipitation* were initially fixed as referring to differences between the periods 1961–1990 and 1901–1930. However, the classification according to changes in climate during individual tree-cohorts' lifetimes was only possible after the sample trees had been cut and their age determined.

At the regional scale, the growth of forests in northern Europe is positively related to temperature, or more precisely to temperature sum above a threshold (e.g. Beuker, 1994). It is tempting to apply these relationships to predict effects of warming over time. However, whether temperature dependence of growth is related to the length of the active growing period, to the rate of gross photosynthesis or to other plant physiological processes can hardly be resolved since temperature affects almost all plant and also many soil processes. The effects air temperature has on photosynthesis and respiration could have been disentangled by separately analysing long-term trends in daily minimum and daily maximum temperatures together with the tree growth development. However, adequate regionalized long-term data on daily extremal temperatures were not available at the time the study was conducted. Besides the direct temperature effects the simultaneous indirect effects of temperature on water availability for plants via evaporation and vapour pressure deficit have also to be considered. Another constraint concerning identifiability of temperature change effects on growth changes in observational field studies results from multi-collinearity with multiple other – potentially growth relevant – quantities like solar radiation, relative air humidity, and precipitation (Cannell, 1999).

In temperate and northern Europe the role of water supply is often considered as not being a major driving factor for tree growth and tree growth variation over time. However, several studies have shown that even under humid conditions, water availability is of primary importance for growth dynamics of trees and forests (Mäkinen et al., 2002 and 2003; Kahle and Spiecker, 1996a and 1996b; Spiecker, 1991a and 1991b).

Analysis of the effect of changes in seasonal climatic water balance on long-term changes in height growth revealed in two cases significant relationships (Figure 14): the starting level of height growth acceleration of the Scots pine and the slope of Δih of the beech sample trees were higher at sites where climatic water balance decreased. These findings do not support the hypotheses (1) of a limitation of height growth of the pine and beech sample trees by climatic water supply in the past, and (2) of a significant role of changes in climatic water balance for the observed height growth changes of spruce.

4.2.6.4 The role of foliar nitrogen nutrition and nitrogen deposition

The magnitude of mean height growth acceleration of beech sample trees with higher foliar nitrogen concentration [N] is more than double that of trees with lower foliar [N] (Figure 11). The same tendency is indicated for pine. During the period 1960–2000 the rate of height growth acceleration decreased with increasing foliar [N] for all three species (Figure 15). The same relation is indicated for total nitrogen deposition (Figure 17).

Nitrogen is the nutrient which plants need in the largest quantities, and in many ecosystems such as temperate and boreal forests its availability is normally a limiting factor for plant production (Vitousek and Howarth, 1991; Oren et al., 2001). The supply of N also regulates many complex interactions within ecosystems, including interactions between different trophic levels (Högberg et al., 2000; Schulze, 2000). The effects of input of N into forest ecosystems via deposition are complex – N addition can result in both nutrient-enrichment (eutrophication) and acidification of the soil. As Kreutzer (1993) pointed out, the response of forest ecosystems to increased N supply varies, mainly depending on the site and stand properties. The previous N status of the system – that is the N pool before major parts of N have been exported from the forest e.g. by timber and wood harvesting and litter removal (Glatzel, 1991; Puhe and Ulrich, 2001; Hunter and Schuck, 2002) – is considered to be the most important factor determining the intrinsic N sensitivity. There is also a noticeable effect of tree species on N input:

evergreen species filter significantly more N (and other pollutants) from the air than deciduous tree species do under similar conditions owing to their often higher leaf-area-index and all-year foliation. Not only N deposition, but also N acquisition, differs between species: for the nutrition of beech ammonium-N plays a more important role than nitrate-N as compared to spruce (Bauer, 1997). Identifiability of the effect of nitrogen nutrition level on growth in addition is complicated because of possible dilution effects, e.g. if needle/leaf mass increased proportionally more than the N content, [N] is decreased simply due to dilution even so total N content might have increased. However, since we only had data on foliar [N] but lacked data on mass of needles/leaves we were not able to disentangle this issue.

In many parts of Europe where there is high N deposition, foliar [N] has increased in recent decades. This has been shown for spruce and pine (Hippeli and Branse, 1992; Spiecker et al., 1992; Heinsdorf, 1993; Bauer et al., 1997; Prietzel et al., 1997; de Vries et al., 2000; Heinsdorf and Branse, 2002) and for beech (Bauer et al., 1997; Flückiger and Braun, 1998; Duquesnay et al., 2000). However, there is no unique and direct link between N deposition and foliar N level of forest trees. Considerable local and regional variations in foliar [N] of spruce and beech were found along a transect extending from northern Sweden to central Italy by Bauer et al. (1997). Evidence from literature suggests that there is also intraspecific genetic variation in potentially adaptive traits related to nutrient acquisition for forest tree species (e.g. Oleksyn et al., 2002); however, its role remains poorly understood.

Kenk and Fischer (1988) compiled results from long-term forest fertilization plots in Germany to quantify the response of forest stands to N fertilization. Over all experiments the application of 100 kg N per hectare resulted in a mean increase in per hectare volume increment of 9% for Norway spruce and 21% for Scots pine, with considerable variation in each case. For both species the effect of fertilization was small both on very poor and on very fertile sites, but was independent of stand age. Also in fertilization experiments in Nordic countries which have recently been reviewed by Näsholm et al. (2000) N fertilization did not always enhance forest growth: no increase in growth was obtained on relatively fertile soils.

The impact of excess N on growth and mortality of Norway spruce and Scots pine has been investigated by several authors (Hofmann et al., 1990; Bergholm et al., 1993; Heinsdorf, 1993; Högberg et al., 2000; Näsholm et al., 2000; Heinsdorf and Branse, 2002; Heinsdorf and Beck, 2003). The applied levels of N input in most cases were extremely high. In addition, because experimental N treatments imposed in instantaneous, stepwise applications may invoke different responses than chronic, gradual changes, the findings from N treatment experiments usually cannot be directly scaled-up to field conditions. No general applicable threshold for N excess exists and the switch from increased to decreased tree growth rate in response to N additions in some experiments highlights the need for long-term studies as a necessary component of future research (Emmett, 1999).

Whether N fertilization affects stem allometric relations of Scots pine has recently been analysed by Anders et al. (2002). No systematic stem form changes could be found for low to medium N deposition rates. However, on sites with very high N deposition loadings both tree height as well as tree diameter growth were significantly reduced. Because tree height growth was more reduced this resulted in more blunt, compact stem forms.

Although additions of N does cause a number of changes in the biochemistry and structure of trees, which can in turn make them more sensitive – e.g. to frost, drought, insect and fungi attack – there are only a few clear indications in the literature of changes in trees' resistance to such stress factors (e.g. Spiecker, 1991a and 1991b; Näsholm et al., 2000).

4.2.6.5 The role of former site index

An inverse relation between the former site index (i.e. the site index of the older sample trees) and change in height growth was found for Scots pine and beech: the larger the sites index of the older trees, the lower the level of change in height growth (Figure 18). For beech and spruce the mean height growth acceleration rate during 1960–2000 declined for samples with higher site index.

The former site index is regarded as a proxy for the *native* site productivity. Whether the former site productivity is indeed reflecting the natural site potential, which is mainly determined by soil and climate factors, or whether it is already the result of human interaction by former land use, can hardly be distinguished. In some rare cases information about former land use may exist (e.g. Glatzel, 1991 and 1999). The most detrimental effect of grazing and litter raking on soil fertility is due to the massive N output associated with these usages. Whether the improvement in site quality is mainly due to 'natural' site recovery from former misuse, or due to enrichment in N via N deposition is also hard to disentangle. However, Hunter and Schuck (2002) note that the effect of former land use should be (i) absent in undisturbed forests, and (ii) spatially heterogeneous, and therefore separable from more evenly distributed agents like N deposition.

4.2.7 Conclusions

Based on growth data from widely scattered plots on the European level we could show that height growth has substantially increased during the last 40 years for the three tree species. Mean annual height growth rates of younger trees for the period 1960–2000 are significantly larger than that of older trees on similar sites when they had the same age. The median increase in annual height increment was +25% for beech and pine, and +23% for spruce. Because of the non-linear relation in conventional yield level models between total wood volume production potential (TVP) and stand height, calculated changes in TVP are even larger than the observed changes in height growth. The model-based results concerning change in TVP between 1960 and 2000 are: +56% for spruce, +45% for pine, and +74% for beech.

Analysis of the development of the height-to-diameter-ratio (h/d-ratio) of the spruce and pine sample trees revealed no difference between the age cohorts neither in the level of the h/d-ratio nor in its long-term development. Together with the finding of increased height growth this provides evidence for isometric changes of growth: increase in height growth was paralleled by an equal increase in diameter growth.

The missing strict orthogonality of the analysed treatment factors *change in seasonal air temperature, change in seasonal precipitation,* and *level of foliar nitrogen nutrition* prevented identification of the independent effect that these variables have on growth change. Especially change in air temperature and level of foliar nitrogen nutrition are mixed so that an evaluation of the specific effects of these factors is hampered. The observed correlation between the different factors' levels could in fact be the result of real interactions between the factors; a hypotheses that has not been tested within the project.

With respect to the effects of changes in the climatic water balance as a proxy for the water supply of the trees we conclude that height growth of the pine and beech sample trees was not severely limited by climatic water supply in the past, and that changes in water balance did not play a significant role for the observed height growth changes of the spruce sample trees.

Overall, the clearest signal is that of foliar nitrogen concentration [N]: beech and pine sample trees with above average foliar [N] in recent years show larger mean height growth acceleration. For spruce the reverse is true. However, the lower foliar [N] is, the larger is the increase in height growth acceleration during 1960–2000 for all three species. With regard to the role of former site index as a proxy for native site productivity it is shown for Scots pine and beech that the higher the site fertility is, the lower is the level of change in height growth.

From the very beginning we hypothesized that it was likely no single factor that caused growth trends on the European level, but that a spatially and temporarily varying combination of, probably species-specific, factors might be responsible for the observed phenomenon (Spiecker et al., 1996; Rehfuess et al., 1999; Hunter and Schuck, 2002). This, together with potential effects of more local factors such as differences in land-use history and history of stand dynamics might explain the large amount of variation in the data on growth changes.

References

Abetz, P., 1976. Beiträge zum Baumwachstum: Der h/d-Wert – mehr als ein Schlankheitsgrad! Forst- und Holzwirt 31: 389–393.

Anders, S., Beck, W., Bolte, A., Hofmann, G., Jenssen, M., Krakau, U.K. and Müller, J. 2002. Ökologie und Vegetation der Wälder Nordostdeutschlands - Einfluss von Niederschlagsarmut und erhöhtem Stickstoffeintrag auf Kiefern-, Eichen- und Buchen-Wald- und Forstökosysteme des nordostdeutschen Tieflandes. Dr. Kessel, Oberwinter. 199 p.

Assmann, E. 1959. Höhenbonität und wirkliche Ertragsleistung. Forstwissenschaftliches Centralblatt 78: 1–20.

Assmann, E. 1961. Waldertragskunde: Organische Produktion, Struktur, Zuwachs und Ertrag von Waldbeständen. Bayerischer Landwirtschaftsverlag, München. 490 p.

Assmann, E. and Franz, F. 1963. Vorläufige Fichten-Ertragstafel für Bayern. Institut für Ertragskunde der Forstlichen Forschungsanstalt München, München. 104 p.

Bauer, G. 1997. Stickstoffhaushalt und Wachstum von Fichten- und Buchenwäldern entlang eines europäischen Nord-Süd-Transektes. Bayreuther Forum Ökologie 53: 168 p.

Bauer, G., Schulze, E.D. and Mund, M. 1997. Nutrient contents and concentrations in relation to growth of Picea abies and Fagus sylvatica along a European transect. Tree Physiology 17: 777–786.

Bergel, D. 1985. Schätzverfahren zur Bestimmung des Ertragsniveaus von Fichtenbeständen. Centralblatt für das gesamte Forstwesen 102: 86–96.

Bergholm, J., Jansson, P.E., Johansson, U., Majdi, H., Nilsson, L.O., Persson, H., Rosengren-Brinck, U. and Wiklund, K. 1993. Air pollution, tree vitality and forest production – The Skogaby project, general description of a field experiment with Norway spruce in south Sweden. In: Nilsson, L.O., Hüttl, R.F., Johansson, U.T. and Mathy, P. (eds.). Nutrient Uptake and Cycling in Forest Ecosystems – Proceedings of a symposium held in Halmstadt, Sweden, 7–10 June 1993. Pp. 69–87.

Beuker, E. 1994. Long-term effects of temperature on the wood production of *Pinus sylvestris* L. and *Picea abies* (L.) Karst. in old provenance experiments. Scandinavian Journal of Forest Research 9: 34–45.

Bradley, R.T., Christie, J.M. and Johnston, D.R. 1966. Forest Management Tables. Forestry Commission Booklet. 218 p.

Cannell, M.G.R. 1999. Relative importance of increasing atmospheric CO_2, N deposition and temperature in promoting European forest growth. Causes and Consequences of Accelerating Tree Growth in Europe. European Forest Institute Proceedings 27: 25–41.

Cannell, M.G.R., Thornley, J.H.M., Mobbs, D.C. and Friend, A.D. 1998. UK conifer forests may be growing faster in response to increased N deposition, atmospheric CO_2 and temperature. Forestry: An International Journal of Forest Research 71: 277–296.

de Vries, W., Reinds, G.J., Dobben, H.v., Zwart, D.d., Aamlid, D., Neville, P., Posch, M., Auée, J., Voogd, J.C.H. and Vel, E.M. 2002. Intensive Monitoring of Forest Ecosystems in Europe – Technical Report 2002. Geneva, UN/ECE, EC, Forest Intensive Monitoring Coordinating Institute (FIMCI). 175 p.

de Vries, W., Reinds, G.J., Kerkvoorde, M.S.v., Hendriks, K.C.M.A., Leeters, E.E.J.M., Gross, K.P., Voogd, J.C.H. and Vel, E.M. 2000. Intensive Monitoring of Forest Ecosystems in Europe – Technical Report 2000. Geneva, UN/ECE, EC, Forest Intensive Monitoring Coordinating Institute (FIMCI). 191 p.

de Vries, W., Reinds, G.J., Salm, C.v.d., Draaijers, G.P.J., Bleeker, A., Erisman, J.W., Auée, J., Gundersen, P., Kristensen, H.L., Dobben, H.v., Zwart, D.d., Derome, J., Voogd, J.C.H. and Vel, E.M. 2001. Intensive Monitoring of Forest Ecosystems in Europe – Technical Report 2001. Geneva, UN/ECE, EC, Forest Intensive Monitoring Coordinating Institute (FIMCI). 177 p.

Duquesnay, A., Dupouey, J.L., Clement, A., Ulrich, E. and Le Tacon, F. 2000. Spatial and temporal variability of foliar mineral concentration in beech (*Fagus sylvatica*) stands in northeastern France. Tree Physiology 20: 13–22.

European Commission (EC) and United Nations Economic Commission for Europe (UN/ECE) (Eds.) 1998. Protection of Forests Against Atmospheric Pollution: European Programme for the Intensive Monitoring of Forest Ecosystems - Basic Documents for the Implementation of the Intensive Monitoring Programme of Forest Ecosystems in Europe. 1.ed: VI/3908/95-EN: 142.

Elfving, B. and Tegnhammar, L. 1996. Trends of tree growth in Swedish forests 1953–1992: an analysis based on sample trees from the National Forest Inventory. Scandinavian Journal of Forest Research 11: 26–37.

Elfving, B., Tegnhammar, L. and Tveite, B. 1996. Studies on growth trends of forests in Sweden and Norway. In: Spiecker, H., Mielikäinen, K., Köhl, M. and Skovsgaard, J.P. (eds.), Growth Trends of European Forests - Studies From 12 Countries. European Forest Institute Research Report 5. Springer, Berlin. Pp. 61–70

Emmett, B.A. 1999. The impact of nitrogen on forest soils and feedbacks on tree growth. Water, Air, and Soil Pollution 116: 65–74.

Flückiger, W. and Braun, S. 1998. Nitrogen deposition in Swiss forests and its possible relevance for leaf nutrient status, parasite attacks and soil acidification. Environmental Pollution 102: 69–76.

Glatzel, G. 1991. The impact of historic land use and modern forestry on nutrient relations of Central European Forest Ecosystems. Fertilizer Research 27: 1–8.

Glatzel, G. 1999. Historic forest use and its possible implications to recently accelerated tree growth in Central Europe. EFI Proceedings 27. European Forest Institute. Pp. 65–74.

Hamilton, G.J. and Christie, J.M. 1971. Forest Managament Tables (Metric). Forestry Commission Booklet. 201 p.

Hasenauer, H., Stampfer, E., Rohrmoser, C. and Sterba, H., 1994. Solitärdimensionen der wichtigsten Baumarten Österreichs. Österreichische Forstzeitung 3: 28–29.

Heinsdorf, D. 1993. The role of nitrogen in declining Scots pine forests (*Pinus sylvestris*) in the lowland of East Germany. Water, Air, and Soil Pollution 69: 21–35.

Heinsdorf, D. and Beck, W. 2003. Langjährige Untersuchungen zur Wirkung hoher N-Zufuhren auf Ernährung und Wachstum eines Kiefernbestandes. Beiträge für Forstwirtschaft und Landschaftsökologie 37: 28–35.

Heinsdorf, D. and Branse, C. 2002. Entwicklung der Nährelementgehalte in den Nadeln von Kiefernbeständen auf charakteristischen pleistozänen Standorten Brandenburgs in den Jahren 1964 bis 1999. Forst und Holz 57: 421–428.

Hippeli, P. and Branse, C. 1992. Veränderungen der Nährelementkonzentrationen in den Nadeln mittelalter Kiefernbestände auf pleistozänen Sandstandorten Brandenburgs in den Jahren 1964 bis 1988. Forstwissenschaftliches Centralblatt 111: 44–60.

Högberg, P., Bengtsson, G., Berggren, D., Högberg, M., Nilsson, I., Nohrstedt, H.Ö., Persson, T. and Sjöberg, M. 2000. How are the nitrogen dynamics of forest soils affected? In: Bertills, U. and Naesholm, T. (eds.). Effects of Nitrogen Deposition on Forest Ecosystems. Berling Skogs, Stockholm. Swedish Environmental Protection Agency. Pp. 29–52.

Hofmann, G., Heinsdorf, D. and Krauss, H.H. 1990. Wirkungen atmogener Stickstoffeinträge auf Produktivität und Stabilität von Kiefern-Forstökosystemen. Beiträge für die Forstwirtschaft 24: 59–73.

Hunter, I. and Schuck, A. 2002. Increasing forest growth in Europe – possible causes and implications for sustainable forest management. Plant Biosystems 136: 133–141.

Kahle, H.P. and Spiecker, H. 1996a. Adaptability of radial growth of Norway spruce to climate variations: results of a site specific dendroecological study in high elevations of the Black Forest (Germany). Radiocarbon: 785–801.

Kahle, H.P. and Spiecker, H. 1996b. Modelling growth-climate relationships of Norway spruces in high elevations of the Black Forest (Germany). EFI Proceedings 5. European Forest Institute. Pp. 205–220.

Kenk, G. and Fischer, H. 1988. Evidence from nitrogen fertilisation in the forests of Germany. Environmental Pollution 54: 199–218.

Kenk, G., Spiecker, H. and Diener, G. 1991. Referenzdaten zum Waldwachstum. KfK-PEF. 59 p.

Kennel, R. 1973. Die Bestimmung des Ertragsniveaus bei der Buche. Forstwissenschaftliches Centralblatt 92: 226–234.

Kiviste, A. 1999. Site index change in the 1950s–1990s according to Estonian forest inventory data. EFI Proceedings 27. European Forest Institute. Pp. 87–100.

Kreutzer, K. 1993. Changes in the role of nitrogen in central European forests. In: Hüttl, R.F. and Mueller-Dombois, D. (eds.). Forest Decline in the Atlantic and Pacific Region. Springer, Berlin. Pp. 82–96.

Lorenz, M. 1995. 10 Jahre ICP-Forests: Von der Kronenzustandserhebung zum intensiven Monitoring. AFZ 50: 1094–1096.

Mäkinen, H., Nöjd, P., Kahle, H.P., Neumann, U., Tveite, B., Mielikäinen, K., Röhle, H. and Spiecker, H. 2002. Radial growth variation of Norway spruce (*Picea abies* (L.) Karst.) across latitudinal and altitudinal gradients in central and northern Europe. Forest Ecology and Management 171: 243–259.

Mäkinen, H., Nöjd, P., Kahle, H.P., Neumann, U., Tveite, B., Mielikäinen, K., Röhle, H. and Spiecker, H. 2003. Large-scale climatic variability and radial increment variation of *Picea abies* (L.) Karst. in central and northern Europe. Trees 17: 173–184.
Ministerium für Ländlichen Raum, Ernährung Landwirtschaft und Forsten Baden-Württemberg and Landesforstverwaltung Baden-Württemberg, 1993. Hilfstabellen für die Forsteinrichtung - Zusammengestellt für den Gebrauch in der Landesforstverwaltung. 188 p.
Mitscherlich, G. 1958. Das Wachstum der Fichte in Baden. Schriftenreihe der Badischen Forstlichen Versuchsanstalt Heft 14: 40 p.
Mitscherlich, G. 1959. Das Fichtenwachstum in Baden-Württemberg im Vergleich zur Ertragstafel von Wiedemann. Allgemeine Forstzeitschrift 14: 521–524.
Mitscherlich, G. 1963. Das Wachstum der Fichte in Europa. AFJZ 134: 29–45, 61.
Moosmayer, H.U. and Ripberger, M. 1990. Überprüfung waldwachstumskundlicher Auswertungen von Standortsgliederungen anhand neuerer Daten aus Versuchsflächen. Mitteilungen des Vereins für Forstliche Standortskunde und Forstpflanzenzüchtung 35: 9–14.
Moosmayer, H.U., Schöpfer, W., Kublin, E. and Tuch, U. 2003. Standörtliche Einflüsse auf Wachstumsänderungen der Fichte in Baden-Württemberg. Mitteilungen des Vereins für Forstliche Standortskunde und Forstpflanzenzüchtung. Pp. 23–30.
Näsholm, T., Nohrstedt, H.Ö., Kårén, O., Kytö, M. and Björkman, C. 2000. How are forest trees affected? In: Bertills,U. and Naesholm,T. (eds.). Effects of Nitrogen Deposition on Forest Ecosystems. Berling Skogs, Stockholm. Swedish Environmental Protection Agency. Pp. 53–75.
Oleksyn, J., Reich, B.R., Zytkowiak, R., Karolewski, P. and Tjoelker, M.G. 2002. Needle nutrients in geographically diverse *Pinus sylvestris* L. populations. Annals of Forest Science 59: 1–18.
Oren, R., Ellsworth, D.S., Johnsen, K.H., Phillips, N., Ewers, B.E., Maier, C., Schäfer, K.V., McCarthy, H., Hendrey, G., McNulty, S.G. and Katul, G.G. 2001. Soil fertility limits carbon sequestration by forest ecosystems in a CO_2-enriched atmosphere. Nature 411: 469–472.
Pretzsch,H. 1992. Zunehmende Unstimmigkeit zwischen erwartetem und wirklichem Wachstum unserer Waldbestände: Konsequenzen für zukünftige ertragskundliche Informationssysteme. Forstwissenschaftliches Centralblatt 111: 366–382.
Pretzsch, H. and Utschig, H. 2000. Wachstumstrends der Fichte in Bayern. Mitteilungen aus der Bayerischen Staatsforstverwaltung 49. 170 p.
Prietzel, J., Kolb, E. and Rehfuess, K.E. 1997. Langzeituntersuchung ehemals streugenutzter Kiefernökosysteme in der Oberpfalz: Veränderungen von bodenchemischen Eigenschaften und der Nährelementversorgung der Bestände. Forstwissenschaftliches Centralblatt 116: 269–290.
Puhe, J. and Ulrich, B. 2001. Global Climate Change and Human Impacts on Forest Ecosystems: Postglacial Development, Present Situation, and Future Trends in Central Europe. Springer, Berlin. Ecological Studies. 592 p.
Rehfuess, K.E., Agren, G.I., Andersson, F., Cannell, M.G.R., Friend, A.D., Hunter, I., Kahle, H.P., Prietzel, J. and Spiecker, H. 1999. Relationships between recent changes of growth and nutrition of Norway spruce, Scots pine, and European beech forests in Europe – RECOGNITION. EFI Working Paper 19. European Forest Institute. 94 p.
Reimeier, S. and Kennel, E. 2001. Zuwachstafeln für ausgewählte Baumarten. Dr. Kessel, Remagen. 93 p.
Schmidt, A. 1973. Ertragsniveau und Standort dargestellt am Beispiel der Kiefer. Forstwissenschaftliches Centralblatt 92: 268–274.
Schober, R. 1975. Ertragstafeln wichtiger Baumarten bei verschiedener Durchforstung. J. D. Sauerländer's Verlag, Frankfurt am Main. 154 p.

Schöpfer, W., Hradetzky, J. and Kublin, E. 1994. Wachstumsänderungen der Fichte in Baden-Württemberg. Forst und Holz 49: 633–644.

Schulze, E.D. 2000. Carbon and Nitrogen Cycling in European Forest Ecosystems. Springer, Berlin. 500 p.

Spiecker, H. 1991a. Growth variation and environmental stresses: Long-term observations on permanent research plots in Southwestern Germany. Water, Air, and Soil Pollution 54: 247–256.

Spiecker, H. 1991b. Liming, nitrogen and phosphorus fertilization and the annual volume increment of Norway spruce stands on long-term permanent plots in Southwestern Germany. Fertilizer Research 27: 87–93.

Spiecker, H., 1992. Which trees represent stand growth? In: Bartholin, T. S., Berglund, B. E., Eckstein, D., and Schweingruber, F. H (eds.). Tree rings and environment. Proceedings of the International Dendrochronological Symposium, Ystad, South-Sweden, 3–9 September 1990. Lundqua-Report 34: 308–312.

Spiecker, H. 1999. Overview of recent growth trends in European forests. Water, Air, and Soil Pollution 116: 33–46.

Spiecker, H., Hildebrand, E.E. and Evers, F.H. 1992. Ernährung und Zuwachs von Fichten (*Picea abies* [L.] Karst.) in einem langfristig beobachteten Düngungsversuch im Buntsandstein-Schwarzwald. Forst und Holz 47: 181–185.

Spiecker, H., Mielikäinen, K., Köhl, M. and Skovsgaard, J.P. (eds.). 1996. Growth Trends in European Forests – Studies From 12 Countries. EFI Research Report 5. Springer, Berlin. 372 p.

Thornthwaite, C.W. and Mather, J.R. 1955. The water balance. Publ. in Climatology – Drexel Inst. of Tech., Lab. of Clim. 8(1). 104 p.

Thornthwaite, C.W. and Mather, J.R. 1957. Instructions and tables for computing potential evapotranspiration and the water balance. Drexel Institute of Technology – Laboratory of Climatology 10(3). 308 p.

Untheim, H. 1996a. Has site productivity changed? A case study in the Eastern Swabian Alb, Germany. In: Spiecker, H., Mielikäinen, K., Köhl, M. and Skovsgaard, J.P. (eds.), Growth Trends in European Forests – Studies From 12 Countries. Springer, Berlin. Pp. 133–147.

Untheim, H. 1996b. Zur Veränderung der Produktivität von Waldstandorten: Untersuchungen zum Höhen- und Volumenwachstum von Fichte (*Picea abies* [L.] Karst.) und Buche (*Fagus sylvatica* L.) auf Standorteinheiten der Ostalb und des Flächenschwarzwaldes. Mitteilungen FVA Baden-Württemberg 198. 239 p.

Vitousek, P.M. and Howarth, R.W. 1991. Nitrogen limitation on land and in the sea: How can it occur? Biogeochemistry 13: 87–115.

Weise, U. 1991. Auf Versuchsflächen im badischen Schwarzwald: Ertragsniveau und Zuwachsgang der Weißtanne. AFZ/Der Wald 46: 192–195.

Wiedemann, E. 1949. Ertragstafeln der wichtigen Holzarten bei verschiedener Durchforstung sowie einiger Mischbestandsformen. Schaper, Hannover. 100 p.

Wiedemann, E. 1948. Die Kiefer. Schaper, Hannover. 337 p.

Wiedemann, E. 1932. Die Rotbuche: 1931. Fortführung des Berichtes von Geheimrat Schwappach 1911 über die preußischen Buchenversuchsflächen. Schaper, Hannover. 191 p.

Wilks, D.S. 1995. Statistical Methods in the Atmospheric Sciences. Academic Press, San Diego. 464 p.

Zar, J.H. 1974. Biostatistical Analysis. Prentice-Hall, Englewood Cliffs. 620p.

4.3 Short-, medium-, and long-term variation in radial growth, and the role of changes in the climatic water balance for the growth of three tree species in Europe

Hans-Peter Kahle[1], Heinrich Spiecker[1], Rüdiger Unseld[1], Pedro-José Pérez-Martínez[1], Jörg Prietzel[2], Karl-Heinz Mellert[2], Ralf Straussberger[2] and Karl-Eugen Rehfuess[2]

[1]*Institute for Forest Growth, University Freiburg, Germany*
[2]*Chair of Soil Science, Technical University Munich, Germany*

4.3.1 Introduction

The aim of this chapter is to analyse and clarify the relationships between water supply, and growth of the three tree species Norway spruce, Scots pine, and beech with special regard to medium- and short-term impacts on growth. Whereas in Chapter 4.2 the emphasis is on height growth, in this chapter the focus is on the analysis of radial growth. From forest growth and dendroecological research studies on trees grown under seasonal climatic conditions it is well known that annual radial growth is usually highly synchronized and closely correlated among neighbouring trees. This highlights the potential of this parameter as an indicator of environmental effects on growth acting on short- to medium-term time-scales (Fritts, 1976; Schweingruber, 1988; Kahle, 1996; Kahle and Spiecker, 1996a). For Norway spruce it has been shown that depressions in radial growth of trees often are closely related to periods of increased tree mortality on the stand and regional level (Spiecker, 1999 and 2000). Variation in radial growth can therefore be regarded as a sensitive measure of tree vitality and indicator of the occurrence and effect of extreme events. Since the level of radial growth is largely affected by stand density (Mitscherlich, 1978), radial growth of trees is not a suitable proxy for forest site productivity (Spiecker et al., 1994). However, here we compare medium-term age-related trends in radial growth of the tree cohorts in order to reveal possible differences in growth that might be indicative of past differences in stand management and stand dynamics in the respective stands.

In earlier studies it has been shown that fluctuations in radial growth of Norway spruce on many sites in Europe are closely linked to variations in water supply (Kahle and Spiecker, 1996a and 1996b; Mäkinen et al., 2002; Spiecker, 1991 and 1995; Neumann and Röhle, 2002). Therefore, in this chapter special emphasis is placed on the role of water supply for the development of radial tree growth during the last century. These relations are analysed in the time as well as in the frequency domain (Brockwell and Davis, 1987; LaMarche, 1974).

Analysis of radial increment contributes to a better understanding of temporal variations in tree vitality, and to a deeper insight into the occurrence of extreme events and their effects on tree growth. Complementary to the analysis of height growth it improves insight into possible confounding effects of age trend, competition, and extreme events on tree growth.

4.3.2 Short-, medium-, and long-term changes in radial growth

In this section changes in radial growth of the sample trees collected for the Present State Analysis (PSA) are analysed with the aims of revealing variations in growth on different time scales and describing their development during the last century.

Retrospective annual radial increment data were derived from stem analysis (Chapter 2.2). For the purpose of this study the time series of annual radial increment were divided into three components according to their characteristic time scales in (1) a long-term component dominated by low-frequency variation (period $1/\lambda \geq 30$ years); (2) a medium-term component characterized by variations with periods in the range 10 yr $\leq 1/\lambda < 30$ yr; and (3) a short-term or high-frequency component of which the inter-annual variability is the dominating feature (1 yr $\leq 1/\lambda < 10$ yr). This decomposition was achieved by fitting in a two-step procedure cubic smoothing splines to the measurement series on an individual tree basis (Van Deusen and Reams, 1993; Mäkinen et al., 2002). In the first step a rigid spline with a 50%-frequency cut-off of 30-years was fitted to the radial increment series. The fitted and plot-wise aggregated spline curves were classified as the 'long-term component'. In the second step a more flexible smoothing spline with a 50%-frequency cut-off of 10-years was fitted to the indices that result from the first step of detrending (ir_{30ind}). These plot-wise aggregated fitted spline curves were classified as the 'medium-term component' (ir_{10pre}), whereas the respective quotients were classified as the 'short-term' or 'high-frequency component' (ir_{10ind}).

Age-related trends in radial growth were investigated in order to reveal possible differences in stand management and stand dynamics in the respective stands in the past. Figure 1 shows the long-term component of mean annual radial increment of the Norway spruce (PCAB), Scots pine (PISY), and beech (FASY) samples plotted versus age. For all three species average annual radial increment (ir) of younger sample trees was significantly larger than that of the older trees during the age period 1 to 50 years ($p > |t| = 0.000$) (PCAB old: ir=2.51 mm, young: ir=2.90 mm; PISY old: ir=2.18 mm, young: ir=2.46 mm; FASY old: ir=2.28 mm, young: ir=2.89 mm). The relative increase referred to the growth rate of the old trees amounts to +15.5% for spruce, +12.8% for pine, and +26.8% for beech.

Culmination of annual radial increment occurred slightly later in Norway spruce (t_{irmax}=13 years) than in the more light-demanding Scots pine (t_{irmax}=7 years). However, within species there is no difference between the age cohorts in this respect. Diameter growth of beech behaved differently than that of the conifers during young developmental stage: it took 30 years or longer until annual radial increment reached a maximum. On average the conifers show an age-related growth pattern typical of trees in even-aged stands, planted on clear-cut areas without canopy trees. The pattern of beech is typical of a shade-tolerant species regenerated

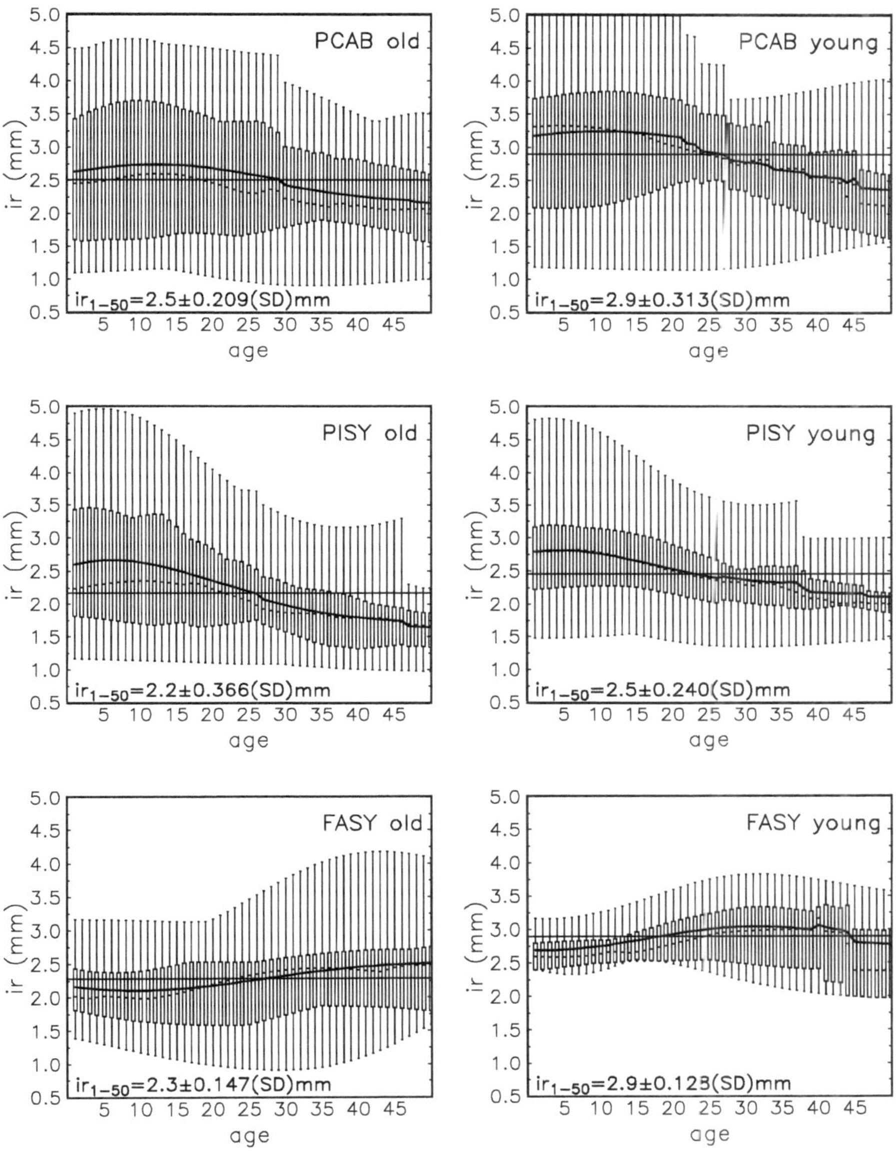

Figure 1. Long-term component of mean annual radial increment (ir_{30pre}) of the Norway spruce (PCAB, n_{Plots}=25), Scots pine (PISY, n_{Plots}=24), and beech (FASY, n_{Plots}=14) sample trees plotted versus cambial age in breast height and stratified into age cohorts. Number of sample trees: n_{PCABo}=110, n_{PCAy}=105, n_{PISYo}=107, n_{PISYy}=106, n_{FASYo}=59, n_{FASYy}=48. Top and bottom of boxes mark the empirical quartile ranges, the absolute limits are taken from the actual limits of the data. Horizontal straight solid line: 50-year average, curved solid line: arithmetic (ensemble) mean. For all three species annual radial increment of younger sample trees during the first 50 years is significantly larger than that of the older sample trees ($P(|t_{\alpha(2),(v1+v2)}| \geq t) < 0.000$).

naturally under the canopy of the mature seed trees. The larger average radial growth rates of the cohorts of younger trees are most probably the result of lower stand densities in the respective stands. However, the similarity of the shape of the radial increment curves between the tree cohorts suggests that the general pattern of stand dynamics during early phases of stand development has not changed between the respective periods.

Medium-term fluctuations of annual radial increment of the sample tree strata versus calendar year can be depicted from Figure 2. A typical feature of these time series is the gradual variation of growth rates between subsequent local minima and maxima recurring in episodic to periodic intervals lasting several years to decades. Radial growth depressions, synchronous among all three species occurred in the mid-1920s, around 1950, and most pronounced in the late 1990s with a local minimum around 1997.

Figure 3 shows short-term variations of annual radial increment of the Norway spruce, Scots pine, and beech sample trees plotted versus calendar year. This high-frequency component of annual radial growth is assumed to be primarily driven by weather and climate variability on time-scales of one to several years. This assumption is supported by the finding that the series are significantly synchronized among cohorts and species (Table 1). The mean inter-series correlations (cross-correlation coefficient, r_{xy}) between young and old trees are highly significant and vary from 0.743 to 0.871, and even between species it ranges from 0.408 to 0.616 ($p < 0.000$). The large extent of synchronous variation among the tree cohorts from various sites in Europe and among the different tree species suggests a strong climatic control over radial growth of the sample trees.

Marked depressions in radial growth occurred for all three species in 1996 and 1997; depressions common to Norway spruce and beech occurred in 1949, 1993, and 1977; species-specific depressions occurred in 1924 for spruce, and in 1941 and 1953 for Scots pine.

4.3.3 Variation in water supply

Time series of the climatic water balance were used as a measure for temporal changes in the atmospheric water supply. For the calculation of the climatic water balance we applied Thornthwaite's formulae (Thornthwaite and Mather, 1955 and 1957) and used spatially interpolated data of monthly mean air temperature and precipitation sum (CRU 0.5 lat./long. gridded monthly climate data). It is well known from literature that Thornthwaite's formulae may produce biased estimates when applied to locations outside the temperate regions for which it has been developed, and that even within the target area seasonally distinct differences to the true values can occur (Schrödter, 1985). However, since we are mainly interested in temporal changes in the amount of water supplied by the atmosphere considering losses due to (potential) evapotranspiration, a systematic bias can be tolerated. The same is true concerning certain special features of forests as vegetation cover in the hydrological cycle which are also not explicitly considered in the simplified formulae (e.g. interception losses, snow storage, etc.).

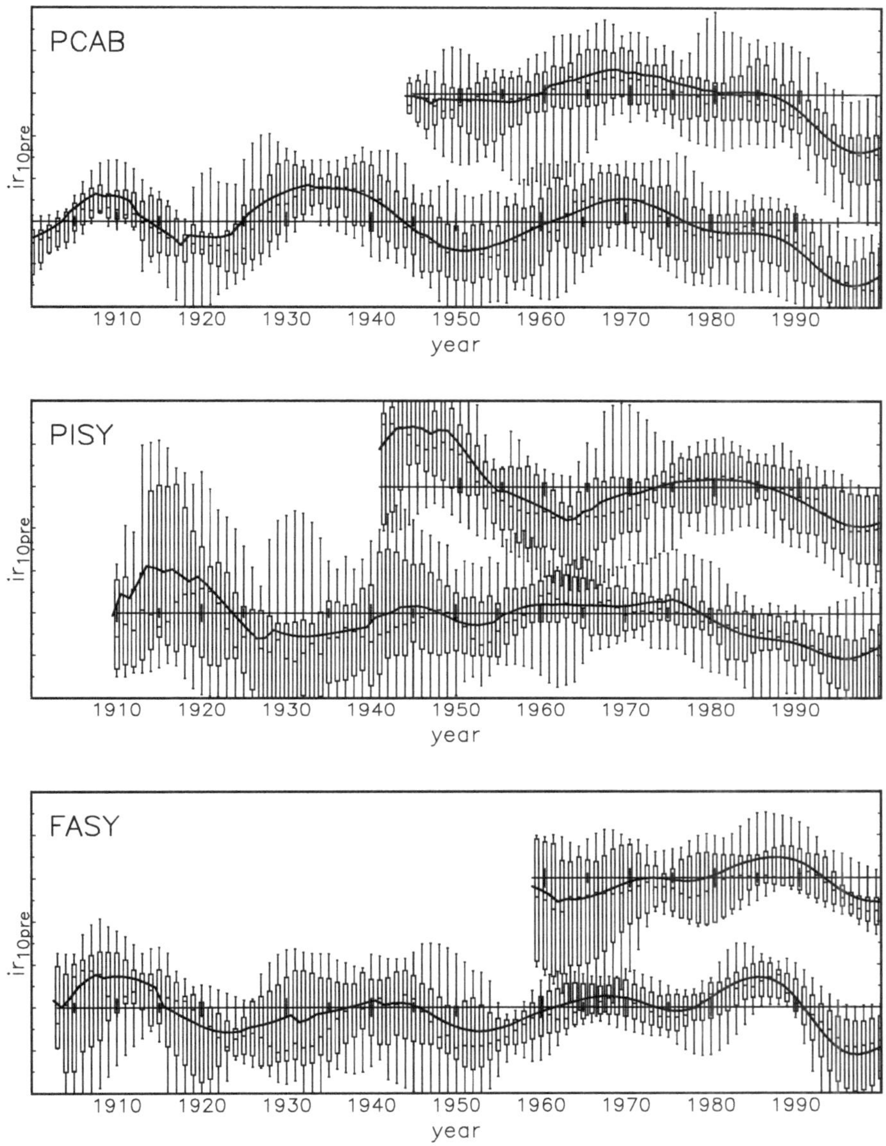

Figure 2. Medium-term component of mean annual radial increment index (ir_{10pre}, dimension: unity), of the Norway spruce (PCAB), Scots pine (PISY), and beech (FASY) sample trees plotted versus calendar year and stratified into tree age cohorts (long series: older trees; short series: younger trees; for reasons of presentation short series are shifted vertically along the y-axis and horizontally by +0.5 years along the x-axis). Horizontal straight solid line: long-term average, curved solid line: arithmetic (ensemble) mean. Only values with a sample depth of $n \geq 5$ observations are included.

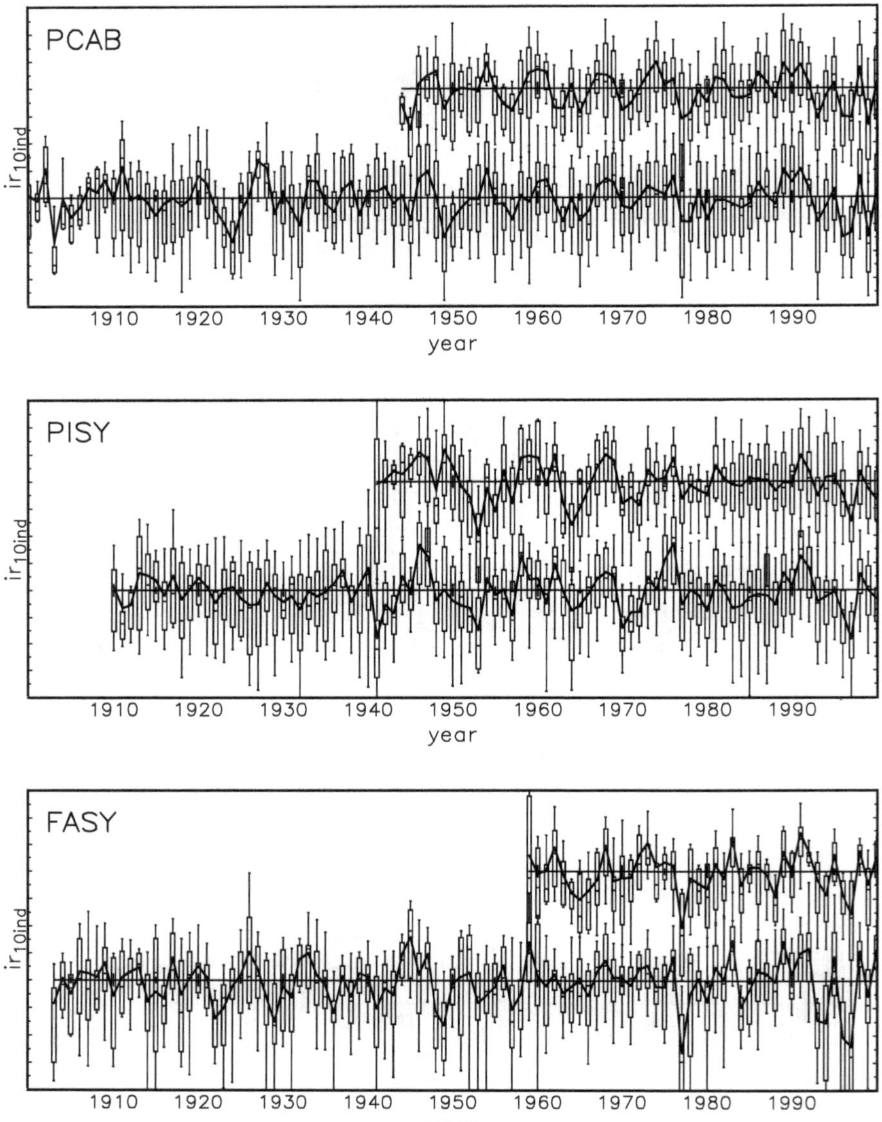

Figure 3. Short-term component of mean annual radial increment index (ir_{10ind}, dimension: unity) indicating high-frequency variation of the Norway spruce (PCAB), Scots pine (PISY), and beech (FASY) sample trees plotted versus calendar year and stratified into tree age cohorts (long series: older trees; short series: younger trees; for reasons of presentation short series are shifted vertically along the y-axis). Horizontal straight solid line: long-term average.

Table 1. Cross-correlation (r_{xy}) between time series of mean annual radial increment (short-term component, see Figure 3) (calculations based on group-wise maximum common overlap periods). All cross-correlation coefficients are significant ($p < 0.000$).

	"old" trees			"young" trees		
	PCAB	PISY	FASY	PCAB	PISY	FASY
	Cross-correlation coefficient r_{xy}					
	1910–2001 (n = 92)					
PCAB old	1	0.408	0.519			
PISY old	–	1	0.421			
FASY old	–	–	1			
	1944–2001 (n = 58)	1941–2001 (n = 61)	1959–2001 (n = 43)	1959–2001 (n = 43)		
PCAB young	0.871	–	–	1	0.596	0.616
PISY young	–	0.743	–	–	1	0.603
FASY young	–	–	0.833	–	–	1

Figure 4 shows maps of absolute deviations of the long-term mean annual and seasonal (May-September) water balance between the periods 1961–1990 and 1931–1960, and the reference period 1901–1930. Overall Europe the annual and seasonal climatic water balances have slightly increased in the period 1961–1990 compared to 1901–1930 (deviation in annual mean: $\Delta cwb_{01\text{-}12}$=+8 mm, deviation in seasonal mean: $\Delta cwb_{05\text{-}09}$=+3 mm). The time series of the mean seasonal climatic water balance at the selected sample sites show distinct medium-term fluctuations; however, the overall time trend is not significantly different from zero (Figure 5). Periods lasting several years with pronounced below average seasonal climatic water balance occurred in the mid-1920s, around 1950, and in the late-1990s.

No overall trend in seasonal climatic water balance can be detected when considering the average of the selected sites. However, there is a large variability in long-term trends between the individual sites. Figure 6 shows the linear trends in deviations of seasonal climatic water balance from its mean during the reference period 1901–1930 at the sampled PSA sites. Compared to the inter-annual and periodic variability, the extent of the long-term trend even at the sites with the most pronounced trend is relatively small.

4.3.4 Relationships between water supply and tree growth

4.3.4.1 Water supply and fluctuations in radial growth

Comparison of the medium-term fluctuations of annual radial increment (Figure 2) with the time series of the mean seasonal climatic water balance (Figure 5) reveals close relationships between both processes: periods lasting several years with pronounced below average seasonal climatic water balance in the mid-1920s,

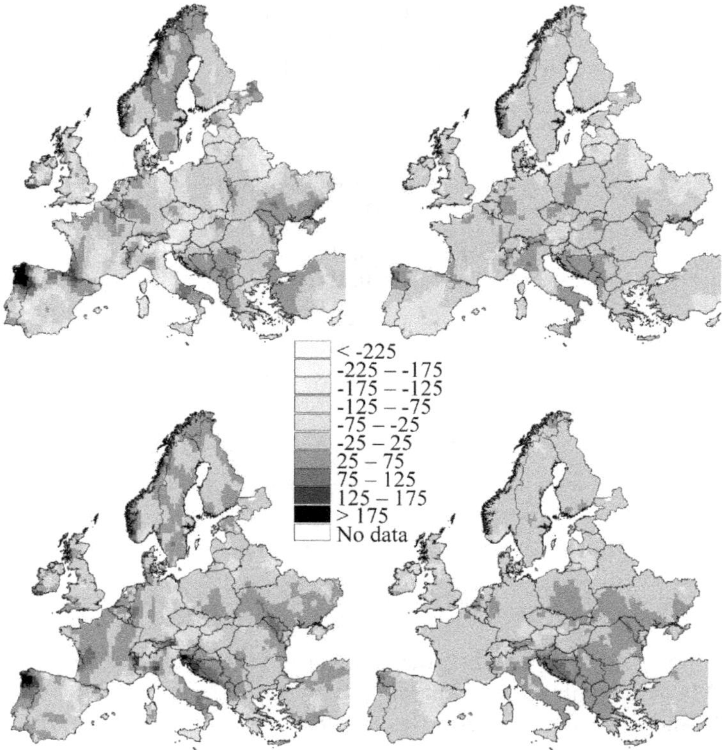

Figure 4. Deviation (mm) of mean annual (figures to the left) and mean seasonal (May–September, figures to the right) water balance between the reference period 1901–1930 and the periods 1961–1990 (figures on top) and 1931–1960 (figures at bottom). Calculation of potential evapotranspiration according to Thornthwaite and Mather (1955) (data source: CRU 0.5). For Europe as a whole, the annual and seasonal climatic water balance has slightly increased in the period 1961–1990 compared to 1901–1930.

around 1950, and in the late-1990s coincide with marked depressions in radial growth which are synchronous for all three species.

Cross-spectral analysis was used in order to investigate the relationship between time series of annual radial increment and seasonal climatic water balance at different frequency ranges (LaMarche, 1974). Analogous to the correlation in the time domain, the coherency $\omega_{x,y}(\lambda)$ is a measure for the similarity of two time series in the frequency domain. The coherency spectrum is the product of the amplitude spectrum and the phase spectrum and indicates the synchronism between two time series at a given frequency (Brockwell and Davis, 1987). The highest frequency λ to be resolved from annual data is $\lambda=1/2$ cycles per year, the theoretical resolvability limit for the low frequency component is $1/n$ cycles per year, where n equals the length of the time series in years.

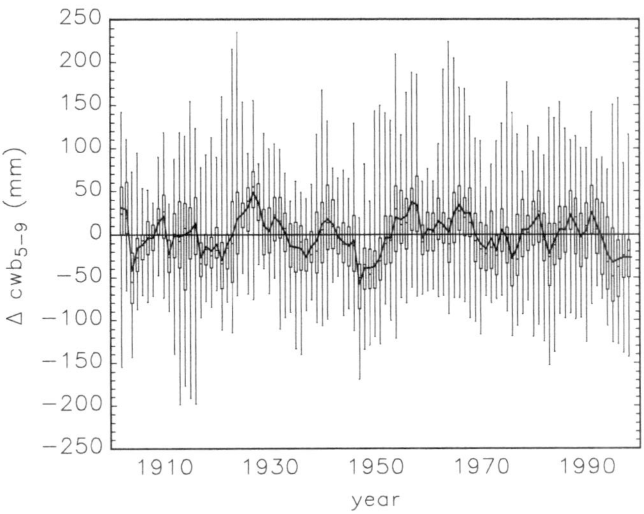

Figure 5. Deviation of seasonal (May–September) climatic water balance ($\Delta\ cwb_{5-9}$) (calculated according to Thornthwaite and Mather (1955)) from the long-term mean during the reference period 1901–1930 at the sampled PSA sites (n=62). The time series are smoothed by applying a causal 5-year running average with linearly declining weights. The top and bottom of the boxes mark the empirical quartile ranges, the absolute maximum and minimum limits are taken from the actual limits of the data. The solid line represents the arithmetic mean. The overall linear long-term trend is not different from zero ($P(|t_{\alpha(2),v-2}| \geq t) > 0.05$).

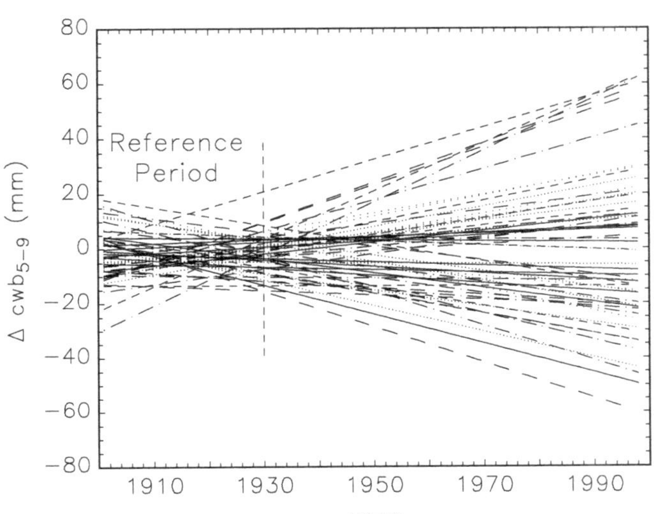

Figure 6. Linear trends of deviations of seasonal (May–September) climatic water balance ($\Delta\ cwb_{5-9}$) from its mean during the reference period 1901–1930 at the sampled PSA sites (n=62). The mean long-term trend is not different from zero ($P(|t_{\alpha(2),v-2}| \geq t) > 0.05$).

Figure 7 shows the coherency between the combined short- and medium-term component of mean annual radial increment (indices of splines with a 50%-frequency cut-off of 30-years: ir_{30ind}) and the seasonal (January to September) climatic water balance (cwb_{01-09}) for the three tree species. Radial growth is significantly linked to changes in climatic water balance at high-, medium-, and low-frequencies (corresponding periods: ~3 years, ~10 years, ~30 years) for all three species. In addition the most prominent peaks in the short- and medium-frequency coherency spectrum are synchronized among the species, highlighting the important role of water supply for radial growth variation. There was more variation in the medium-frequency range for Scots pine and Norway spruce than for beech.

4.3.4.2 Water supply and sensitivity in radial growth

In Figure 8 mean sensitivity values of the annual radial increment time series of the three species are plotted versus absolute values of the long-term mean seasonal climatic water balance. The mean sensitivity of radial increment (ms_{ir}) is defined as the mean percentage change from each measured annual radial increment value in a time series to the next (Fritts, 1976). The variability among the sample sites with respect to long-term mean seasonal climatic water balance (cwb_{5-9}) is larger for spruce than for the other species (spruce: cwb_{5-9}=–18 mm, range R=105 mm; pine: cwb_{5-9}=–42 mm, R=50 mm; beech: cwb_{5-9}=–17 mm, R=70 mm).

Despite the large differences between the sample sites with respect to long-term mean climatic water balance, the sensitivity in radial increment is independent of the mean level of the climatic water balance. For all species the mean sensitivity tended to be higher for lower mean seasonal climatic water balances; however, in no case is the slope of the regression line significantly different from zero. Older trees tend to react more sensitive than younger trees.

4.3.5 Summary and conclusions

Analysis of the short-, medium-, and long-term variations in radial growth, and the role of changes in the climatic water balance for the growth of spruce, pine and beech in Europe resulted in the following findings:

- The average annual radial growth rate has significantly increased: mean radial increment of the cohort of younger trees was significantly larger than that of the older trees for all three species (age period 1 to 50 years). Compared to the average growth rate of the reference trees the growth acceleration was +15.5% for spruce, +12.8% for pine, and +26.8% for beech.
- The overall trend of the radial growth curves of the two age cohorts is similar. Hence no indications for major differences in the applied stand management principles between the tree cohorts are evident.
- The radial increment index series are significantly inter-correlated. Comparison between the detrended radial increment series reveals a significant degree of synchronicity in growth fluctuations between tree cohorts, between sites and between tree species.

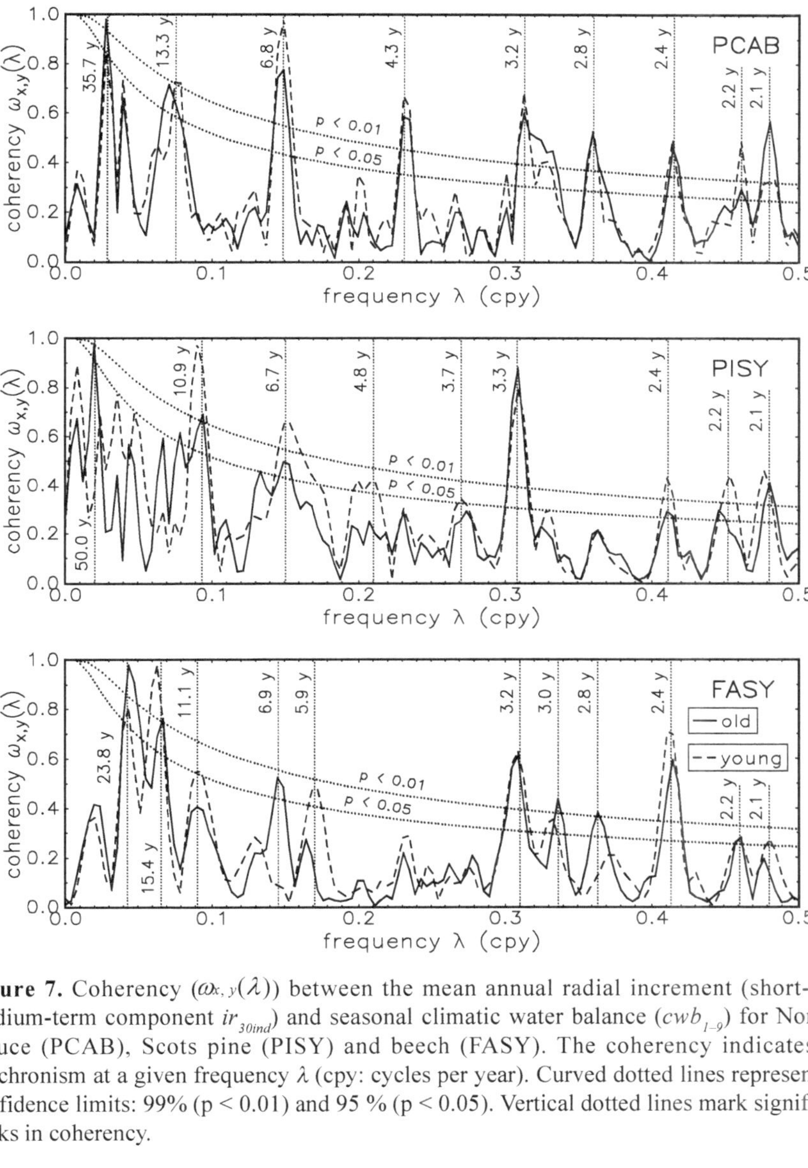

Figure 7. Coherency ($\omega_{x,y}(\lambda)$) between the mean annual radial increment (short- and medium-term component ir_{30ind}) and seasonal climatic water balance (cwb_{1-9}) for Norway spruce (PCAB), Scots pine (PISY) and beech (FASY). The coherency indicates the synchronism at a given frequency λ (cpy: cycles per year). Curved dotted lines represent the confidence limits: 99% ($p < 0.01$) and 95 % ($p < 0.05$). Vertical dotted lines mark significant peaks in coherency.

- Coherency analyses of bivariate time series reveal close direct relations between variations in climatic water balance and radial growth of all sample tree strata at high and medium frequencies.
- Based on the CRU 0.5 lat./long. gridded monthly climate data overall Europe the mean climatic water balance has slightly increased in the period 1961–1990 compared to 1901–1930: +8 mm in annual mean, and +3 mm in seasonal (May–Sept.) mean.

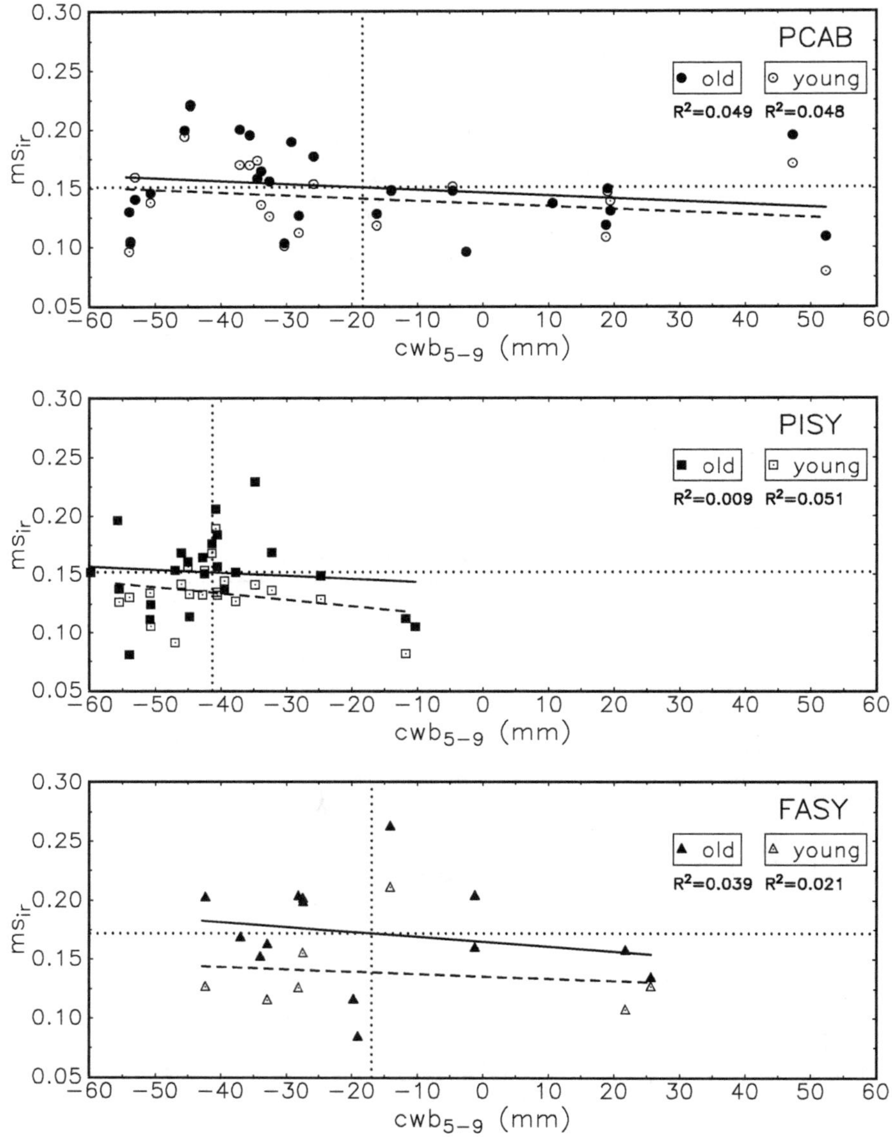

Figure 8. Mean sensitivity of annual radial increment (ms_{ir}) of the Norway spruce (PCAB), Scots pine (PISY), and beech (FASY) sample trees plotted versus long-term mean seasonal (May–September) climatic water balance ($cwb_{5\text{-}9}$) (common overlap period: old trees 1907–1998; young trees: 1956–1998). In no case the slope of the linear regression model (solid lines: old trees; dashed lines: young trees) is significantly different from zero ($P(|t_{\alpha(2),v\text{-}2}| \geq t) > 0.05$). Horizontal dotted lines: overall mean sensitivity. Vertical dotted lines: overall mean seasonal climatic water balance.

- Among the sampling sites there is a considerable variability in the long-term trend of the annual and seasonal climatic water balance. Over all samples no significant long-term trend is found during the period 1901–1998.
- Despite the large scatter of the sample sites with respect to long-term mean climatic water balance, the sensitivity of radial increment is independent of the mean level of the climatic water balance.

Based on these results achieved the following conclusions are drawn:

- Under the assumption that tree height and form factor remained unchanged the described acceleration in mean diameter growth compared to the reference growth would result in a mean increase of single tree volume increment of +33% for spruce, +27% for pine, and +61% for beech. The assumption of constant height and form factor is rather conservative since not only diameter but also height growth was significantly increased (see Chapter 4.2). However, the calculated increase in volume increment refers to an individual tree basis and cannot be directly scaled up to the stand level. The considerable acceleration in diameter growth of the younger sample trees is most probably the effect of a lower stand density in the stands from which these were sampled.
- The finding that only a small increase in climatic water balance over Europe can be found, and that this increase is not reflected by the average over the sample sites suggests that water supply as expressed by the climatic water balance is unlikely to be the major driver of the observed long-term increase in forest site productivity over Europe.
- Independent of the baseline level of climatic water supply, variations in climatic water balance have a strong effect on diameter growth of trees and hence on volume production of forests particularly on short- to medium-term time scales.

References

Brockwell, P.J. and Davis, R.A. 1987. Time Series: Theory and Methods. Springer, New York. 519 p.

Fritts, H.C. 1976. Tree Rings and Climate. Academic Press, London. 567 p.

Kahle, H.P. 1996. Modelling the dynamics of growth-climate relationships of Norway spruce and silver fir in high elevations of the Black Forest. Proceedings IUFRO, Section 4.01: 93–103.

Kahle, H.P. and Spiecker, H. 1996a. Adaptability of radial growth of Norway spruce to climate variations: results of a site specific dendroecological study in high elevations of the Black Forest (Germany). Radiocarbon: 785–801.

Kahle, H.P. and Spiecker, H. 1996b. Modelling growth–climate relationships of Norway spruces in high elevations of the Black Forest (Germany). In: Päivinen, R., Roihuvuo, L., and Siitonen, M. (eds.) Large-Scale Forestry Scenario Models: Experiences and Requirements. EFI Proceedings 5. European Forest Institute. Pp. 205–220.

LaMarche, V.C.J. 1974. Frequency-dependent relationships between tree-ring series along an ecological gradient and some dendroclimatic implications. Tree-Ring Bulletin 34: 1–20.

Mäkinen, H., Nöjd, P., Kahle, H.P., Neumann, U., Tveite, B., Mielikäinen, K., Röhle, H. and Spiecker, H. 2002. Radial growth variation of Norway spruce (*Picea abies* (L.) Karst.) across latitudinal and altitudinal gradients in central and northern Europe. Forest Ecology and Management 171: 243–259.

Mitscherlich, G. 1978. Wald, Wachstum und Umwelt. 1. Bd.: Form und Wachstum von Baum und Bestand. Sauerländer's Verlag, Frankfurt. 144 p.

Neumann, U. and Röhle, H. 2002. Radial- und Höhenzuwachsentwicklung der Fichte (*Picea abies*) in den sächsischen Mittelgebirgen. Forstarchiv 73: 171–178.

Schrödter, H. 1985. Verdunstung – Anwendungsorientierte Messverfahren und Bestimmungsmethoden. Springer, Berlin. 186 p.

Schweingruber, F.H. 1988. Tree Rings: Basics and Applications of Dendrochronology. Kluwer Academic Publishers, Dordrecht. 276 p.

Spiecker, H. 1991. Growth variation and environmental stresses: long-term observations on permanent research plots in Southwestern Germany. Water, Air, and Soil Pollution 54: 247–256.

Spiecker, H. 1995. Growth dynamics in a changing environment – long-term observations. Plant and Soil 168–169: 555–561.

Spiecker, H. 1999. Overview of recent growth trends in European forests. Water, Air, and Soil Pollution 116: 33–46.

Spiecker, H. 2000. The growth of Norway spruce (*Picea abies* [L.] Karst.) in Europe within and beyond its natural range. In: Hasenauer, H. Forest Ecosystem Restoration – Ecological and Economical Impacts of Restoration Processes in Secondary Coniferous Forests, Proceedings of the International Conference held in Vienna, Austria 10–12 April, 2000. Pp. 247–256.

Spiecker, H., Mielikäinen, K., Köhl, M. and Untheim, H. 1994. Growth trends of European Forests – Has site productivity changed? EFI Working Paper 4. European Forest Institute. 68 p.

Thornthwaite, C.W. and Mather, J.R. 1955. The water balance. Publ. in Climatology – Drexel Inst. of Tech., Lab. of Clim. 8(1). 104 p.

Thornthwaite, C.W. and Mather, J.R. 1957. Instructions and tables for computing potential evapotranspiration and the water balance. Drexel Institute of Technology – Laboratory of Climatology. Vol 10, No 3: 308 p.

Van Deusen, P.C. and Reams, G.A. 1993. Frequency domain tree-ring standardization. Forest Science 39: 55–65.

4.4 Evaluation of past and future changes in European forest growth by means of four process-based models

Marcel van Oijen[1], Göran I. Ågren[2], Oleg Chertov[3], Seppo Kellomäki[4], Alex Komarov[3], Deena Mobbs[1] and Maureen Murray[1]

[1]*Centre for Ecology and Hydrology, Edinburgh Research Station, UK*
[2]*Swedish University of Agricultural Sciences, Uppsala, Sweden*
[3]*European Forest Institute*
[4]*University of Joensuu, Faculty of Forest Sciences, Joensuu, Finland*

4.4.1 Introduction

This chapter reports on the results obtained by process-based modelling in Recognition. Results from empirical-statistical studies on forest growth determining factors are reported in Chapters 4.1, 4.2 and 4.3, and explicit comparisons between the different methods are made in Chapter 5.1.

One means of analysing causes of growth change is by using a process-based model to reproduce the observations, followed by a sensitivity analysis of the model to identify the major drivers of change. A possible problem with this approach is that a process-based model may reproduce observations closely, but not in a unique manner. Other, differently structured models may also be able to account for the observations, while showing different results during the subsequent sensitivity analysis. This problem of model identifiability is more serious when the competing models are parameter-rich, allowing each of them to be fit to the data by calibration of parameter values. However, the use of process-based models seems unavoidable when insufficient data are available from multi-factorial experiments to identify key causal factors directly. Moreover, such models may be the best tools available when predicting growth under conditions not previously observed, such as further changes in climate and atmospheric CO_2.

The twin questions of explaining past changes in European forest growth and predicting future changes were examined in Recognition (Chapter 1). Process-based forest ecosystem modelling was used in the project. The problem of model identifiability was addressed by including four very differently structured models in the project (Chapter 3.1), testing them against data collected in the project (Chapter 3.2), and finally by analysing differences between the models in terms of underlying processes, reported here.

Four process-based models were used: EFIMOD, EFM, FinnFor and Q. The models simulate flows of resources (C, N, water) into and out of the forest system, but in very different ways (Chapter 3.1). We will show the answers that the models give to the project questions, and identify the key areas of uncertainty – i.e. those processes and resource use components for which the models show different sensitivities to environmental change.

4.4.2 Methodology

4.4.2.1 Models

Four process-based models were used to simulate forest growth: Q, EFIMOD, FinnFor and EFM. The models differed greatly in structure and complexity. The two simplest models, Q and EFIMOD (44 and 119 parameters, respectively) simulated forest primary productivity as determined mostly by nitrogen uptake from the soil, whereas the two more complex models FinnFor (249 parameters) and EFM (261 parameters) simulated primary productivity as the consequence of interactions between photosynthetic carbon assimilation, nitrogen and water uptake. In all four models, the growth processes were affected by nitrogen deposition, radiation and temperature, but only FinnFor and EFM included sensitivity to CO_2, water availability, wind speed and humidity. The models are described in detail in Chapter 3.1.

4.4.2.2 Environmental conditions

Each model simulated forest growth for the same set of 22 sites across Europe: 13 with Scots pine and 9 with Norway spruce (Figure 1). The average temperature of the sites during the period 1975–1990 ranged from –0.6 to 10.0°C, and rainfall from 403 to 1169 mm yr^{-1}. Soil carbon and nitrogen contents varied by nearly an order of magnitude. All simulations reported here were for unthinned stands. Further details of site conditions are presented in Chapter 3.2.

Three time-periods of forest growth were simulated for each of the sites: 1920–2000, 1960–2040 and 2000–2080. The first period was used to analyse past forest growth, and the second and third period were used to predict future growth. For the whole period from 1920 to 2080, scenarios of nitrogen deposition, atmospheric CO_2 and the weather variables were constructed from observations extrapolated by means of environmental models (Chapter 3.2). Only one Europe-wide scenario was constructed for CO_2, but for nitrogen deposition and weather, 22 site-specific scenarios were derived. Simulations were run using these scenarios of actual and expected change, but also using modified scenarios in which one or more of the environmental factors were kept at their initial values (base level scenarios). For example, whereas the actual scenario of CO_2 progressed from 302 μmol mol^{-1} in 1920 to 370 μmol mol^{-1} in 2000, the base level scenario for CO_2 in the first time-period was a constant 302 μmol mol^{-1}, and the base level scenario for the third time-period was a constant 370 μmol mol^{-1}. For climate variables, base level scenarios consisted of repeating the first 8 years of each time-period ten times, to maintain inter-annual variation. For all six weather variables, spatial variability (standard deviation of year-averaged weather on the 22 sites) was larger than temporal variability (standard deviation of site-averaged weather over years).

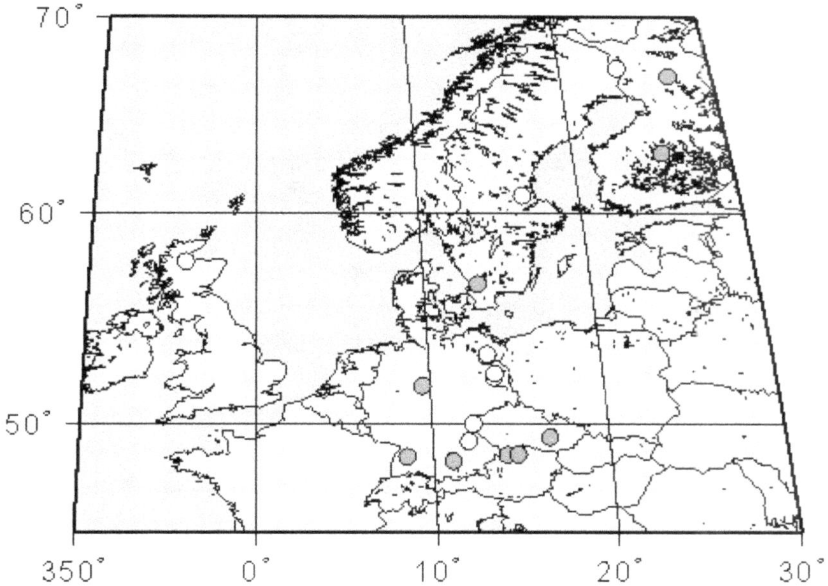

Figure 1. Distribution of the 22 sites used for process-based modelling in Recognition. Dark circles: Norway spruce (n = 9); Light circles: Scots pine (n = 13).

The environmental conditions on the 22 sites and over the three time-periods are summarised in Table 1. Average rate of N-deposition was 4.2 ± 1.5 (SD) kg N ha^{-1} yr^{-1} in 1920. Compared to this average initial base level, N-deposition increased on most sites, such that average N-deposition over the whole period 1920–2000 was 10.5 ± 5.2 kg N ha^{-1}, i.e. an increase of about 150%. The increase was most pronounced at lower latitudes. By 1975–1990, N-deposition exceeded 18 kg N ha^{-1} yr^{-1} on all sites in Germany, Austria and Czechia whereas it remained less than 13 kg N ha^{-1} yr^{-1} on all sites in the UK, Sweden and Finland (Chapter 3.2). The other variables changed much less. The average atmospheric CO_2 concentration in the first time-period was 7.3% higher than its 1920 base level, and most weather variables changed by less than 1%. Temperature increased 0.52 ± 0.24°C from its 1920–1927 base level, but this cannot sensibly be expressed as a percentage of a base level.

During the second and third time-periods, further changes in N deposition were expected to be much smaller than between 1920 and 2000, with deposition even decreasing at most sites during the first decade of the 21st century (Table 1, Chapter 3.2). Changes in CO_2 and most climatic variables, on the other hand, were accelerating.

Table 1. Environmental variables. Averages and standard deviations (SD) for 22 sites and three time-periods. "Base levels" are averages for the first eight years of each eighty-year period. "Changes" refer to the differences between the eighty-year average and the base level.

		Time period 1920–2000		1960–2040		2000–2080	
		Average	SD	Average	SD	Average	SD
N-deposition	*Base level (kg N ha^{-1} yr^{-1})*	4.2	1.46	11.3	5.82	12.3	6.26
	Change (kg N ha^{-1} yr^{-1})	6.3	3.74	1.2	0.73	–1.4	0.94
	Change (%)	148.6	88.8	10.9	6.5	–11.0	7.7
CO_2	*Base level (µmol mol^{-1})*	302	0	316	0	370	0
	Change (µmol mol^{-1})	22	0	63	0	110	0
	Change (%)	7.3	0.0	19.9	0.0	29.7	0.0
Temperature	*Base level (°C)*	6.9	0.65	7.2	0.65	8.0	0.69
	Change (°C)	0.52	0.24	1.20	0.19	1.67	0.26
Precipitation	*Base level (mm yr^{-1})*	597	153	592	154	597	153
	Change (mm yr^{-1})	5.0	16.7	–0.8	25.0	–4.9	33.6
	Change (%)	0.8	2.8	–0.1	4.2	–0.8	5.6
Global radiation	*Base level (MJ m^{-2} d^{-1})*	9.6	0.43	9.6	0.43	9.6	0.43
	Change (MJ m^{-2} d^{-1})	–0.01	0.12	0.26	0.27	0.45	0.38
	Change (%)	–0.1	1.3	2.7	2.8	4.6	4.0
Humidity	*Base level (% RH)*	77.2	1.4	77.0	1.4	76.2	1.4
	Change (% RH)	–0.5	0.3	–1.4	0.7	–2.1	1.1
	Change (%)	–0.7	0.3	–1.9	0.9	–2.8	1.4
Wind speed	*Base level (m s^{-1})*	2.99	0.44	3.00	0.44	3.01	0.44
	Change (m s^{-1})	0.01	0.01	0.01	0.01	0.01	0.02
	Change (%)	0.4	0.3	0.3	0.5	0.3	0.7

4.4.2.3 Distribution of the two tree species

The 9 spruce sites were somewhat wetter than the 13 pine sites (average rainfall of 673 and 544 mm yr^{-1}, respectively) but otherwise weather conditions were similar. The sites differed greatly in soil fertility, with the spruce sites having on average 5300 kg total nitrogen ha^{-1}, and the pine sites only 2500 kg N ha^{-1} (Chapter 3.2).

4.4.2.4 Forest growth

For each time-period, we used two simulated measures of forest growth rate: (i) average net primary productivity (NPP; t dry matter ha^{-1} yr^{-1}); and (ii) final tree height at the end of each period (m). The underlying physiological causes of growth change were analysed by quantifying the rates at which the trees acquired different resources, and the efficiency with which they used the resources to sustain NPP. Six resource use components were defined as follows:

$$NPP = LU \times LUE = NU \times NUE = WU \times WUE \quad (1)$$

Where LU is rate of absorption of photosynthetically active radiation (MJ PAR m^{-2} d^{-1}), LUE is light-use efficiency (g dry matter MJ^{-1} PAR), NU is gross nitrogen uptake (kg N ha^{-1} yr^{-1}), NUE is nitrogen-use efficiency (g dry matter g^{-1} N), WU is water uptake or transpiration (mm yr^{-1}), and WUE is water-use efficiency (g dry matter kg^{-1} water). Note that we have given units for the resource use components that are in common use, rather than the ones that would make Eq. 1 dimensionally correct. Each resource use component was calculated as an average over the 80-yr duration of a time-period. The models Q and EFIMOD did not simulate tree water relations and, therefore, did not calculate WU and WUE.

4.4.2.5 Analysis of simulation output

In most cases, simulated NPP and height were higher in the actual environmental scenarios than in the base level scenarios. The effects of individual environmental factors on forest growth during a time-period were accordingly quantified as the percentage increase in NPP and height resulting from replacing the general base level scenario (i.e. the scenario in which all of N-deposition, CO_2 and climate variables were kept at their initial values) with actual scenarios.

Standard deviations of simulated growth and resource use components were calculated from the output of the four different models to identify major areas of uncertainty.

4.4.3 Simulations of European forest growth between 1920 and 2000

In this section, we present the results of applying the models to the first question of Recognition: explaining observed changes in European forest growth during the 20th century. The simulations for the first time-period (1920–2000) address the question.

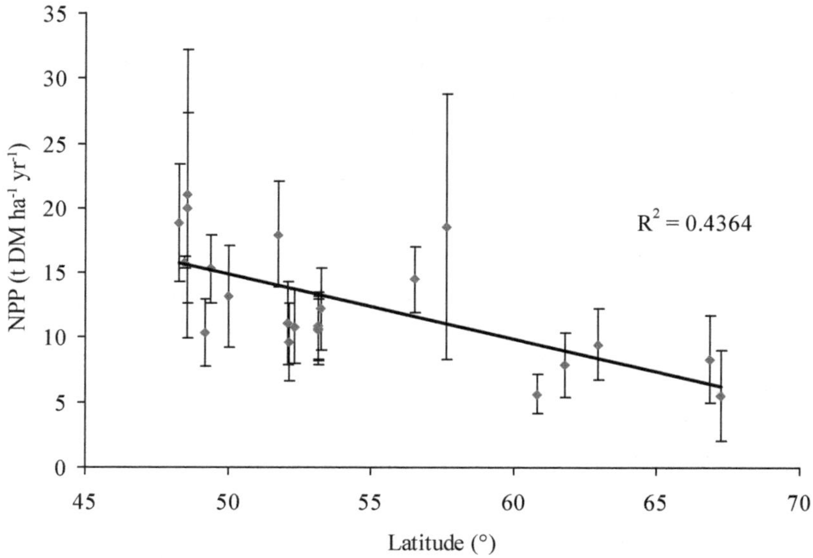

Figure 2. Rates of net primary productivity (NPP, t DM ha^{-1} yr^{-1}) as simulated by the process-based models EFM, EFIMOD, FinnFor and Q for 22 sites (Figure 1) over the period 1920–2000. Points are the four-model means; error bars show ± the standard deviation.

4.4.3.1 NPP and height

For the general base level scenario, which held N-deposition, CO_2 and climate at the levels of 1920 or 1920–1927, the models simulated an average NPP of 12.6 ± 2.3 (SD) tons of dry matter ha^{-1} yr^{-1}. (Table 2). NPP was negatively correlated with latitude (r^2=0.44; Figure 2). The lower NPP at high latitudes was the result of climate differences (temperature had a negative correlation with latitude, r^2=0.71) and, to a lesser extent, of differences in soil fertility (total soil nitrogen had a negative correlation with latitude, r^2=0.07, and without the exceptionally fertile northern Scottish site, r^2 was 0.16). All four models had similar average NPP and latitudinal effects, but for individual sites the models differed greatly (Figure 2).

Simulations using the actual growing conditions, with the environmental variables all changing, yielded values of NPP that were on average 8.4% higher than the base level runs (Table 2). Considerable variation between the models in their estimate of average NPP-change was expressed in a standard deviation of 4.2% (Table 2). When averaged over the sites, model Q showed relatively low sensitivity to environmental change (Figure 3).

The effect of environmental change not only varied between the models but also between the sites (Figure 3). NPP-change varied over the 22 investigated sites from about –5% to +20% according to all four models (Figure 3). Sites high in total soil nitrogen had generally smaller changes in NPP (r^2 for this negative correlation was 0.46).

Compared to NPP, height growth was much less affected by environmental change. Whereas the models showed site-averaged NPP-increases of 8.4 ± 4.2%, final height was increased by only 1.9 ± 2.0% (Table 2).

4.4.3.2 Tree species

Average base level NPP of spruce was higher (15.7 t DM ha^{-1} yr^{-1}) than NPP of pine (10.5 t DM ha^{-1} yr^{-1}). Average response of spruce NPP to overall environmental change was 6.8% and that of pine 9.5%, with effects of changing N-deposition of 4.1% and 7.0%, respectively. In contrast, spruce NPP increased more than pine NPP in response to increasing atmospheric CO_2 concentration: NPP change of 4.1% for spruce compared to 2.6% for pine.

4.4.3.3 Drivers of change

In most cases the effects of the three environmental change factors were additive (Figure 3). The models concurred in identifying changes in N-deposition as the major factor of environmental change at most sites (Figure 3). The changes in N-deposition were not found to be equally important on all sites. The models EFM, FinnFor and EFIMOD showed smaller effects of changing N-deposition at high latitudes, and at sites high in initial total soil nitrogen content.

4.4.3.4 Resource use components

The largest simulated change in rate of acquisition or efficiency of use of any resource was the 23 ± 19% increase in gross N-uptake rate (Table 2), which was largely the direct result of increasing N-deposition.

The different environmental drivers affected the trees via different mechanisms, as was evident from how they differed in their effects on the resource use components. Increased uptake of nitrogen due to increasing N-deposition reduced NUE greatly, but increased LUE and WUE. Increasing CO_2 increased all of LUE, NUE and WUE, but had little effect on the rates at which resources were acquired. Effects of climate change on resource use components were not apparent in the simulations of the first time-period.

We can compare the uncertainties in the simulations of the different resource use components by comparing the standard deviations of the components, and of their response to change, with the base level values. The largest degree of uncertainty seems to be attached to nitrogen uptake and NUE (Table 2). The base level values of N-uptake rate were 51, 97, 28 and 51 kg N ha^{-1} yr^{-1} for EFIMOD, EFM, FinnFor and Q, respectively, and the corresponding changes in these rates due to the combined action of the three environmental drivers were 7, 32, 46 and 7%. For each model, most of the change in N-uptake rate was due to enhanced N-deposition, with only minor effects of increasing CO_2 and climate change. The base level values for NUE were 279, 192, 331 and 247 kg DM kg^{-1} N for EFIMOD, EFM, FinnFor and Q, respectively. Environmental change affected NUE less than N uptake, by 2, –14, –20 and –3% for the four respective models, with again most of the effect being caused by enhanced N-deposition rather than the other two drivers.

Table 2. Simulations of conifer forest growth and resource use. Averages and standard deviations (SD) for four models and three time-periods. For all values, n=4 except for height (not calculated by model Q; n=3), water use and water use efficiency (not calculated by models EFIMOD and Q; n=2) and for the changes caused by CO_2 (not calculated by models EFIMOD and Q; n=2).

		Time period					
		1920–2000		1960–2040		2000–2080	
		Average	SD	Average	SD	Average	SD
Net Primary Productivity	*Base level (t DM ha^{-1} yr^{-1})*	*12.6*	*2.3*	*13.9*	*2.9*	*14.9*	*3.2*
	Change (%)	8.4	4.2	8.8	6.8	9.0	6.6
	Change caused by CO_2 (%)	3.2	2.5	7.8	4.0	9.4	3.3
	Change caused by Climate (%)	0.9	0.9	2.1	1.7	4.9	4.4
	Change caused by N (%)	5.8	2.3	2.4	2.0	–1.2	0.8
Height	*Base level (m)*	*18.3*	*4.1*	*18.7*	*4.4*	*19.3*	*4.1*
	Change (%)	1.9	2.0	3.5	2.9	4.4	4.1
	Change caused by CO_2 (%)	0.7	1.8	1.9	3.6	2.7	4.8
	Change caused by Climate (%)	0.4	0.8	1.4	0.7	3.2	1.6
	Change caused by N (%)	1.0	1.4	0.7	0.8	–0.4	0.5
Light absorption	*Base level (MJ PAR m^{-2} yr^{-1})*	*1245*	*150*	*1293*	*127*	*1309*	*115*
	Change (%)	2.4	1.0	2.1	1.2	1.2	2.0
	Change caused by CO_2 (%)	0.3	0.9	0.5	1.7	0.3	2.0
	Change caused by Climate (%)	–0.3	0.2	0.5	0.7	1.6	1.4
	Change caused by N (%)	2.5	1.3	1.3	1.3	–0.6	0.5
Gross N uptake	*Base level (kg N ha^{-1} yr^{-1})*	*56.7*	*28.9*	*72.6*	*45.0*	*78.0*	*50.5*
	Change (%)	22.8	19.4	9.1	6.9	2.0	5.0
	Change caused by CO_2 (%)	–1.1	0.8	–1.4	1.0	–1.9	1.6
	Change caused by Climate (%)	1.3	1.6	2.4	2.6	6.7	5.6
	Change caused by N (%)	22.4	17.8	7.2	5.2	–3.6	2.8

Table 2. continued.

		Time period					
		1920–2000		1960–2040		2000–2080	
		Average	SD	Average	SD	Average	SD
Water uptake	*Base level (mm yr^{-1})*	*277*	*3*	*273*	*1*	*283*	*10*
	Change (%)	1.2	3.6	1.4	2.5	3.5	1.1
	Change caused by CO_2 (%)	0.2	0.7	0.0	1.4	–0.4	1.9
	Change caused by Climate (%)	1.0	1.8	1.6	0.8	3.9	0.3
	Change caused by N (%)	0.1	1.2	–0.2	0.4	0.0	0.1
Light Use Efficiency (LUE)	*Base level (g DM MJ^{-1} PAR)*	*1.03*	*0.27*	*1.08*	*0.28*	*1.15*	*0.31*
	Change (%)	6.0	4.4	6.6	5.4	7.4	5.1
	Change caused by CO_2 (%)	3.0	1.6	7.2	2.3	9.1	1.2
	Change caused by Climate (%)	1.1	0.8	1.6	1.1	3.2	3.8
	Change caused by N (%)	3.4	2.7	1.2	1.5	–0.7	0.8
Nitrogen Use Efficiency (NUE)	*Base level (g DM g^{-1} N)*	*262*	*58*	*233*	*57*	*237*	*57*
	Change (%)	–8.6	9.9	0.5	3.7	7.3	8.1
	Change caused by CO_2 (%)	4.4	1.8	9.4	2.9	11.7	1.3
	Change caused by Climate (%)	–0.3	1.0	0.0	1.6	–1.4	2.2
	Change caused by N (%)	–10.9	11.0	–4.0	3.6	2.6	2.8
Water Use Efficiency (WUE)	*Base level (g DM kg^{-1} H_2O)*	*4.2*	*0.9*	*4.8*	*1.4*	*5.2*	*1.7*
	Change (%)	10.1	2.8	12.7	1.5	10.3	0.3
	Change caused by CO_2 (%)	3.0	1.8	7.8	2.5	9.9	1.2
	Change caused by Climate (%)	0.6	1.7	1.1	0.9	1.0	2.5
	Change caused by N (%)	6.2	2.6	2.9	3.2	–1.5	0.8

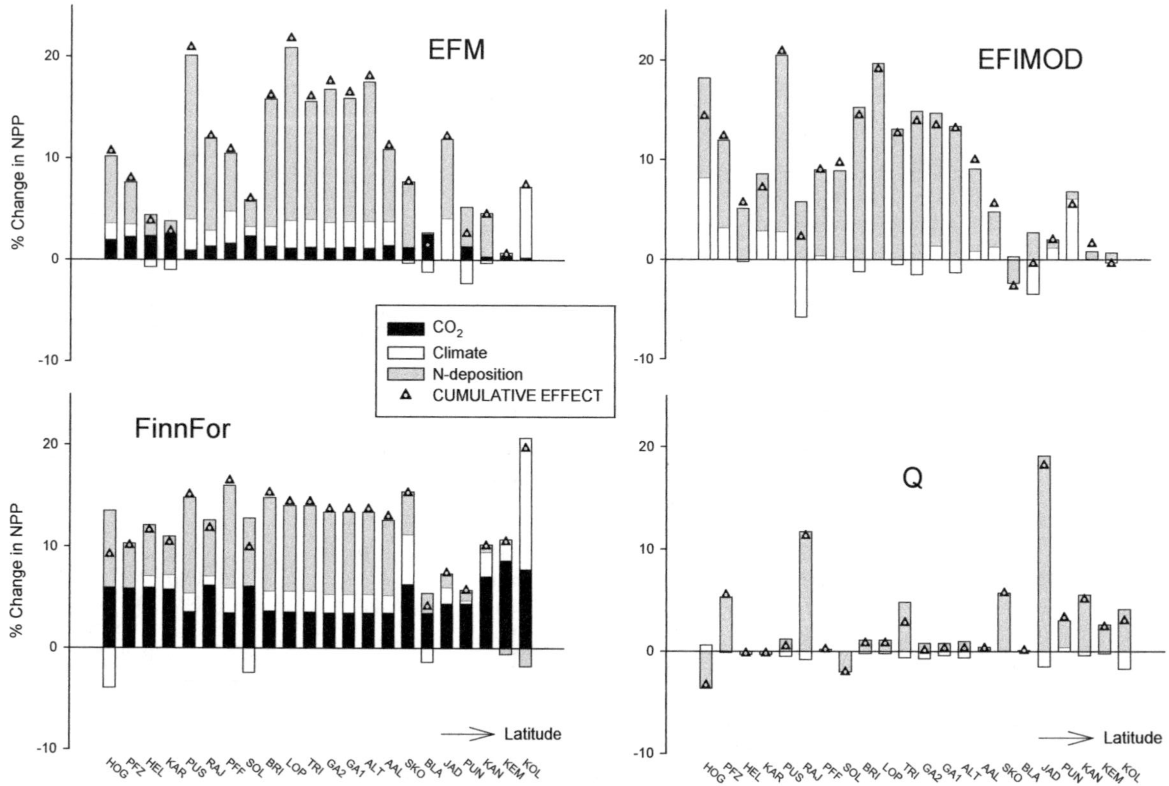

Figure 3. Changes (%) in forest net primary productivity (NPP) as calculated by the process-based models EFM, EFIMOD, FinnFor and Q for 22 sites (Figure 1) over the period 1920–2000. The changes are calculated relative to the base level scenario runs with no environmental change over the 80-yr period. The differently shaded parts of the columns indicate the single-factor effects of changes in either atmospheric CO_2, climate or nitrogen deposition. The cumulative effects of all three types of environmental change are indicated by triangular symbols. The effects of the three factors are additive when the cumulative effect equals the sum of the three single-factor effects, i.e. when the triangles are positioned at the top of the columns. Note that the effect of elevated CO_2 was only calculated for EFM and FinnFor.

4.4.4 Simulations of European forest growth: 1960–2040 and 2000–2080

In this section, we present the results of applying the models to the second question of Recognition: predicting future changes in European forest growth. Simulations for the second time-period (1960–2040) and third time-period (2000–2080) address this question.

4.4.4.1 NPP and height

The models showed that further forest growth rate increases may be expected in the future. The simulations predict increases in NPP during the second and third time-periods, which are of similar magnitude (about 9%) as in the first time-period (Table 2). Increases in height growth rate are expected to be larger in future than in the past, but still less than increases in NPP.

4.4.4.2 Tree species

Whereas in the first period, pine NPP had responded slightly more than spruce NPP to environmental change, the difference disappeared in the second period (both species increasing their NPP by 8.8% on average), and during the third period, spruce even benefited more than pine from continued environmental change (NPP-increase spruce: 11.4%; pine: 7.2%).

4.4.4.3 Drivers of change

The changes in forest growth during the second and third time-periods are attributed to different causes than in the first time-period. Increasing CO_2 replaces changing N-deposition as the major driver of change in NPP, whereas height growth is increasingly stimulated by climate change. However, it should be remembered that N-deposition changes much less during the second and third periods than during the first. The level of N-deposition during the later periods remained sufficiently high to maintain the increase in production effected over the first period, but it did not drive significant further change.

4.4.4.4 Resource use components

During the second and third time-periods, WUE was the resource use component that changed most, largely in response to increasing atmospheric CO_2 concentration. As for uncertainties in the components, NUE remains a key factor for which the models disagree (Table 2). On average though, NUE no longer decreased as it had done during the first time-period, and it even increased during the third period.

4.4.5 Discussion

4.4.5.1 NPP and height

Simulated NPP was negatively correlated with latitude (r^2=0.44; Figure 2). This compares well with the negative correlation (r^2=0.55) reported by Valentini et al. (2000) for the relationship between latitude and NEE (Net Ecosystem Exchange = NPP – soil respiration), based on eddy covariance measurements in European forests between 40°N and 65°N. NEE is directly measurable with the eddy covariance technique, but it differs from NPP, and forest biomass increase provides a better estimate of NPP. Schulze (1982) reviewed the literature on forest growth rates and found an average value of 16 t DM ha^{-1} yr^{-1} for temperate evergreen coniferous forest. The 22 sites we simulated included slower growing boreal sites (Figure 1), which may explain our lower value for average simulated NPP of around 14 t DM ha^{-1} yr^{-1} under actual conditions during the period 1920–2000. NPP is the primary output variable of the process-based models, whereas height growth is calculated from biomass increase using either sophisticated tapering functions (EFIMOD and FinnFor), a cylindrical stem shape (EFM), or it is not calculated at all (Q). There have been no measurements of NPP on the 22 sites to provide a direct comparison with our simulations. Height growth measurements were available, and the models accounted for 18–51% of variation in measured height growth (Chapter 3.2).

Spiecker et al. (1996) collected a large number of studies that showed forest growth rates in many parts of Europe have increased during the 20th century. Their study was summarised by Hunter and Schuck (2002), who stated that volume growth rate currently is about 50% higher than 50 years ago, and height growth rate is about 10–20% higher. Consistent with the findings of Spiecker et al. (1996), our simulations showed increases in NPP and height growth due to changes in N-deposition, atmospheric CO_2 and climate. Furthermore, height growth responded less than NPP to environmental change in all three time-periods. The relatively strong response of NPP and volume growth compared to height growth can be explained if the trees did not markedly alter their shape in response to environmental change, because isometric growth means that volume scales with height cubed. The findings are also consistent with observations by Tissue et al. (1996), who found that, after 19 months of elevated CO_2, only the biomass (and not stem height or diameter) of loblolly pine (*Pinus taeda*) showed a statistically significant increase.

The reported increases of 50% in volume growth rate seem larger than the increases of about 9% in NPP per time-period reported here. However, the percentages are based on different comparisons. The values we calculated for increasing NPP refer to increases of average productivity over a whole 80-yr period, whereas the volume growth measurements are reported for shorter periods. Because our scenarios for environmental change were always identical with the base level scenarios at the beginning of each run, differences between scenarios only developed gradually during the runs. Final values of *current* NPP were thus increased by a multiple of the 9% reported for *average* NPP. Furthermore, additional analysis of the EFIMOD model showed that growing stock increases more than proportionately for a given increase in NPP. Remaining differences between observations and simulations may be the result of errors in the empirical estimates of volume growth (Hunter and Schuck, 2002), and of effects on forest growth rate of changes in forest management, a factor we have not addressed here.

4.4.5.2 Tree species

Each model had different sets of parameter values for Norway spruce and Scots pine, with the differences being derived from independent calibrations on spruce and pine experiments (Chapter 3.2). However, most of the differences between the two species reported here, concerning base level NPP and sensitivity of NPP to environmental change, could be explained by differing site conditions rather than differing physiological characteristics. Spruce NPP was generally higher because of higher soil fertility, and it increased less during the first period because nitrogen was then the major driver of change (see section 4.4.5.3), and the high soil nitrogen contents reduced the effect of extra N-deposition. The high soil nitrogen contents allowed tree growth on the spruce sites to benefit more from increasing atmospheric CO_2 concentration, but experiments under controlled conditions have shown that spruce physiology can be stimulated more by elevated CO_2 than that of pine (Yeatman, 1970, as cited by Wullschleger et al., 1997).

During the second and third time-periods, the response of spruce to environmental change no longer lagged behind that of pine. In fact, growing conditions for spruce started to improve faster than those for pine because environmental change was becoming dominated by changes in CO_2 and climate, rather than by changes in N-deposition, and the fertile spruce sites allowed these factors to greatly stimulate NPP.

4.4.5.3 Drivers of change

The models all identified increasing nitrogen deposition as the major cause of NPP increase in the 20th century (Figure 3, Table 2). This finding is partly the result of the large increase in N-deposition rate during that period (nearly 150%, Table 1). Note, however, that at the given rates, N-deposition may still have provided an order of magnitude less mineral nitrogen than soil mineralisation (Nadelhoffer, 2001). The sensitivity to nitrogen deposition does imply that the trees were generally not growing on nitrogen-saturated soils. Most simulations showed a smaller response to changes in nitrogen deposition at higher latitudes, consistent with the relatively small increases of nitrogen deposition itself on the Scandinavian sites, and with the findings of Spiecker et al. (1996) that Scandinavian forests did not follow the general trend of increases in growth rate.

With regard to the smaller effect on NPP of climate change relative to increased N-deposition, one caveat should be made: only two of the four models simulated water relations (EFM and FinnFor, see Chapter 3.2). However, the average annual increase in rainfall intensity over the period 1901–1995 was less than 0.003 mm d^{-1} for all sites, and less than 0.001 mm d^{-1} for 12 out of the 22 studied sites (Chapter 3.2). These small increases were, according to both EFM and FinnFor, insufficient to account for significant increases in NPP, temperature being the dominant climate change variable. Moreover, the absence of water relations in the other two models did not automatically increase their sensitivity to N – that being lowest in model Q (Figure 3).

Our simulations also showed that the relative importance of the different changing environmental factors might change over time. Changes in CO_2 and climate are predicted to become increasingly important in the next two time-periods, whereas changes in nitrogen deposition will become less significant. This finding suggests that if the goal is prediction for the 21st century, there is some danger in calibrating forest models on observations made during the 20th century only, because the calibrated models may underestimate sensitivity to CO_2 and climate change – i.e. those factors that only become prominent later. The opposite problem may happen when models that are calibrated on past observations do not include the N-cycle at all. The absence of N-effects may then erroneously be compensated for by increasing the models' sensitivity to CO_2 and climate. A further source of potential error in this respect is that instead of simulating reality, which is an ongoing slow increase in atmospheric CO_2 concentration, the experimental database consists mostly of short-term trials with large-jump increases in CO_2.

In the simulations, the effects of the environmental factors were largely additive. This is in contrast to the findings of Tissue et al. (1993) who found downregulation of photosynthesis in loblolly pine at elevated CO_2 only under low N conditions, and those of Eamus and Jarvis (1989) and Murray et al. (2000) who found no positive effect on trees of elevated CO_2 at nutrient limitation. Oren et al. (2001) found, in a FACE experiment, that carbon increment in woody tissue of maturing loblolly pine trees gradually lost sensitivity to elevated CO_2, but that the CO_2 effect was resumed after nitrogen fertilisation. However, the literature on the interactive effects of CO_2 and nutrition on trees is still inconclusive. Luxmoore et al. (1986), Norby et al. (1986) and O'Neill et al. (1987) found positive CO_2 effects on conifers growing at low nutrient availability in field experiments. In our simulations, the trees responded to increased N availability, so there must have been some nitrogen limitation, but still the effects of N and CO_2 were mostly additive. Any interaction may have been obscured in the simulations by the fact that most growth effects were quite small (< 10%). Factorial experiments typically have larger differences between treatments.

The relatively small response to increasing atmospheric CO_2 concentration during the first time-period, as seen in the simulations with models FinnFor and EFM, was not unexpected. Wullschleger et al. (1997) reviewed the literature, covering 398 measurements, and found that the average tree growth response to CO_2 doubling was 32%. This compares to a 3.2% increase in NPP caused by a 7% increase in average atmospheric CO_2 concentration in our simulations. In their review of the literature, Graumlich and Brubaker (1995) concluded that there is no direct evidence yet for a growth increase in natural forests by elevated CO_2. Also, Eamus and Jarvis (1989) found coniferous species to be less responsive to elevated CO_2 than deciduous species.

The simulations showed that climate effects might become increasingly significant, especially at high latitudes. Further analysis with one of the models (EFM, results not shown) showed that changes in temperature were probably the most influential aspect of climate change, the other weather variables having less impact. However, our study did not include drought-prone southerly sites, nor did we include drought-sensitive deciduous tree species. The climate change effect on forest growth was not mechanistically analysed, but it may have had complicated causes as the models realistically include effects of temperature on tree phenology (apparent in the length of the growing season), tree physiology, and on soil transformation processes.

4.4.5.4 Resource use components

For all three periods, simulated rates of water uptake rate were low, around 280 mm yr^{-1} – i.e. less than 50% of average yearly precipitation (Tables 1 and 2). This corresponds well with the general observation that conifers have low transpiration rates (Woodward, 1995). Also, the predicted increases in WUE during the second and third time-periods seem consistent with the literature: Field et al. (1995) found, for 23 tree species, that CO_2 doubling reduced stomatal conductance on average by 23%. The stomatal closure will have reduced water loss more than CO_2 uptake, thereby increasing WUE. The models further predicted that increasing atmospheric CO_2 concentration will increase NUE, and increasing nitrogen deposition will increase both light absorption and LUE, consistent with the findings for conifers in a review by Cannell (1989). We have defined NUE as the ratio of NPP and gross nitrogen uptake (Equation 1, section 4.4.2.4), which makes simulations of NUE difficult to verify empirically as gross uptake of N is difficult to measure accurately. However, the simulated stimulation of NUE by CO_2 seems to be consistent with observations of reduced tree nitrogen concentrations at elevated CO_2 (e.g. Norby et al., 1986). We found no stimulation of nitrogen uptake by CO_2, in contrast to some observations that elevated CO_2 may stimulate fine root growth and thereby the capacity for nutrient uptake (King, 1993).

4.4.5.5 Uncertainties

Uncertainty is evident where the models yield highly different results. For the first time-period (1920–2000), the coefficient of variation (CV=SD/Average) between the models of the base levels of growth variables and resource use components was much smaller than the CV of the changes in these variables (Table 2). This difference reflects, besides structural differences between the models, the often limited scope of model calibration and testing in forest research: we had some data to calibrate our models to common growing conditions (Chapter 3.2), but mostly lacked data to calibrate the sensitivity of our models to environmental change. Uncertainty increased further when the models were applied to prediction. The CV of changes during the periods 1960–2040 and 2000–2080 were larger than those during the first time-period. Our capability for prediction is less than that for explanation.

Changing nitrogen deposition was identified as the single factor having the largest impact on forest growth rate during the 20th century. This result was quite robust, as all four models identified this factor. However, the simulation results for the absolute level of N-uptake had a larger coefficient of variation than for any of the other resource use components (Table 2). Also, besides the absolute rates, the changes in rate of acquisition and use efficiency of N showed a larger CV than those of light and water. The relatively high base level of gross N-uptake rate of the EFM (97 kg N ha^{-1} yr^{-1} for the period 1920–2000) compared to the other three models (28–51 kg N ha^{-1} yr^{-1}) was due to it being the only model that simulates N-efflux from roots. Net uptake of nitrogen thus differed less than gross uptake. Environmental change, in particular enhanced N-deposition, increased N-uptake more in EFM and FinnFor (29–44%) than in EFIMOD and Q (8%) because the former two models are mainly carbon balance driven and allow variation in the N-concentration of all plant organs (Chapter 3.1). Such structural differences between the models also explain why enhanced N-deposition had almost no effect on the NUE in EFIMOD and Q, whereas NUE was decreased significantly (-14 to -20%) by high N-availability in EFM and FinnFor. Clearly, these large differences in model behaviour and model structure require empirical testing.

4.4.6 Conclusions

1. All four process-based models used in this study identified increasing nitrogen deposition as the major cause of observed changes in European forest growth during the 20th century.
2. Further changes in forest growth are more likely to be caused by increasing atmospheric CO_2 concentration and, especially in northern latitudes, by increasing temperature.
3. There are uncertainties, evident in diverging model results, regarding the response of nitrogen uptake and nitrogen use efficiency to environmental change, that need to be resolved in experiments and models.

References

Cannell, M.G.R. 1989. Physiological basis of wood production: A review. Scand. J. For. Res. 4: 459–490.

Eamus, D. and Jarvis, P.G. 1989. The direct effects of increase in the global atmospheric CO_2 concentration on natural and commercial temperate forest. Adv. Ecol. Res. 19: 1–55.

Field, C.B., Jackson, R.B. and Mooney, H.A. 1995. Stomatal responses to increased CO2: implications from the plant to the global scale. Plant, Cell and Env. 18: 1214–1225.

Graumlich, L.J. and Brubaker, L.B. 1995. Long-term records of growth and distribution of conifers: integration of paleoecology and physiological ecology. In: Smith, W.K. and Hinckley, T.M. (eds.). Ecophysiology of Coniferous Forests. Academic Press, San Diego. Pp. 37–62.

Hunter, I. and Schuck, A. 2002. Increasing forest growth in Europe – possible causes and implications for sustainable forest management. Plant Biosystems 136: 133–141.

King, D.A. 1993. A model analysis of the influence of root and foliage allocation on forest production and competition between trees. Tree Physiol. 12: 119–135.

Luxmoore, R.J., O'Neill, E.G., Ells, J.M. and Rogers, H.H. 1986. Nutrient uptake and growth response of Virginia pine to elevated atmospheric carbon dioxide. J. Env. Qual. 15: 244–251.

Murray, M.B., Smith, R.I., Friend, A. and Jarvis, P.G. 2000. Effect of elevated [CO_2] and varying nutrient application rates on physiology and biomass accumulation of Sitka spruce (*Picea sitchensis*). Tree Physiology 20: 421–434.

Nadelhoffer, K.J. 2001. The impacts of nitrogen deposition on forest ecosystems. In: Follett, R.F. and Hatfield, J.L. (eds.). Nitrogen in the Environment: Sources, Problems and Management. Elsevier. Pp. 311–331.

Norby, R.J., Pastor, J. and Melillo, J.M. 1986. Carbon-nitrogen interactions in CO_2-enriched white oak: physiological and long-term perspectives. Tree Physiol. 2: 233–241.

O'Neill, E.G., Luxmoore, R.J. and Norby, R.J. 1987. Elevated atmospheric CO_2 effects on seedling growth, nutrient uptake, and rhizosphere bacterial populations of Liriodendron tulipifera L. Plant and Soil 104: 3–11.

Oren, R., Ellsworth, D.S., Johnsen, K.H., Phillips, N., Ewers, B.E., Maier, C., Schäfer, K.V.R., McCarthy, H., Hendrey, G., McNulty, S.G. and Katul, G.G. 2001. Soil fertility limits carbon sequestration by forest ecosystems in a CO_2-enriched atmosphere. Nature 411: 469–472.

Schulze, E.D. 1982. Plant life forms and their carbon, water and nutrient relations. Encycl. Plant Physiol., New Ser. 12B: 615–676.

Spiecker, H., Mielikäinen, K., Köhl, M. and Skovsgaard, J.P. (eds.) 1996. Growth Trends in European Forests. EFI Research Report 5. Springer.

Tissue, D.T., Thomas, R.B. and Strain, B.R. 1993. Long-term effects of elevated CO_2 and nutrients on photosynthesis and Rubisco in loblolly pine seedlings. Plant, Cell and Env. 16: 859–865.

Tissue, D.T., Thomas, R.B. and Strain, B.R. 1996. Growth and photosynthesis of loblolly pine (Pinus taeda) after exposure to elevated CO_2 for 19 months in the field. Tree Physiol. 16: 49–59.

Valentini, R. and twenty-nine co-authors 2000. Respiration as the main determinant of carbon balance in European forests. Nature 404: 861–865.

Woodward, F.I. 1995. Ecophysiological controls of conifer distributions. In: Smith, W.K. and Hinckley, T.M. (eds.), Ecophysiology of Coniferous Forests. Academic Press, San Diego. Pp. 79–94.

Wullschleger, S.D., Norby, R.J. and Gunderson, C.A. 1997. Forest trees and their response to atmospheric carbon dioxide enrichment: A compilation of results. In: "Advances in Carbon Dioxide Effects Research", ASA Special Publication no. 61: 79–100.

Yeatman, C.W. 1970. CO_2 enriched air increased growth of conifer seedlings. For. Chron., June.Pp. 229–230.

5. Combined Approach

5.1 A comparison of empirical and process-based modelling methods for analysing changes in European forest growth

Marcel van Oijen[1], Göran I. Ågren[2], Oleg Chertov[3], Hans-Peter Kahle[4], Seppo Kellomäki[5], Alex Komarov[3], Karl Mellert[6], Jörg Prietzel[6], Heinrich Spiecker[4] and Ralf Straussberger[6]

[1]*Centre for Ecology and Hydrology, Edinburgh Research Station, UK*
[2]*Swedish University of Agricultural Sciences, Uppsala, Sweden*
[3]*European Forest Institute*
[4]*Institute for Forest Growth, University Freiburg, Germany*
[5]*University of Joensuu, Joensuu, Finland*
[6]*Chair of Soil Science, Technical University Munich, Germany*

5.1.1 Introduction

Empirical and process-based models are different tools with which ecosystem dynamics can be analysed. Process-based models differ from empirical models in that they represent a natural system at two hierarchical levels at least: that of the overall system itself and that of the underlying processes (Van Oijen, 2002). Process-based models of tree growth typically calculate growth by integrating processes of resource acquisition and conversion into biomass. In contrast, empirical models relate a variable of interest, e.g. tree growth rate, directly to possible explanatory variables such as weather variables. Both types of method have been applied in Recognition (Chapter 1), in an effort to determine the causes of observed growth changes in European forests. The observations, concerning tree volume and height growth, suggest that European forests have started to grow faster during the 20th century (Spiecker et al., 1996; Hunter and Schuck, 2002). Recognition assembled data concerning the possible causes of the growth changes (i.e. changes in nitrogen deposition, climate and atmospheric CO_2), and applied models to identify the key variables. Predictions concerning the possible continuation of growth change were made as well.

Empirical and process-based models differ in the kind of variables that they are capable of analysing, the sensitivity with which they can detect effects on forest growth, and the sources of uncertainty affecting the results. The combined use of both approaches in the project was motivated by the idea that the methods would complement each other. The different strengths and weaknesses of the methods, and their complementarity, will be analysed in this chapter. Comparisons between empirical and process-based models in forestry have been made before by various authors (e.g. Sharpe, 1990; Mohren and Burkhart, 1994; Vanclay, 1994; Amaro and Tomé, 1999). These comparisons have tended to be philosophical in outlook, the examined models being taken from the literature and not applied to the same problem. An exception has been the study of Courbaud (2000) who derived empirical and process-based models from yield tables and investigated their dynamical behaviour under various hypothetical conditions. In Recognition, different models were developed independently, but then applied to the same problem of explaining observed forest growth changes across Europe. This permits a thorough and realistic evaluation of their behaviour.

In the project, two empirical modelling approaches were employed: the Historical Development Investigation (HDI) and the Present State Analysis (PSA) (Chapters 2.1 and 2.2). Both methods analysed the role of changes in temperature, precipitation and foliar nutrient concentration on the growth of Norway spruce, Scots pine and European beech at a number of sites across Europe. Height growth was related to these environmental factors by means of statistical methods – i.e. methods based on multiple linear regression. The approaches differed in the particular equations used, with respect to the sources of reference forest growth data and in the diversity and length of the foliar nutrient time-series.

Four process-based forest ecosystem models were used in the project: EFIMOD, EFM, FinnFor and Q (Chapters 3.1 and 3.2). The models simulated flows of resources (C, N, water) into and out of the forest system, and the consequences for productivity, but at different levels of complexity (see 5.1.2). In contrast to the empirical models, two of the process-based models also included effects of atmospheric CO_2 concentration, but nutrient relations were limited to nitrogen.

Both empirical and process-based modelling approaches were used to analyse past changes in forest growth, but only the latter method was used for prediction as well. We will mention some of these predictions in this chapter, but will concentrate on the results for past growth where the methods and results can be compared. Our major aim is evaluating project methodology; an in-depth discussion of the results themselves is presented in Chapter 5.2.

5.1.2 Modelling methods

In this section, we will briefly describe the different modelling methods that were applied in Recognition to quantify the effects of different environmental factors on European forest growth. The project comprised both empirical and process-based modelling methods, and for both approaches a number of different models were developed and applied. A common feature for all models was that they were applied to a limited number of forest sites across Europe. No upscaling to regional or continental forest production was attempted. In the following, we will first describe the sites to which the models were applied, and the environmental conditions at those sites. Thereafter, the modelling methods themselves will be described in somewhat more detail.

5.1.2.1 Site conditions – soil, CO_2, climate and N-deposition

All four process-based models were used to simulate forest growth at 22 sites across Europe: 13 covered with Scots pine and 9 covered with Norway spruce.

Latitude varied from 48°N to 67°N. The average air temperature of the sites over the years 1975–1990 ranged from –0.6 to 10.0°C, and rainfall from 403 to 1169 mm yr^{-1}. The spruce sites had more rainfall on average than the pine sites (673 and 544 mm yr^{-1}, respectively) but otherwise the ranges of weather conditions were similar. Soil carbon and nitrogen contents varied by nearly an order of magnitude between sites, with the spruce sites having on average 5300 kg total N ha^{-1} and the pine sites only 2500 kg N ha^{-1} (Chapter 3.2).

To investigate the effects of past changes in environmental conditions on growth, the process-based models simulated forest growth on the 22 sites from 1920 to 2000. Therefore, for this whole growth period, scenarios (time-series) of gradual change in nitrogen deposition, atmospheric CO_2 and the weather variables were constructed from regional observations (Chapter 3.2). Only one Europe-wide scenario was constructed for CO_2, progressing from 302 μmol mol^{-1} in 1920 to 370 μmol mol^{-1} in 2000. For nitrogen deposition and weather, 22 site-specific scenarios were derived from measured data. Average rate of N-deposition increased from 4.2 ± 1.5 (SD) kg N ha^{-1} yr^{-1} in 1920 to 17.7 ± 9.3 kg N ha^{-1} yr^{-1} in 1980 and down to 12.5 ± 6.4 (SD) kg N ha^{-1} yr^{-1} in 2000. Deposition increased more at the Central European sites than in Scandinavia and the UK (Chapter 3.2). Between 1920 and 2000, most weather variables changed by less than 1%. Average annual temperature increased 0.52 ± 0.24 °C.

To predict current and future changes in forest growth, the process-based models were run for the periods 1960–2040 and 2000–2080 as well, and environmental scenarios were constructed for those periods (Chapter 3.2). However, as the focus of the present chapter is on the comparison between process-based models and empirical models, and as the latter were only applied to 20th century growth, we will not give details of the simulations for current and future conditions here.

The sites used in the two empirical modelling approaches HDI and PSA overlapped largely with the sites used for process-based modelling, described above, and the ranges of environmental conditions were similar. HDI was applied to 12 sites covered with Scots pine and 7 sites covered with Norway spruce, whereas PSA used 19 pine sites and 17 spruce sites (Chapter 4.1). The two empirical methods were also applied to a small number of sites with European beech (*Fagus sylvatica* (L.)), but beech was not part of the process-based modelling study and will not be discussed here.

To ensure consistency between methods, process-based and empirical modelling used the same sources of information to quantify site-specific time courses of weather and nitrogen deposition. However, time-series with higher temporal resolution (daily or monthly values) were derived for the process-based modelling than for the empirical modelling (monthly, seasonal and annual values). The approaches also differed in that the process-based models simulated growth on all sites from 1920–2000, whereas periods of investigation were not standardised for HDI and PSA. Both empirical approaches operated on actual observations of tree growth rather than on simulated growth, so the periods of investigation were determined by site-specific dates of tree planting and measurement series. Further details of site conditions have been presented in Chapter 3.2.

5.1.2.2 Empirical modelling: HDI and PSA

The Present State Analysis (PSA) and the Historical Development Investigation (HDI) were used to analyse forest growth. The PSA used a two-stands approach to quantify changes in forest growth. At each investigated site, data were gathered from neighbouring young and old stands, and stem analysis (Chapter 2.2) was used to determine the time course of past height growth for both stands, from their respective dates of planting to the date of sampling. The average age difference between young and old stands across sites was about 40 years. The extent to which tree growth rate had changed over time was quantified by comparing the patterns of annual height growth of both stands at the same age. The site-specific patterns were analysed by linear regression on concurrent changes in temperature, precipitation and foliar nitrogen concentration.

In contrast to the PSA, the HDI was a single-stand approach, i.e. it used data from only one forest stand per site. Whereas the PSA used neighbouring older stands as a reference against which to quantify growth rate changes, HDI used empirical height growth curves from past experience in the same region, expressed as yield tables or site index curves (Spiecker, 1999; Chapter 2.2). This approach reduced the number of stems required for analysis, but suffered from the fact that yield tables are unreliable references for the detection of site-specific height growth trends, as they represent regional averages and growth trends of certain time periods (Spiecker, 1999; Chapter 5.2). On the other hand, problems inherent in the two-stands approach, i.e. uncertainty about the extent to which neighbouring stands will indeed have had common land-use and forest management histories, were thus avoided. After growth rates were normalised by dividing by the reference growth rate, they were analysed in the HDI by multiple linear regression:

$$ih_{ref} = a_0 + \sum_{i=1}^{6} a_i x_i$$

Where ih_{ref} is the ratio of observed annual height increment and reference height increment for trees of the same age (%), the a_i are regression constants, $\{x_1, x_2, x_3\}$ are precipitation, air temperature and foliar nitrogen concentration or nitrogen deposition for the present year and $\{x_4, x_5, x_6\}$ are the moving averages of the same variables over the last 5 years (Chapter 4.1). Besides foliar nitrogen, other nutrients (P, Mg, Ca, K) were included in earlier versions of the regression equations, but were never found to account for a significant proportion of observed variation in growth rate and were therefore discarded (Chapter 4.1). For further details of HDI and PSA, see Chapters 2.1, 2.2, 4.1 and 4.2.

5.1.2.3 Process-based modelling

The four process-based models Q, EFIMOD, FinnFor and EFM differed in their structure. The two simplest models, Q and EFIMOD (44 and 119 parameters, respectively) simulated forest primary productivity as determined mostly by nitrogen uptake from the soil, whereas the two more complex models FinnFor (249 parameters) and EFM (261 parameters) simulated primary productivity as the consequence of interactions between photosynthetic carbon assimilation, nitrogen and water uptake. In all four models, the growth processes were affected by nitrogen deposition, radiation and temperature, but only FinnFor and EFM included sensitivity to CO_2, water availability, wind speed and humidity.

Simulations were carried out for unthinned stands at each of the 22 sites used for process-based modelling. The models produced different output variables but all of them simulated net primary productivity of the forests (NPP; t dry matter ha^{-1} yr^{-1}), and the analysis therefore focused on this variable (Chapter 4.4).

Growth between the years 1920 and 2000 was simulated for each site using multiple environmental scenarios. A base level scenario was defined in which CO_2 and N-deposition were kept at their 1920 levels, and weather conditions were maintained at the 1920–1927 levels, in an 8-year cycle. Simulations using these site-specific base level scenarios provided reference forest growth rates, comparable to the reference growth rates from old stands and yield tables in PSA and HDI. Scenarios in which one or more of the environmental scenarios changed as observed until 2000 were used to quantify the contribution of the different environmental variables to growth change, and their interactions (Chapter 4.4).

Using the base level scenarios, the models simulated an average NPP of 12.6 ± 2.3 (SD) tons of dry matter ha^{-1} yr^{-1} across the sites (Chapter 4.4). NPP was negatively correlated with latitude (r^2 = 0.44) (Figure 1) because of climate differences (temperature had a negative correlation with latitude, r^2 = 0.71) and, to lesser extent, because of differences in soil fertility (total soil nitrogen had a negative correlation with latitude, r^2 = 0.16, excluding the exceptionally fertile northern Scottish site).

The underlying physiological causes of growth change were analysed by quantifying resource use components, i.e. the rates at which the trees intercepted or acquired light, water and nitrogen, and the efficiencies with which these resources were used to sustain NPP.

For further details of the process-based modelling, see Chapters 3.1, 3.2 and 4.4.

5.1.3 Comparison of the results

The results of applying the empirical modelling approaches have been presented in Chapters 4.1, 4.2, and 4.3, and the results of the process-based modelling have been presented in Chapter 4.4. In this section, we summarise these results and compare the outcomes of the different modelling approaches. Note that the results of the empirical approaches describe height growth – with the exception of section 4.2.3.2 –, whereas the presented results of the process-based modelling describe NPP.

5.1.3.1 Growth change caused by all factors combined

The results of the process-based models are summarised in Figure 1. The top left panel shows the average NPP that the four models predicted for 22 sites across Europe using each site's base level scenario. Sites are ordered by latitude. The two other panels in the top row show the same model outputs plotted against average air temperature and soil nitrogen content at the sites. These results constitute the reference growth rates, and the panels in the second row show how much simulated

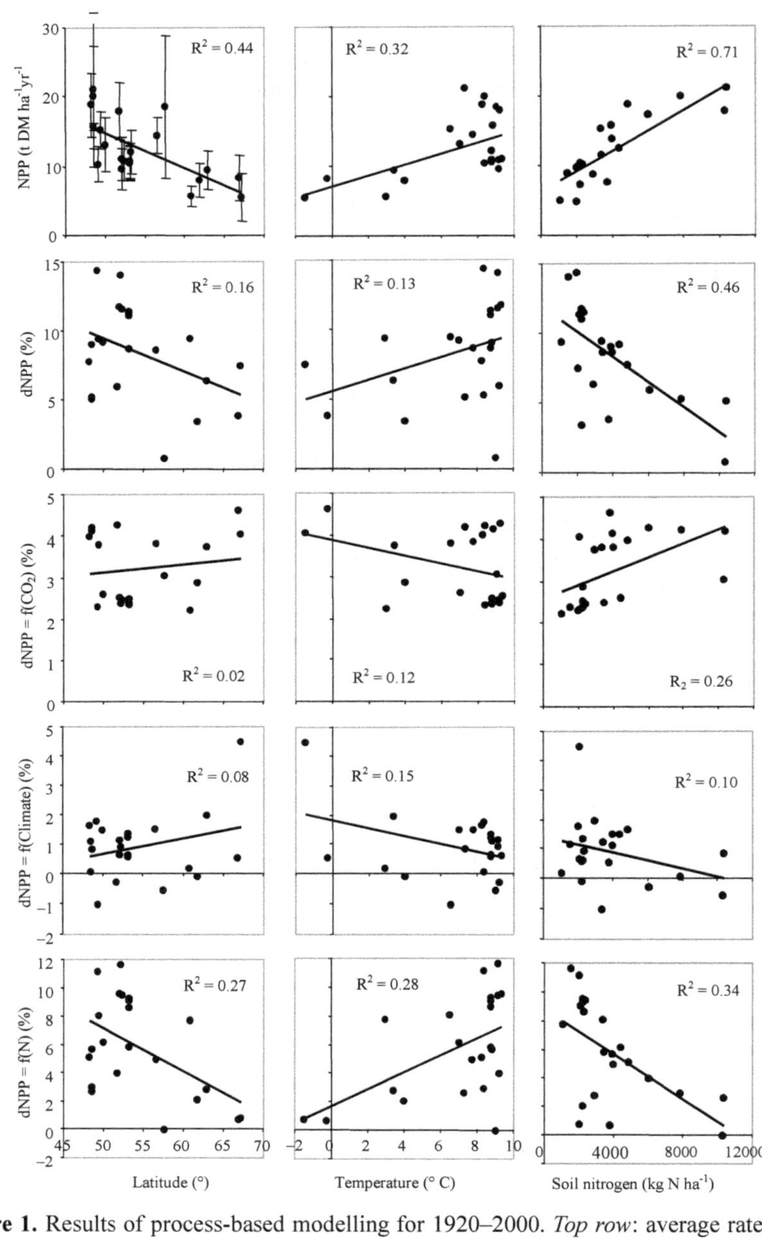

Figure 1. Results of process-based modelling for 1920–2000. *Top row*: average rates of net primary productivity (NPP, t dry matter ha^{-1} yr^{-1}) as simulated for 22 sites across Europe. Error bars indicate standard deviations between the four process-based models used: EFIMOD, EFM, FinnFor and Q. Sites and models are described in detail in Chapter 3.2. Error bars are only shown in the top-left panel for clarity, similar coefficients of variation apply throughout. *Second row*: simulated relative changes in NPP because of combined changes in atmospheric CO_2 concentration, climate and N-deposition. *Rows three, four and five*: single-factor effects of changing CO_2, climate or N-deposition, respectively, with each row showing the % change in average NPP for 1920-2000 if only one factor were changing during this period The three *columns* show identical results, but ordered according to latitude, average annual temperature and soil nitrogen content of the different sites.

NPP changed compared to these references (%) because of the combined effect of changes in CO_2, climate and N-deposition between 1920 and 2000. Average simulated NPP increased on all 22 sites, by between 1 to 15%, with a slightly positive correlation with site temperature, and negative correlations with latitude and site fertility (Figure 1). The average response of NPP to the environmental changes was slightly higher on the pine sites (9.5%) than on the sites with spruce (6.8%). The HDI showed increased growth rates on the lower latitude (Central European) pine sites, but not elsewhere, nor on the spruce sites. The PSA showed increases of mean annual height increment over the period 1960 to 2000 in both pine (25%) and spruce (23%) but no significant relation between height growth acceleration and latitude (Chapter 4.2 and 5.2). Based on conventional yield level models the observed average increase in stand height between 1960 and 2000 results in an average increase in total wood volume production potential of + 56% for spruce, and + 45% for pine (Chapter 4.2). These values seem higher than the 1-15% increase in NPP accounted for by the process-based models but the result cannot be compared rigorously because they refer to different variables, time-periods and management regimes (Chapter 4.4). The process–based models predicted that forest growth rate will keep on increasing in the 21st century (Chapter 4.4).

5.1.3.2 CO_2

The effect of atmospheric CO_2 concentration ($[CO_2]_a$) on forest NPP was included in two of the four process-based models: EFM and FinnFor. According to these models, the increase in $[CO_2]_a$ between 1920 and 2000 accounted for only a small part of the increase in NPP: between 2 and 5% for most of the 22 sites (Figure 1, third row). Spruce responded slightly stronger than pine to increasing $[CO_2]_a$: 4.1% compared to 2.6%. Increasing $[CO_2]_a$ had little effect on the rates at which light, water and nitrogen were acquired by the trees but did increase the efficiency with which these resources were converted into biomass. The empirical models did not address the effect of $[CO_2]_a$ on tree growth. The process-based models predicted that increasing $[CO_2]_a$ will become an increasingly important driver of forest growth rate increase in the 21st century.

5.1.3.3 Climate

The process-based models did not show any important effects of climate change on NPP, except for the most northerly site (Figure 1, fourth row). HDI and PSA investigated the effects of changes of air temperature and precipitation on tree height growth and also did not reveal important climate change effects. However, the process-based models predicted that climate change would become an increasingly important driver of forest growth rate increase in the 21st century, especially on the Scandinavian sites.

5.1.3.4 Nutrients

The process-based models identified changes in N-deposition as the major factor of environmental change at most sites. This is evident, for example, in Figure 1 where the changes in NPP because of changes in N-deposition alone (bottom row) almost equal the changes in NPP because of changes in all factors combined (second row). Effects of changing N-deposition were smaller at high latitudes, and at sites high in initial soil nitrogen content. The average response of spruce NPP to changing N-deposition was 6.8% and that of pine was 9.5%. The stimulating effect of N-deposition operated through an increase in gross N-uptake rate of 23% averaged across sites, but the effect was counteracted by a reduction in nitrogen use efficiency of 9% (Chapter 4.4). HDI and PSA also identified changes in N-availability as the major cause of forest growth change in the 20th century. Other nutrients were not found to be important, nor were any significant interactions between nutrition and climate found. Nitrogen deposition is expected to decrease or stabilise in the 21st century, so the process-based models predicted that future changes in N-deposition will have minor impact on forest growth.

5.1.4 Discussion

5.1.4.1 Uncertainties in empirical and process-based modelling

All models are abstractions of the real world. In the process-based models, tree metabolism and interactions with the environment were represented by differential equations. In the empirical models, the relationships between tree size or growth rate and environmental factors were represented by linear regression equations. Both empirical and process-based models produced results which showed that forest growth rates increased on many sites across Europe. The status of these results was not the same. The empirical methods established changes in growth rate directly from *tree* data, whereas the process-based models inferred changes in growth rates from *environmental* data and their simulated effects on trees. Therefore, only the empirical models provided statistically significant evidence for changes in forest growth. However, the project was not aimed at establishing growth change as such – that had been done before (Spiecker et al., 1996) – but at identifying the underlying causal factors. In this respect, the fact that both modelling approaches found increased growth rates (albeit with exceptions for the empirical methods, see Chapter 4.1) suggested that the process-based models were not falsified by the data, and could be used as instruments for further analysis.

The process-based models tried to represent the complex functioning of the forests, whereas the empirical models tried to identify patterns within the data directly. Both approaches provided uncertain information. For process-based modelling, there were three sources of uncertainty: (1) imprecisely known inputs (i.e. climate and soil information); (2) imprecisely known parameter values; and (3) uncertainty regarding the realism of model structure itself. For the empirical approaches, the sources of uncertainty were: (1) the reliability and representativeness of the data on environmental conditions, recent forest growth and past (reference) growth; and (2) the possible over-simplification involved in expressing ecological relationships in a limited number of linear regression equations. Empirical model uncertainty was increased by the fact that the distribution of environmental factors over sites was not orthogonal, so confounding of factor effects could not be excluded. This was not a problem for the process-based models, which could define any full-factorial set of combinations of environmental factors. A positive consequence of the simpler empirical approach was that, once the type of regression equations was decided upon, techniques like least-squares estimation guaranteed that a unique solution was found, whereas model structure was never uniquely identified in the process-based approach.

The process-based modelling uncertainties were addressed in the project by the use of the four different models (Chapter 3.1), testing them against data collected in the project (Chapter 3.2), and analysing differences between the models in terms of underlying processes (Chapter 4.4). The process-based models differed much less in their output for the base levels of growth and use of resources (especially uptake and use of N) than for the changes in these variables (Chapter 4.4). Besides reflecting structural differences between the models, this reflected the poor availability of data for calibrating the sensitivity of the models to environmental change. Uncertainty, expressed as differences between model outputs, increased further when the models were applied to prediction: the capability for robust prediction apparently being less than that for explanation (Chapter 4.4).

5.1.4.2 Applicability of empirical and process-based modelling to different environmental factors

In contrast to the process-based models, the empirical models were unable to directly address the effect of increasing $[CO_2]_a$ on tree growth, as there was no variation for this factor between sites, and the temporal variation had been too gradual. In Chapter 2.1 it is argued that, if increasing $[CO_2]_a$ had been an important stimulant of European forest growth, increased tree growth would have been expected across all sites. As this was not the case, any effects of CO_2 may have been small. This argument was corroborated by the process-based models. The simulated effects of increasing atmospheric CO_2 concentration were indeed more uniform across sites than those of changes in climate and N-deposition (Figure 1), but the effects of increasing CO_2 on NPP were generally small, between 2 and 4%, well below the detection threshold of the empirical approaches.

The contributions of changes in climate to forest growth change were examined by both the empirical and the process-based modelling approaches. Both approaches could be applied because long-term Europe-wide climate data were available for use in the project, the same data being used by the different approaches (Chapters 2.1, 2.2 and 3.2). No significant effects concerning growth acceleration were found by the empirical approaches (Chapters 4.1 and 4.2), and the effects shown by the process-based models were small (Figure 1). For assessing the role of climate change in forest growth, each method seemed valid, but using both together enhanced the capacity to interpret the results. Although no statistically significant climate change effect was found by the empirical models, the results of the process-based models suggested that effects were present albeit very small. However, it is clear that the results of the process-based modelling concerning climate change effects should be treated with caution given the small data support.

Both the empirical and process-based modelling approaches identified changes in nitrogen availability as having had a large impact on forest growth rate during the 20th century. The different approaches arrived at this common conclusion using very different sources of information about nitrogen availability. The process-based modelling made direct use of data on nitrogen deposition, whereas the empirical approaches used either nitrogen deposition or the proxy of foliar nitrogen concentration. Foliar nitrogen concentration may, however, be a good proxy for the target variable N-deposition, as shown in Chapter 4.2. However, both data sources had their disadvantages. N-deposition data in the process-based modelling approach were not taken from on-site measurements but from spatial interpolation models (Chapter 3.2), which added uncertainty to the estimates. On the other hand, the use of foliar N concentration as an independent variable in the analysis implied replacing an environmental variable (N-deposition) with a plant variable, even though it is the effects of environmental change on plants that we are aiming to evaluate.

The HDI used long time-series of foliar nitrogen concentration or nitrogen deposition data, whereas the PSA relied on recent data only. Process-based modelling and HDI could thus directly examine the role of changes in nutrition over time, whereas PSA could only infer that *relative* changes in N-availability as a result of deposition will have been high on those sites where foliar nitrogen content was found to be low. This indirect method of estimating nitrogen availability adds to the uncertainty of the PSA results concerning nitrogen.

The models were consistent in identifying higher increases in forest growth rate on soils relatively poor in soil N (process-based models, Figure 1) or in trees with relatively low foliar N concentration (empirical models). The different methods thus supported each other.

5.1.4.3 Interactions between factors

None of the approaches found clear interactions between changes in CO_2, climate or N-deposition concerning their effects on forest growth. Even if these factors did interact, identifying the interactions would be extremely difficult for the empirical approaches given the relatively small magnitude of changes in forest growth, the small number of sites, and the inherent error in measurements. In principle, process-based modelling may be a more useful tool to identify potential interactions. For each of the 22 considered sites, all four process-based models were run with each combination of constant or changing CO_2, climate and N-deposition, but no consistent, strong interactions were found (Chapter 4.4).

Evidence for interactions between the factors was found in experiments where the factors were changed by large amounts. Eamus and Jarvis (1989) and Murray et al. (2000) found no positive effect on trees of elevated CO_2 at nutrient limitation. In a FACE (Free Air CO_2 Enrichment) experiment, Oren et al. (2001) found that carbon increment in woody tissue of maturing loblolly pine trees gradually lost sensitivity to elevated CO_2, but that the CO_2 effect was resumed after nitrogen fertilisation. However, the sensitivity of such experiments to interactions is bought at the price of applying unrealistically large or sudden changes in environmental conditions.

5.1.4.4 Applicability of empirical and process-based modelling to different sites and species

Each process-based model had different sets of parameter values for Norway spruce and Scots pine, with the differences being derived from independent calibrations on spruce and pine experiments (Chapter 3.2). However, differences between simulations for the two species, concerning sensitivity of NPP to environmental change, were mostly caused by differences in site conditions rather than in physiological characteristics. Simulated spruce NPP increased less than that of pine because the spruce sites were more fertile, thereby reducing the effect of extra N-deposition. This finding was corroborated by the HDI, which also found strongest effects of nitrogen availability on Central European pine sites. Similar results were obtained with the PSA, which also detected a larger increase in tree height growth of pine than of spruce whereas average site index of pine was lower than of spruce.

The relatively high level of soil fertility on the spruce sites allowed simulated spruce growth to benefit strongly from increasing atmospheric CO_2 concentration, but for this phenomenon no empirical corroboration was possible because of the inaccessibility of CO_2 data to the HDI and PSA analyses. Figure 1 shows various examples of changes in NPP resulting from interacting site factors, as revealed by the process-based models. Process-based models offer a unique tool to disentangle such effects on tree growth of correlated environmental factors (site, fertility) and of interacting effects of factors (N-availability, CO_2).

Process-based modelling obviously fails as an analytic tool when growth response to environmental change is predominantly caused by factors not incorporated in the models. Concerning mineral nutrients only the effect of nitrogen was investigated using the process-based models. Fortunately, HDI was able to assess the availability of P, K, Mg and Ca, using time-series of foliar concentrations, and found no evidence of nutrient limitations other than for nitrogen (Chapter 4.1).

5.1.4.5 Empirical and process-based models as tools for explanation

The first goal of Recognition was to explain observed changes in past European forest growth. The project results show that both empirical and process-based modelling were able to account for the changes, but in different ways and to different extents. The empirical models correlated observed tree growth data to specific factors, such as nitrogen availability. Correlation is not the same as causation, so one may consider this achievement a description of facts rather than an explanation. However, the fact that many plausible factors could be excluded suggests that the key elements for explanation were found, but not yet embedded in a mechanistic framework. The process-based models did not address the growth data directly – and even focused on NPP rather than height growth – but were able to show mechanistically how forest ecosystem processes could have responded to the various environmental changes. They showed how the different environmental drivers affected the trees via different mechanisms, with specific effects on the acquisition rates and use efficiencies of the various resources (Chapter 4.4). However, we cannot state that one single mechanistic explanation, represented in the process-based models, was identified, because the observations could be accounted for by the differently structured process-based models used in the project (Chapter 4.4).

We conclude that empirical models provide incomplete explanations and process-based models provide uncertain ones, making them complementary tools for scientific analysis.

5.1.4.6 Empirical and process-based models as tools for prediction

The second goal of Recognition was to predict future forest growth. As stated above, the empirical models were developed and used to analyse past growth data, not to provide predictions. The process-based models, on the other hand, were used for prediction as well as explanation. This division of tasks was chosen because empirical regression models are derived using data from specific conditions, and should not be extrapolated to new conditions. The limited use of empirical methods for prediction was recognised from the outset in Recognition, the project having been initiated to explain the fact that site productivity has increased on many sites in Europe (Spiecker et al., 1996). The deficiencies of yield tables and other empirical management models under conditions of global change are known, and have stimulated the use of process-based models as tools in forest management (Johnsen et al., 2001; Mäkelä and Landsberg, 2000). However, it should be noted that process-based models can be used to extrapolate when applied to future, new conditions, but the extrapolation involved in such models is hidden in their parameterisation and structure. In process-based modelling, current or past physiological data are used to quantify parameter values and identify essential feedback processes, and these may become invalid or incomplete when applied to trees growing under different conditions. However, the fact remains that process-based models are the only tools with which various 'what-if' questions can be explored.

One of the major results of the predictive use of the process-based models in project Recognition, was the finding that the key environmental drivers of forest growth change are likely to change in the future. Whereas N-deposition constituted the major factor causing change in forest growth in the 20th century, changes in CO_2 and climate are likely to become the most important changes in the next century (Chapter 4.4). The effect of these latter drivers is expected to be largest on fertile sites; so, in contrast to the growth increases of the 20th century, spruce may benefit more than pine. Although we cannot verify these predictions at present, they do suggest that predictive use of empirical models (like the HDI and PSA), which excluded changes in climate and CO_2 as being insignificant, would have been incorrect.

Another form of empirical extrapolation, i.e. using data from experiments in which trees were subjected to expected future environmental conditions, would also have been problematic as such experiments typically are short-term trials with unrealistic large-jump changes in conditions.

We conclude that, for prediction, process-based modelling may remain the least problematic choice because it can show how future effects of environmental drivers may differ from their present effects. However, prediction by means of process-based models should be exercised with caution because models still vary widely in their sensitivity to environmental factors (Van Oene and Ågren, 1995; Chapter 4.4).

5.1.5 Conclusions

Empirical modelling is a method in which observational data are analysed directly, with fewer assumptions about forest ecosystem functioning than are required by process-based modelling. It thus is the best tool for identifying the variables that should be part of an explanation of changes in forest growth. Process-based models, on the other hand, embody more assumptions, and require parameterisation of more parameters – all adding to output uncertainty – but this allows them to disentangle the various correlations and interactions in the forest ecosystem, and thereby provide mechanistic explanation of growth change. We conclude that using both approaches together may be the best way to provide reliable explanations, well supported by the data and mechanistically understood. If the combined approach shows consistent results, as it did in Recognition, the process-based models can be used to explore possible future forest growth changes.

References

Amaro, A. and Tomé, M. (eds.) 1999. Empirical and Process Based Models for Forest Tree and Stand Growth Simulation. Ed. Salamandra, Lisbon: 591 p.

Courbaud, B. 2000. Comparing light interception with stand basal area for predicting tree growth. Tree Physiology 20: 407–414.

Eamus, D. and Jarvis, P.G. 1989. The direct effects of increase in the global atmospheric CO_2 concentration on natural and commercial temperate forest. Advances in Ecological Research 19: 1–55.

Hunter, I. and Schuck, A. 2002. Increasing forest growth in Europe – possible causes and implications for sustainable forest management. Plant Biosystems 136: 133–141.

Johnsen, K., Samuelson, L., Teskey, R., McNulty, S. and Fox, T. 2001. Process models as tools in forestry research and management. Forest Science 47: 2–8.

Mäkelä, A. and Landsberg, J.J. 2000. Process-based models for forest management – A collection of peer reviewed papers from among those presented at the conference "Process-Based Models for Forest Management" held at Saariselkä, Finland, August 1998. Tree Physiology 20: 289–419.

Mohren, G.M.J. and Burkhart, H.E. 1994. Contrasts between biologically-based process models and management-oriented growth and yield models. Forest Ecology and Management 69: 1–5.

Murray, M.B., Smith, R.I., Friend, A. and Jarvis, P.G. 2000. Effect of elevated [CO_2] and varying nutrient application rates on physiology and biomass accumulation of Sitka spruce (*Picea sitchensis*). Tree Physiology 20: 421–434.

Oren, R., Ellsworth, D.S., Johnsen, K.H., Phillips, N., Ewers, B.E., Maier, C., Schäfer, K.V.R., McCarthy, H., Hendrey, G., McNulty, S.G. and Katul, G.G. 2001. Soil fertility limits carbon sequestration by forest ecosystems in a CO_2-enriched atmosphere. Nature 411: 469–472.

Sharpe, P.J.H. 1990. Forest modeling approaches: compromises between generality and precision. In: Dixon, R.K., Meldahl, R.S., Ruark, G.A. and Warren, W.G. (eds.), Process-based Modelling of Forest Growth Responses to Environmental Stress. Timber Press, Portland: 180–190.

Spiecker, H., Mielikäinen, K., Köhl, M. and Skovsgaard, J.P. (eds.) 1996. Growth Trends in European Forests. EFI Research Report 5. Springer: 372 p.

Spiecker, H. 1999. Forest growth. In: Rehfuess, K.E. , Ågren, G.I., Andersson, F., Cannell, M.G.R., Friend, A., Hunter, I., Kahle, H.P., Prietzel, J. and Spiecker, H. (eds.). Relationships Between Recent Changes of Growth and Nutrition of Norway Spruce, Scots Pine, and European Beech Forests in Europe – RECOGNITION. EFI Working Paper 19. European Forest Institute. Pp. 8–17.

Vanclay, J.K. 1994. Modelling Forest Growth and Yield. CAB International, Wallingford: xvii + 312 p.

Van Oene, H. and Ågren, G.I. 1995. Complexity versus simplicity in modelling acid deposition effects on forest growth. Ecological Bulletins 44: 352–262.

Van Oijen, M. 2002. On the use of specific publication criteria for papers on process-based modelling in plant science. Field Crops Research 74: 197-205.

5.2 Analysis of the relative importance of nutrients, climatic factors, and CO_2 as causes of observed changes in growth

Hans-Peter Kahle[1], Marcel van Oijen[2], Heinrich Spiecker[1], Göran I. Ågren[3], Oleg G. Chertov[4], Seppo Kellomäki[5], Karl-Heinz Mellert[6], Pedro-José Pérez-Martínez[1], Jörg Prietzel[6], Ralf Straussberger[6], Karl-Eugen Rehfuess[5] and Rüdiger Unseld[1]

[1]*Institute for Forest Growth, University Freiburg, Germany*
[2]*Centre for Ecology and Hydrology, Edinburgh Research Station, UK*
[3]*Swedish University of Agricultural Sciences, Uppsala, Sweden*
[4]*European Forest Institute*
[5]*University of Joensuu, Joensuu, Finland*
[6]*Chair of Soil Science, Technical University Munich, Germany*

5.2.1 Introduction

In the Recognition project empirical and process-based models have been applied to investigate causes of changes in growth of forests across Europe (Chapter 1). Three tree species were investigated: Norway spruce (*Picea abies* (L.) Karst.), Scots pine (*Pinus sylvestris* L.), and European beech (*Fagus sylvatica* L.). The main objective of the empirical modelling approach was the identification and quantitative statistical description of relationships between changes in forest growth as response variable, and changes in climate and foliar nutrition as effect variables. Process-based models were used to explain past and to predict future changes in forest growth.

For the statistical approach two types of modelling were employed: the Historical Development Investigation (HDI) and the Present State Analysis (PSA) (Chapter 2.1). Emphasis in the HDI was on the statistical analysis of temporal variation in multiple data series referring to long-term experimental research plots, whereas the PSA concentrated on the factorial analysis of an observational growth data set and its spatial variation. The two statistical modelling approaches are complementary and closely linked by the common target variable *change in growth,* which by definition operates over time. However, whereas the variable *time* is explicitly included in the HDI models, model types used in PSA are in addition to growth versus time models either cross-sectional models with varying factor levels or spatially explicit models (Chapter 2.2). The four forest ecosystem models which were used in the process-based modelling were developed independently and differ in the level of complexity at which processes have been simulated (Chapter 3.1).

Results of the HDI are presented in Chapter 4.1, of the PSA in Chapter 4.2 and 4.3, and of the process-based modelling approach in Chapter 4.4. A comparison of the empirical and process-based modelling methodology with regard to the results achieved in both approaches is given in Chapter 5.1. The aim of this chapter is to summarize the results of the different approaches with special emphasis on the relative importance of nutrients, climatic factors and CO_2 as causes of observed changes in growth.

5.2.2 Databases

For the empirical modelling approach retrospective data on forest growth, experimental (HDI) or survey data (PSA) on foliar nutrition, and spatially modelled weather (air temperature and precipitation) and nitrogen deposition data have been analysed for a number of research sites across Europe:

- **HDI** was applied to 12 sites with Scots pine and 7 sites with Norway spruce. For an additional 16 Scots pine sites and 16 Norway spruce sites only time series of foliar nutrition have been analysed. Suitable data on forest growth have not been available for these sites.
- **PSA** used data on 19 sites with Scots pine, 17 sites with Norway spruce and 9 sites with European beech.
- The **process-based models** simulated forest growth at 22 sites across Europe: 13 with Scots pine and 9 with Norway spruce. These sites largely overlap with the sites used for the HDI.

5.2.3 Data analysis and modelling methods

A detailed description of the methods applied in the empirical approaches is presented in Chapter 2.2, and of the process-based modelling methods in Chapter 3.1. An in-depth comparison of the empirical and the process-based modelling methodology is given in Chapter 5.1. Only the most important characteristics of the different approaches are recalled here.

5.2.3.1 Methods of empirical modelling

The statistical models applied in the empirical modelling approaches are multiple linear regression models either of the type of the classical linear model (HDI and PSA) or of the generalized linear model type (only HDI). Methods of time series statistics (HDI), combined time-series and cross-sectional analysis methods and geostatistics (PSA) have been employed for modelling temporal, factorial and/or spatial relationships in the empirical approach. The response variable in the empirical modelling is change in tree height growth. The HDI made use of site index curves as references to determine changes in height growth, whereas in the PSA height growth of younger trees was referred to the height growth that currently older trees – growing on similar sites in close proximity to the younger trees – had at the same age. The predictor variables in the regression equations were (i) monthly or seasonal aggregates of air temperature and precipitation either in measurement units or as anomalies (deviations from long-term mean); (ii) derivations thereof (e.g. climatic water balance), and (iii) foliar nutrition levels; and (iv) nitrogen (N-)deposition rates.

5.2.3.2 Methods of process-based modelling

In two of the process-based models, Q and EFIMOD, simulated forest net primary productivity is determined primarily by N-uptake from the soil, whereas the other two more complex models, FinnFor and EFM, simulated primary productivity as the result of interactions between photosynthetic carbon assimilation, N-uptake and water-uptake. In all four models the growth processes were affected by N-deposition, radiation and temperature, but only FinnFor and EFM included sensitivity to CO_2, water availability, wind speed and humidity. The analysis of the results of the process-based modelling focused on the simulated net primary productivity (NPP). Reference growth rates for the determination of changes in forest site productivity have been provided by simulations according to the site-specific base level scenarios which held N-deposition and CO_2 at the levels of 1920 and climate at the baseline level of the period 1920–1927.

5.2.4 Relative importance of causative factors according to the Historical Development Investigation (HDI)

In the regression analyses of the HDI the following factors were considered as independent variables to model change in tree height growth over time: foliar nutrition (N, P, K, Ca, Mg); air temperature; precipitation; and water supply (climatic water balance calculated using Thornthwaite's formulae). The respective reference periods were: annual; growing season (month 5–9, May–September); and early growing season (month 4–6, April–June).

5.2.4.1 Changes in air temperature and precipitation

During the period 1960–2000 mean annual air temperature increased at almost all sites investigated in the HDI. Mean growing season and mean early growing season air temperature in central Europe and the UK exhibited positive trends at nearly all sites, whereas mean growing season air temperature remained unchanged on the Fennoscandian Peninsula (Sweden and Finland). Only during the early growing season the trends were (at nearly all sites) slightly positive. Annual precipitation remained unchanged in central Europe, but decreased at some sites in northern Europe. Regarding growing season and early growing season precipitation slight decreases in central Europe prevail; no consistent pattern exists in northern Europe.

5.2.4.2 Changes in nutrition of Scots pine

Due to the coincidence of poor (often anthropogenically devastated) initial site conditions and relatively high atmospheric N-deposition rates during recent decades, the N-nutrition of the Scots pine stands in central Europe in most cases has improved considerably and often significantly. Also the foliar levels of nutrients other than N - P, K, Mg, and Ca - increased in many cases, but usually to a smaller extent. This resulted in a widespread increase of the ratios of foliar N over other macro-nutrients. A continuation of this trend could mean that in the long run, nitrogen will be replaced as a limiting nutrient for growth by some other macro-nutrient. The observed trends indicate that the changes in the nutritional status of Scots pine stands in central Europe are mainly driven by the interaction of elevated atmospheric N-input and natural recovery of the sites from former devastative land-use practices such as litter raking and grazing.

In northern Europe and the UK the situation of nutritional trends for the investigated Scots pine stands is much more heterogeneous. A rise in nutrition as induced by elevated atmospheric N-deposition and by site recovery may have occurred at some sites; yet N seems to be far less important as a potential driving factor for long-term growth changes of Scots pine stands in Fennoscandia and the UK than in central Europe.

Six of nine stands in central Europe experienced acceleration in height growth during the second part of the 20th century, which in five cases was significant. In contrast to central Europe, trends in height growth of Scots pine stands in Fennoscandia for the periods of foliar nutrient monitoring were all negative.

Changes in air temperature and precipitation could statistically not explain height growth acceleration. However, an acceleration of tree height growth was in all cases associated with an increase in foliar N-nutrition. It seems that the extent of the acceleration in height growth, as induced by increased N supply, is controlled by water availability.

5.2.4.3 Changes in nutrition of Norway spruce

In contrast to Scots pine, most investigated Norway spruce stands in central Europe were well supplied with N and P, and also sufficiently supplied with Ca, Mg, and K. Even though foliar N-concentration of the analysed Norway spruce stands slightly decreased during the period of investigation, foliar N-content increased due to increased needle masses. A similar pattern was observed for P. For northern Europe trends in N- and P-nutrition were heterogeneous. Most investigated Norway spruce stands in central and northern Europe did not show any significant trend of foliar Mg-concentration. Increasing N/P-, N/Mg-, and N/Ca-ratios indicate that increased nitrogen supply is not accompanied by an equivalent increase in the supply of other essential nutrients. The more abundant supply of nitrogen can in certain circumstances induce deficiencies of other nutrients with the risk of nutrient imbalances.

In central Europe, three of four investigated Norway spruce stands grew significantly faster in height than expected according to yield tables. The relative importance of different factors (air temperature, precipitation, N-nutrition) which are significantly related to height growth acceleration varied among the studied sites.

5.2.4.4 Changes in nutrition of European beech

The N-nutritional status of all except one studied European beech stands improved considerably during the monitoring period; however, P-nutrition and at most sites also the supply of Ca and Mg decreased. The trends of the N/P-, N/Mg-, and N/Ca-ratios in almost all cases were strongly positive. This suggests that the potential nutritional imbalances induced or amplified by increasing N-levels in the foliage could – among the tree species considered – occur first in European beech.

5.2.5 Relative importance of causative factors according to the Present State Analysis (PSA)

In the statistical analyses of the PSA the following factors were considered as independent variables to model change in tree height growth: change in seasonal air temperature, change in seasonal precipitation, change in seasonal water balance, level of foliar nitrogen nutrition, and level of total (reduced and oxidized) nitrogen deposition.

The methods of detection and quantification of changes in height growth differed between the HDI and the PSA. Whereas in the HDI, height growth of the subject trees was related to site index curves from yield tables, in the PSA height growth of younger trees was referred to that of currently older trees growing on similar sites at the same age. Data sampling in the PSA was based on a quasi-experimental factorial design. Because the treatment factors were not strictly orthogonal to each other, detection and identifiability of factor specific effects on changes in height growth were limited. In addition to the analysis of height growth, radial growth of the sample trees has been analysed; first, to reveal variations in an additional growth aspect, second, to detect potential differences in stand dynamics or management regimes between the sample tree cohorts which could violate the suitability of the cohort comparison approach for the purpose of this study.

5.2.5.1 Mean changes in height growth of Norway spruce, Scots pine and European beech

During the period 1960–2000 height growth rates were significantly increased for all investigated species. The average percentage acceleration in height growth ($\Delta ih_{1960-2000}$) was 25.1% for European beech, 24.8% for Scots pine, and 22.8% for Norway spruce. The level of change in height growth is relatively stable over the whole observation period. However, a small yet significant increase in height growth acceleration was found for Norway spruce ($\Delta ih_{1960-2000}$: +0.26% yr^{-1}) and European beech ($\Delta ih_{1960-2000}$: +0.22% yr^{-1}).

The results concerning trends in tree height growth presented here are well in agreement to the findings of Spiecker et al. (1996a) that on many sites across Europe forests grew faster in recent decades. In the meanwhile a growing body of evidence for trend-like changes in forest growth has been assembled on different spatial and temporal scales and by applying different methods (e.g. Keeling et al., 1996; Myneni et al., 1997 and 2001; Andersson et al., 1998; Mäkipää et al., 1999;

Joyce and Nungesser, 2000; Pretzsch and Utschig, 2000; Reimeier and Kennel, 2001; Anders et al., 2002; Kenk, 2002).

5.2.5.2 Changes in height growth in relation to changes in air temperature and precipitation

Mean height growth acceleration did not differ significantly between the two temperature-change and precipitation-change strata (increase or decrease in temperature and precipitation respectively) for any of the three species. For beech the starting level of height growth acceleration is positively related whereas the slope is negatively related to the change in mean seasonal temperature. No significant effect of change in seasonal precipitation on height growth acceleration could be inferred from the data. Neither change in temperature nor change in precipitation accounted for a significant part of the change in height growth acceleration for spruce and pine from 1960–2000.

5.2.5.3 Changes in height growth in relation to variations in climatic water balance

Changes in climatic water balance (calculated according to Thornthwaite and Mather, 1955 and 1957) and level of height growth acceleration are inversely related for the Scots pine sample: height growth increased more at sites where climatic water supply decreased. The same applies for the slope of Δih of beech: the larger the decrease in climatic water balance the larger the increase in slope.

The mean climatic water balance overall Europe has slightly increased in the period 1961–1990 compared to 1901–1930 (+8 mm in annual mean, and +3 mm in May-September mean) (data source: CRU 0.5 lat./long. gridded monthly climate data). There is considerable variation in the long-term trend of the annual and seasonal climatic water balance among the different sample sites, but in the mean over all sample sites no significant long-term change is found for the period 1901–1998.

5.2.5.4 Changes in height growth in relation to levels of foliar nitrogen nutrition and nitrogen deposition

The mean height growth acceleration rate of the European beech sample trees with higher nitrogen (N) nutrition is more than two times larger than of those with lower foliar N concentration [N]. The same tendency holds for Scots pine, whereas for Norway spruce the opposite is indicated: sample trees with lower N nutrition at the end of the observation period show consistently larger mean height growth acceleration rates. None of the regression models of change in height growth versus level of foliar [N] is significant. However, for all species the trend in height growth acceleration was inversely associated to foliar [N].

At sites with high N deposition rates the starting level of height growth acceleration also was increased. However, during the period 1960–2000 height growth acceleration rates decreased with increasing N deposition rates.

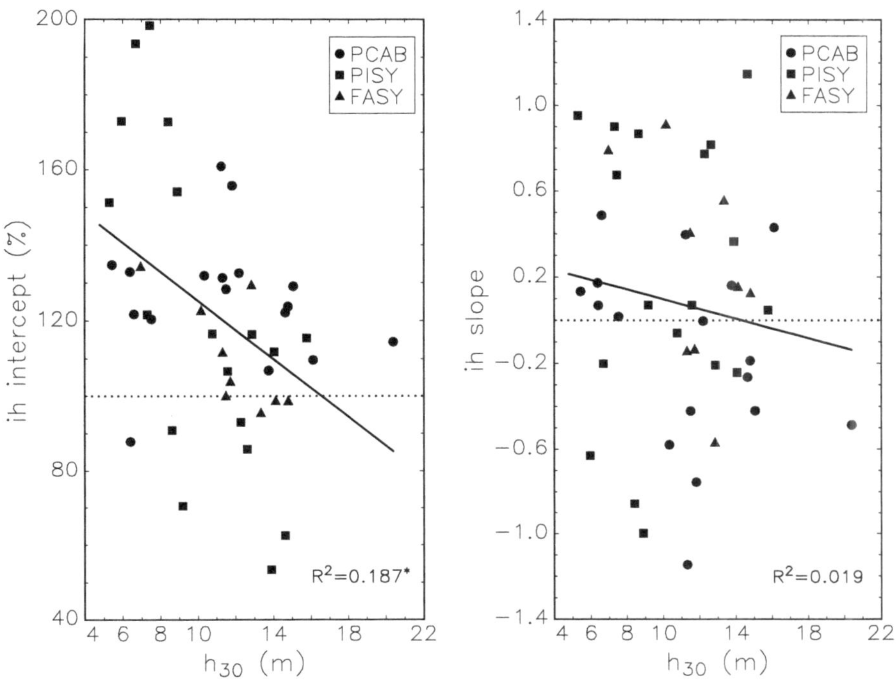

Figure1. Intercepts (b_0, left) and slopes (b_1, right) of linear regression models of Δih (%) versus time t ($\Delta ih(\%) = b_0 + b_1 t$) for Norway spruce (PCAB), Scots pine (PISY) and European beech (FASY) plotted versus site index of older trees (h_{30}: average tree height at age in breast height $t_{1.3}$ = 30 years). Lines are fitted by simple linear regression. The all-species model for the intercept is significant ($P(|t_{\alpha(2),42-2}| \geq t) = 0.0060$), the species specific models (not shown) are not significant.

5.2.5.5 Changes in height growth in relation to site index

Site index of the old trees is regarded as an indicator of the native site productivity. The mean level of change in height growth of Norway spruce, Scots pine, and European beech is significantly related to the former site index: the higher the site index the lower the level of change in height growth (Figure 1).

5.2.5.6 Spatial patterns of changes in height growth

In none of the three species latitude or longitude alone is significantly related to tree height growth acceleration. However, a spatially explicit statistical model has been parameterized which accounts for a significant amount of the spatial variability in changes of mean height growth rate of Norway spruce and Scots pine across parts of central and northern Europe. Model estimates are based on collocated co-Kriging, where mean level of foliar [N] was included as auxiliary predictive variable. According to the model estimates, the largest increases in height growth occurred in an area extending from central Norway in the north-east, over eastern Germany, western Poland and the Czech Republic to Slovakia in the south-east.

5.2.5.7 Changes in radial growth and their relation to variations in climatic water balance

Average annual radial growth rates of the sample trees of all three species have significantly increased in recent decades. For the age period 1 to 50 years increase in average radial growth rate relative to that of the reference trees was +15.5% for spruce, +12.8% for pine, and +26.8% for beech.

Significant direct relationships between variations in radial growth and climatic water balance were found for all three species. These relations are expressed over all time scales considered (short-, medium- and long-term), with most pronounced effects on the high- and medium-frequency scale. The large extent of synchronous variation among the tree cohorts from various sites in Europe and among the different tree species points to a strong climatic control over radial growth of the sample trees.

5.2.5.8 Changes in the height-to-diameter-ratio

Analysis of the development of the height-to-diameter-ratio (h/d-ratio) of the spruce and pine sample trees revealed no difference between the age-cohorts neither in the level of the h/d-ratio nor in its long-term development. Together with the finding of increased height growth this provides evidence for isometric changes of growth: increase in height growth was paralleled by a proportional increase in diameter growth.

5.2.6 Relative importance of causative factors according to the process-based modelling

The input variables that were varied in the process-based modelling are: (daily) minimum and maximum air temperature (all models); N-deposition (all models); precipitation (EFIMOD, EFM, FinnFor), air humidity (EFM, FinnFor), radiation (EFIMOD, EFM, FinnFor), wind (EFM, FinnFor), and CO_2 (EFM, FinnFor) (see Table 3.1 in Chapter 3.1). However, only three factors causing growth change were analysed: CO_2, climate and N-deposition. All weather variables were lumped under the ‘climate change’ factor.

5.2.6.1 Simulated effects of CO_2, climate and N-deposition on net primary productivity

Simulations were run for the periods 1920–2000, 1960–2040 and 2000–2080. The simulated effects of CO_2, climate and N-deposition on net primary productivity (NPP), were quantified as the percentage increase in NPP that results from replacing the general base level scenario (i.e. the scenario in which all

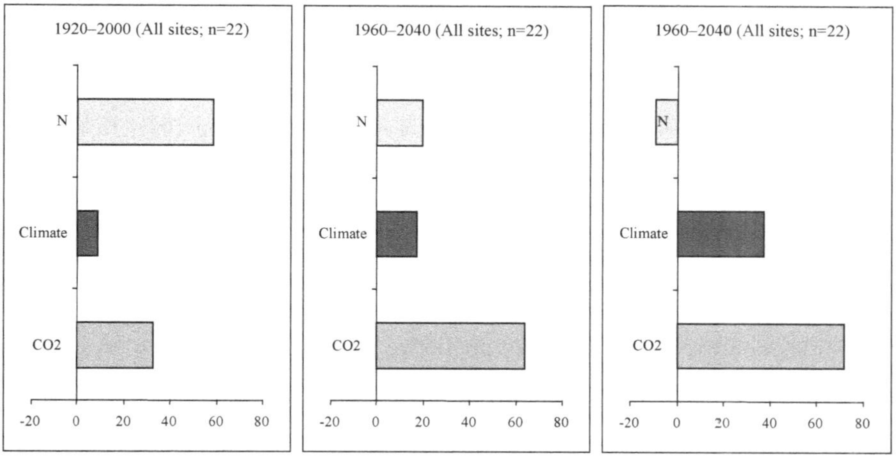

Figure 2. Simulations of conifer forest growth and resource use (averages for four models) for three time periods. Bar-charts show the relative magnitude of the three effects (CO_2, climate, N-deposition) on net primary productivity. The effects are not completely independent. The figures refer to the total effect of changing CO_2, climate and N. The data for CO_2 refers to the average CO_2-effect of only two models (EFM and FinnFor).

environmental variables were kept at their initial values) one-by-one with actual scenarios. Results are presented in this way in Figure 2. On average over all models and for both coniferous species, changing N-deposition was most important in the first period (N-deposition accounted for 58.5% of the change in NPP), followed by CO_2 (32.5%) and climate (9.0%). The relative magnitude of the factor specific effects changed gradually to the third period where CO_2 (71.9%), followed by climate (37.3%) are the predominant factors. In the last period N-deposition even exaggerates a negative effect on changes in NPP (–9.2%). The simulations showed that effects of climate might become increasingly significant, especially at northerly latitudes. In order to elucidate the effects of the climate change factor some extra simulations have been run with one of the models (EFM), which showed that the relatively small climate change effect on NPP was largely caused by changes in temperature.

5.2.6.2 Effect of changes in N-deposition on NPP

The growth enhancing effect of N-deposition operated through an increase in gross N-uptake rate of 23% averaged across sites. However, the net effect was counteracted by a reduction in simulated nitrogen use efficiency of 9%.

On the Scots pine sites the role of N in the first simulation period is more pronounced (65.6%) than on the Norway spruce sites (46.1%). In the third period the limitation by N is also stronger on the Scots pine sites (–15.5%) than on the Norway spruce sites (–3.3%).

5.2.7 Discussion

5.2.7.1 Representativity and uncertainty in empirical modelling

Empirical modelling is data-adaptive statistical modelling. The strength of the statistical modelling is the ability to accurately fit statistical models to the observational data. The main weaknesses are its high demand for extensive field data and its limited potential to establish and prove cause-effect relationships (Holland, 1986; Schreuder and Thomas, 1991). Depending on the type and quality of the data and the model structure the degree of generality of statistical models is usually low. Hence the model structure and outputs of statistical models are not *a priori* valid in other contexts, and their predictive power is closely limited to and by the conditions reflected in the data used for calibration.

Specific aspects of the statistical modelling concerning uncertainties and representativity of the data, the model parameters, and the validity of the models and the implications on the results achieved by the HDI and the PSA are discussed and evaluated in the following section.

Critical discussion and evaluation of the data used in the empirical modelling

- The major drawback in the empirical modelling approaches is the limited size and quality of the database with respect to (i) the total number of observations, (ii) the numbers of replications, (iii) the spatial and temporal coverage of the data, as well as (iv) the lack of a sound sampling and experimental design covering all relevant assessments. Together these drawbacks do not permit statistical inference and extrapolation of the results to larger units. Both, the experimental data used for HDI as well as the monitoring data from the EC-UN/ECE Level II monitoring plots were selected on a preferential basis with no explicit indication of the target population of concern and with no explicit indication of the model linking the sample plots/sites to the target population (Ferretti and Chiarucci, 2003).
- Most forest stands studied in the HDI comprise control plots of long-term (old) fertilisation experiments. These experiments had often been preferentially installed at particularly poor sites and for tree species that were usually cultivated in such situations (Scots pine >> Norway spruce > European beech). With regard to effects of alterations in nutrient supply on forest growth, findings derived from those plots might be biased since responsiveness of tree and stand growth might be overestimated.
- In some cases the data series are unbalanced, i.e. comprise different numbers of observations, different times of commencement and/or duration of the periods as well as different observation intervals e.g. of foliar nutrient assessments. Shifts in the aggregated series, which only result from varying numbers of observations, could be misinterpreted as signals. To avoid this, subsets of the data with a common-overlap were used wherever possible.
- For many studied stands, data on 'dry mass of needles or leaves' (e.g. 100-needle weight) were not available. In these cases the discrimination of changes in the supply of a nutrient from dilution or concentration effects was not possible. However, for an in-depth analysis of cause-effect relations with regard to physiological processes this distinction is essential (Landsberg, 1986).

- For most sample sites no data on environmental factors – e.g. air temperature, N-deposition – from on-site measurements were available. Data either from nearby measurement stations or spatially interpolated or otherwise modelled data had to be used as substitutes. The error that is involved herein cannot directly be assessed.

Critical discussion and evaluation of the variables used in the empirical modelling

- With regard to the nutritional status of stands and trees, only macro-nutrients were considered in the analyses. The nutritional status in relation to micro-nutrients was not taken into account. In certain, however rather extreme, circumstances micro-nutrients may exert a decisive effect on tree growth (Fiedler et al., 1973).
- The characteristic time-scale in the statistical modelling was one year. The minimum aggregation interval of the available weather data was one month. Effects of weather events that are likely to affect tree growth but which operate on smaller time scales (e.g. early and late frost events, storms) were not considered in the analyses. The time scales of the growth response to this type of event are typically short- to medium-term. However, changes in the frequency and intensity of these often extreme events could induce effects which, when accumulated, result in trend-like changes in growth.
- Changes in height increment are determined as deviations between observed and expected height increment. For the HDI the retrospectively derived height increment of the sample trees has been compared with the expected height increment according to site index curves from yield tables. The appropriate site index has been determined by the age-height relationship given in the reference year 1960. The yield tables used have been selected following the suggestions made by growth and yield experts familiar with the local growth conditions. Since it was not possible within the project to verify the specific suitability of the applied site index curves, uncertainty remains whether observed changes in height increment really are caused by changes in growth or if they are due to the comparison with inappropriate site index curves.
- In the PSA height growth of older trees has been used as reference for expected height growth. This height growth reference is much closer to the local growth conditions than the site index curves from yield tables as used in the HDI. However, there is another potential source of bias: the stand components which have been used for the height growth comparison might differ between the younger and the older tree cohorts. Evidently, those trees which – at a certain point in time – belong to the stand component of top height trees might not have belonged to the same stand component over their whole lifetime, but may have joined the group of the e.g. 100 thickest trees per hectare by emerging from a less dominant, smaller diameter group of trees through accelerated growth (“social climber”). If the relation between height and diameter in the sampled stands follows typical stand-height-curves, then those trees that are smaller in diameter are also expected to be shorter in height. The retrospectively derived height growth development of trees currently belonging to the stand component of top height trees therefore underestimates the real stand top height in the past. Consequently, if this reconstructed stand top height is used as a reference for the top height of currently young stands, the

acceleration in height growth is overestimated (*cf.* Magnussen and Penner, 1996). The risk of overestimation increases with increasing time interval between the date of tree sampling and the date to which the reconstructed height growth is referred. However, if we consider the current or periodic annual height increment instead of the average annual height increment (tree height divided by tree age) – as was done in the PSA – the situation is different. During the transition phase when smaller diameter trees speeded up their growth rates in diameter and height – relative to the stand component of the 100 thickest trees per hectare – to become members of this stratum then a given acceleration in height growth will be underestimated. The effect of overestimation as described above is also counteracted by the fact that the older trees themselves may have already experienced acceleration in height growth. As a result the full extent of height growth acceleration might by this method most probably be underestimated (Sterba, 1996). Since we could not detect differences in the mean development of the height-to-diameter-ratios (tree height divided by tree diameter in breast height) between the old and young age cohorts for the spruce and pine sample trees, we consider the risk of overestimation of the height growth acceleration due to the effect of "social climbing" in the spruce and pine reference data as small.

Critical evaluation of the statistical inference from empirical models

- The empirical modelling approaches are built upon observational studies (Chapter 2.2), with the consequence that it is only possible to infer associative relationships between the responses and the treatments since the effect of the explanatory variables on the response variable might be confounded with the effect of other treatment variables or with other but omitted effect variables.

5.2.7.2 Representativity and uncertainty in process-based modelling

The high degree of generality is the key advantage of the process-based modelling. Results of the process-based modelling are regarded as valid as long as no sites with other limiting factors as those incorporated in the models are considered. In this study no sites with clear limitations of P, K, Mg or Ca were found.

The climate change effect on forest growth was not mechanistically analysed, but it may have had complicated causes as the models realistically include effects of temperature on tree phenology (apparent in the length of the growing season), tree physiology, and on soil transformation processes. Further aspects of uncertainty in process-based modelling are discussed in Chapter 5.1

5.2.7.3 Review of the relative importance of nutrients, climatic factors, and CO_2 as causes of observed changes in growth

The discussion about and search for possible causes of growth changes has motivated the Recognition project (Rehfuess et al., 1999). The role of changes in climate, nutrition and deposition on the productivity of forest ecosystems is the subject of ongoing debate (e.g. Binkley and Högberg, 1997; Mickler and Fox,

1997; Andersson et al., 1998; Nadelhoffer et al., 1999; Birdsey et al., 2000; Caspersen et al., 2000; Joyce and Birdsey, 2000; Joyce and Nungesser, 2000; Schimel et al., 2000; Schulze, 2000; Oren et al., 2001; Puhe and Ulrich, 2001; Janssens et al., 2003; Valentini et al., 2003), and research is addressing these questions on different temporal and spatial scales using a variety of different parameters and assessment methods.

In the following paragraph studies are discussed which focussed on changes in forest growth over large spatial scales. Contrary to the Recognition study the effects of changes in site productivity could not be discriminated from other causes of forest growth changes, like e.g. changes in forest management and forest composition.

A multi-decadal increase in growth activity of the northern hemispheric vegetation cover is described by Keeling et al. (1996) who analyzed time series of atmospheric CO_2 measurements. Myneni et al. (1997) and Zhou et al. (2001) analyzed satellite data on reflectance characteristics of the vegetation cover and found clear indications of increased plant growth in the northern high latitudes since 1981. Concerning putative causal factors for the observed changes in growth of forests in Europe, Russia and America, Myneni et al. (2001) hypothesize that "... *increased fires and infestations in Canada, forest cultivation in the Nordic countries, declining harvests in Russia, fire suppression and forest regrowth in the USA, and longer growing seasons from warming are possible reasons for why some forests are storing carbon and others are losing it*". Caspersen et al. (2000) highlight the role of land-use history to carbon accumulation in US forests, whereas Nemani et al. (2003) hypothesize that the enhanced plant growth in the northern mid- and high-latitudes is mainly caused by recent climatic changes. Roderick et al. (2001) attribute the growth stimulating effect which was observed following the Mt. Pinatubo (Philippines) eruption in June 1991 in part to the enhancement of the diffuse fraction caused by volcanic aerosols that have reduced the volume of shade within vegetation canopies. The rate of increase in canopy photosynthesis depends on how far the system is from light-saturation. In order to evaluate the relevance of this mechanism for long-term changes in forest growth, further analyses are necessary.

Karnosky (2003) reviewed results from research into the impacts of elevated atmospheric CO_2 on forest trees and forest ecosystems. It still remains difficult to predict future forest growth and productivity under elevated atmospheric CO_2, since tree and forest responses are as yet largely uncertain

> "*... due to differences in responsiveness by species, genotype, and functional group, and the complex interactions of elevated atmospheric CO_2 with soil fertility, drought, pests, and co-occurring atmospheric pollutants such as nitrogen deposition and O_3. Furthermore, it is impossible to predict ecosystem-level responses based on short-term studies of young trees grown without interacting stresses and in small spaces without the element of competition*" (Karnosky, 2003: 161).

The finding that maximum rates of forest growth acceleration occur on medium and poor sites is not in agreement with the assumption of a CO_2-fertilisation effect as a cause of the increase in forest productivity. Improvement of forest tree growth due to higher atmospheric CO_2 concentrations should be expected predominantly under conditions of favourable nutrient supply, not under conditions of nutrient shortage. Concerning the role of CO_2, Cannell (1999) concluded that tropical forests are likely to be greater carbon sinks than high latitude forests, because photosynthesis is more responsive to elevated CO_2 at high temperatures.

Studies on the role of Nitrogen for forest growth changes
Nadelhoffer et al. (1999), based on the studies of stoichiometric and metabolic limitations on carbon/nutrient ratios in forest ecosystems, concluded that "*Nitrogen deposition makes a minor contribution to carbon sequestration in temperate forests*" (Nadelhoffer et al., 1999: 145). In contrast, Puhe and Ulrich (2001), referring to the growth estimates of Kuusela (1994) concluded that

> "*...the additional forest growth potential due to atmospheric nitrogen input in central Europe could significantly exceed the estimated increase in mean net annual volume increment between 1950 and 1990 of 2.1 m^3 ha^{-1}. Under favourable conditions, it may even approach the magnitude of the estimated current mean net annual volume increment of forests in central Europe of 5.8 m^3 ha^{-1}*" (Puhe and Ulrich, 2001: 176).

Also others – e.g. Hofmann et al. (1990), Binkley and Högberg (1997), Mäkipää et al. (1999) – give evidence for the decisive role of N for changes in forest growth and site productivity. Based on extensive data collected at the EC–UN/ECE ICP Forests Level I and Level II monitoring plots de Vries et al. (2003) modelled the effects of N deposition on the carbon sequestration of forests on the European level. According to this study, N deposition accounts for 10–20% of the increase in standing biomass during the period 1960–2000.

The conclusions concerning the role of N deposition differ between studies with different scopes and perspectives; however, differences largely diminish when one considers that the sequence of events that occur when N inputs have increased is highly complicated. Many of the processes are intricately related and in addition operate over different time scales. The impact of the effects of airborne N deposition depends on: (i) the dose, duration and total amount of the inputs; (ii) the form of the N input; (iii) the intrinsic sensitivity of the plant species present; and (iv) the abiotic conditions in the ecosystem, with respect to acid neutralizing capacity, soil nutrient availability, and soil factors which influence the nitrification potential and N immobilization rate (Bobbink and Lamers, 2002). The last two items can be influenced by both past and present land use, and by management. As a consequence, high variation in sensitivity to N is not surprising, and in fact is to be expected between different ecosystems.

Growth depressions caused by over-utilisation of forest sites can mainly be ascribed to the decoupling of the nitrogen cycle (Fiedler, 2001; Puhe and Ulrich, 2001). These effects drive the transition from mull to moder humus forms and from high to low elasticity of the forest ecosystems. The importance of former land use and soil recovery for the observed changes in forest growth is discussed in detail in Hunter and Schuck (2002).

Detrimental effects of excessive N applications on nutrition and growth of Scots pine stands based on long-term data from experimental fertilization plots are described by Heinsdorf and Beck (2003). Severe damages and increased mortality have been observed at foliar N concentrations exceeding 30 mg g^{-1}. However, only a few years after N applications had stopped, foliar [N] decreased to pre-treatment levels and height and diameter growth recovered.

Discussions of the factors considered most important are given in the recent review studies by Cannell (1999) and by Hunter and Schuck (2002). Cannell (1999: 37) concluded that "... *there has been a positive, non-zero effect of increasing* CO_2 *on the growth of European forest this century, possibly increasing NPP by as much as 10–15%. ... increasing temperatures may have had a smaller effect than might be deduced from site studies*". Hence, since neither N deposition nor CO_2 fertilization and temperature increase alone or in combination could fully explain the observed increase in forest growth, also changes in land use practices and improved forest management are considered to be of relevance. Spiecker et al. (1996b: 369) pointed out that the phenomenon is complex since "... *the general increasing growth trend may have been caused by one factor, a factor combination or by regionally changing factors which finally had similar effects on growth ... [and] the significance of each factor possibly varies in space and time*".

This study gives some empirical evidence for the prominent role of increasing N nutrition as the key driver of forest site productivity increases. However, to which part the increase in N nutrition is driven by recovery of soil from former often devastative land use, and to which extent it is attributed to increased N deposition has not been clarified. The process based modelling gives clear evidence that N played the most important role for forest site productivity changes in the past, but in the future climate change and especially CO_2 fertilization will be of increasing importance.

References

Anders, S., Beck, W., Bolte, A., Hofmann, G., Jenssen, M., Krakau, U.K. and Müller, J. 2002. Ökologie und Vegetation der Wälder Nordostdeutschlands – Einfluss von Niederschlagsarmut und erhöhtem Stickstoffeintrag auf Kiefern-, Eichen- und Buchen-Wald- und Forstökosysteme des nordostdeutschen Tieflandes. Dr. Kessel, Oberwinter. 199 p.

Andersson, F., Braekke, F.H. and Hallbäcken, L. 1998. Nutrition and growth of Norway spruce forests in a Nordic climatic deposition gradient. TemaNord 566. 209 p.

Binkley, D. and Högberg, P. 1997. Does atmospheric deposition of nitrogen threaten Swedish forests? Forest Ecology and Management 92: 119–152.

Birdsey, R.A., Mickler, R.A., Hom, J. and Heath, L.S. 2000. Summary of prospective global change impacts on northern U.S. forest ecosystems. In: Mickler, R.A., Birdsey, R.A. and Hom, J. (eds.). Responses of Northern U.S. Forests to Environmental Change. Springer, New York. Pp. 543–568.

Bobbink, R. and Lamers, L.P.M. 2002. Effects of increased nitrogen deposition. In: Bell, J.N.B. and Treshow, M. (eds.). Air Pollution and Plant Life. John Wiley & Sons, Chichester. Pp. 201–235.

Cannell, M.G.R. 1999. Relative importance of increasing atmospheric CO_2, N deposition and temperature in promoting European forest growth. In: Karjalainen, T., Spiecker, H. and Laroussinie, O. (eds.). Causes and Consequences of Accelerating Tree Growth in Europe. EFI Proceedings 27. European Forest Institute. Pp. 25–41.

Caspersen, J.P., Pacala, S.W., Jenkins, J.C., Hurtt, G.C., Moorcroft, P.R. and Birdsey, R.A. 2000. Contributions of land-use history to carbon accumulation in U.S. forests. Science 290: 1148–1151.

de Vries, W., Reinds, G. J., Posch, M., Sanz, M. J., Krause, G. H. M., Calatayud, V., Renaud, J. P., Dupouey, J. L., Sterba, H., Vel, E. M., Dobbertin, M., Gundersen, P. and Voogd, J. C. H. 2003. Intensive Monitoring of Forest Ecosystems in Europe – Technical Report 2003. 162 p.

Ferretti, M. and Chiarucci, A. 2003. Design concepts adopted in long-term forest monitoring programs in Europe – problems for the future? The Science of the total Environment 310: 171–178.

Fiedler, H.J. 2001. Böden und Bodenfunktionen in Ökosystemen, Landschaften und Ballungsgebieten. Expert-Verlag, Renningen-Malmsheim. Forum Eipos 7. 598 p.

Fiedler, H.J., Nebe, W. and Hoffmann, F. 1973. Forstliche Pflanzenernährung und Düngung. Gustav Fischer, Stuttgart. 481 p.

Heinsdorf, D. and Beck, W. 2003. Langjährige Untersuchungen zur Wirkung hoher N-Zufuhren auf Ernährung und Wachstum eines Kiefernbestandes. Beiträge für Forstwirtschaft und Landschaftsökologie 37: 28–35.

Hofmann, G., Heinsdorf, D. and Krauss, H.H. 1990. Wirkungen atmogener Stickstoffeinträge auf Produktivität und Stabilität von Kiefern-Forstökosystemen. Beiträge für die Forstwirtschaft 24: 59–73.

Holland, P.W. 1986. Statistics and causal inference. Journal of the American Statistical Association 81: 945–960.

Hunter, I. and Schuck, A. 2002. Increasing forest growth in Europe – possible causes and implications for sustainable forest management. Plant Biosystems 136: 133–141.

Janssens, I.A., Dore, S., Epron, D., Lankreijer, H., Buchmann, N., Longdoz, B., Brossaud, J. and Montagnani, L. 2003. Climatic influences on seasonal and spatial differences in soil CO_2 efflux. In: Valentini, R. (ed.). Fluxes of Carbon, Water and Energy of European Forests. Springer, Berlin: 233–253.

Joyce, L.A. and Birdsey, R. (eds.) 2000. The impact of climate change on America's forests: A technical document supporting the 2000 USDA Forest Service RPA Assessment. RMRS-GTR-59. 132 p.

Joyce, L.A. and Nungesser, M. 2000. Ecosystem productivity and the impact of climate change. In: Joyce, L.A. and Birdsey, R. (eds.). The impact of climate change on America's forests: A technical document supporting the 2000 USDA Forest Service RPA Assessment. RMRS-GTR-59. Pp. 46–68.

Karnosky, D.F. 2003. Impacts of elevated atmospheric CO_2 on forest trees and forest ecosystems: knowledge gaps. Environment International 29: 161–169.

Keeling, C.D., Chin, F.S. and Whorf, T.P. 1996. Increased activity of northern vegetation inferred from atmospheric CO_2 measurements. Nature 382: 146–149.

Kenk, G. 2002. Variation der periodischen Zuwächse von Fichten und Buchen in Baden-Württemberg. AFZ/Der Wald 57: 300–302.

Kuusela, K. 1994. Forest Resources in Europe 1950–1990. European Forest Institute Research Report 1. Cambridge University Press, Cambridge. 154 p.

Landsberg, J.J. 1986. Physiological Ecology of Forest Production. Academic Press, London. 198 p.

Magnussen, S. and Penner, M. 1996. Recovering time trends in dominant height from stem analysis. Canadian Journal of Forest Research 26: 9–22.

Mäkipää, R., Karjalainen, T., Pussinen, A. and Kellomäki, S. 1999. Effects of climate change and nitrogen deposition on the carbon sequestration of a forest ecosystem in the boreal zone. Canadian Journal of Forest Research 29: 1490–1501.

Mickler, R.A. and Fox, S. 1997. The Productivity and Sustainability of Southern Forest Ecosystems in a Changing Environment. Springer, New York. 892 p.

Myneni, R.B., Dong, J., Tucker, C.J., Kaufmann, R.K., Kauppi, P.E., Liski, J., Zhou, L., Alexeyev, V. and Hughes, M.K. 2001. A large carbon sink in the woody biomass of Northern forests. Proceedings of the National Academy of Sciences of the United States of America 98: 14784–14789.

Myneni, R.B., Keeling, C.D., Tucker, C.J., Asrar, G. and Nemani, R.R. 1997. Increased plant growth in the northern high latitudes from 1981 to 1991. Nature 386: 698–702.

Nadelhoffer, K.J., Emmett, B.A., Gundersen, P., Kjonaas, O.J., Koopmans, C.J., Schleppi, P., Tietema, A. and Wright, R. 1999. Nitrogen deposition makes a minor contribution to carbon sequestration in temperate forests. Nature 398: 145–148.

Nemani, R.R., Keeling, C.D., Hashimoto, H., Jolly, W.M., Piper, S.C., Tucker, C.J., Myneni, R.B. and Running, S.W. 2003. Climate-driven increases in global terrestrial net primary production from 1982 to 1999. Science 300: 1560–1563.

Oren, R., Ellsworth, D.S., Johnsen, K.H., Phillips, N., Ewers, B.E., Maier, C., Schäfer, K.V., McCarthy, H., Hendrey, G., McNulty, S.G. and Katul, G.G. 2001. Soil fertility limits carbon sequestration by forest ecosystems in a CO_2-enriched atmosphere. Nature 411: 469–472.

Pretzsch, H. and Utschig, H. 2000. Wachstumstrends der Fichte in Bayern. Mitteilungen aus der Bayerischen Staatsforstverwaltung 49. 170 p.

Puhe, J. and Ulrich, B. 2001. Global Climate Change and Human Impacts on Forest Ecosystems: Postglacial Development, Present Situation, and Future Trends in Central Europe. Springer, Berlin. 592 p.

Rehfuess, K.E., Ågren, G.I., Andersson, F., Cannell, M.G.R., Friend, A.D., Hunter, I., Kahle, H.P., Prietzel, J. and Spiecker, H. 1999. Relationships between recent changes of growth and nutrition of Norway spruce, Scots pine, and European beech forests in Europe – RECOGNITION. European Forest Institute Working Paper 19. 94 p.

Reimeier, S. and Kennel, E. 2001. Zuwachstafeln für ausgewählte Baumarten. Dr. Kessel, Remagen. 93 p.

Roderick, M.L., Farquhar, G.D., Berry, S.L. and Noble, I.R. 2001. On the direct effect of clouds and atmospheric particles on the productivity and structure of vegetation. Oecologia 129: 21–30.

Schimel, D., Melillo, J., Hanqin, T., McGuire, A.D., Kicklighter, D., Kittel, T., Rosenbloom, N., Running, S. et al. 2000. Contribution of increasing CO_2 and climate to carbon storage by ecosystems in the United States. Science 287: 2004–2006.

Schreuder, H.T. and Thomas, C.E. 1991. Establishing cause-effect relationships using forest survey data. Forest Science 37: 1497–1512.

Schulze, E.D. (ed.) 2000. Carbon and Nitrogen Cycling in European Forest Ecosystems. Springer, Berlin. 500 p.

Spiecker, H., Mielikäinen, K., Köhl, M., and Skovsgaard, J.P. (eds.) 1996a. Growth Trends in European Forests – Studies From 12 Countries. Springer, Berlin. 372 p.

Spiecker, H., Mielikäinen, K., Köhl, M., and Skovsgaard, J.P. 1996b. Conclusions and summary. In: Spiecker, H., Mielikäinen, K., Köhl, M., Skovsgaard, J.P. (eds.). Growth Trends in European Forests – Studies From 12 Countries. EFI Research Report 5. Springer, Berlin. Pp. 369–372.

Sterba, H. 1996. Possibilities to insert site productivity changes into growth models. Conference on Effects of Environmental Factors on Tree and Stand Growth, Proceedings IUFRO Conference, Berggiesshübel near Dresden, September 23–27, 1996. Pp. 243–250.

Thornthwaite, C.W. and Mather, J.R. 1955. The water balance. Publ. in Climatology – Drexel Inst. of Tech., Lab. of Clim. 8(1). 104 p.

Thornthwaite, C.W. and Mather, J.R. 1957. Instructions and tables for computing potential evapotranspiration and the water balance. Drexel Institute of Technology – Laboratory of Climatology. Vol 10, No 3: 308 p.

Valentini, R., Matteucci, G., Dolman, A.J., Schulze, E.D. and Jarvis, P.G. 2003. The carbon sink strength of forests in Europe: a synthesis of results. In: Valentini, R. (Ed.), Fluxes of Carbon, Water and Energy of European Forests. Springer, Berlin. Pp. 225–232.
Zhou, L., Tucker, C.J., Kaufmann, R.K., Slayback, D., Shabanov, N.V. and Myneni, R.B. 2001. Variations in northern vegetation activity inferred from satellite data of vegetation index during 1981 to 1999. Journal of Geophysical Research 106: 20069–20083.

5.3 Analysis of the long-term consequences for sustainability of observed growth changes of the European forests

Göran I. Ågren[1], Oleg Chertov[2], Hans-Peter Kahle[3], Seppo Kellomäki[4], Alex Komarov[2], Karl-Heinz Mellert[5], Marcel van Oijen[6], Jörg Prietzel[5], Heinrich Spiecker[3], Ralf Straussberger[5] and Karl-Eugen Rehfuess[5]

[1] *Swedish University of Agricultural Sciences, Uppsala, Sweden*
[2] *European Forest Institute*
[3] *Institute for Forest Growth, University Freiburg, Germany*
[4] *University of Joensuu, Faculty of Forest Sciences, Joensuu, Finland*
[6] *Centre for Ecology and Hydrology, Edinburgh Research Station, UK*
[5] *Chair of Soil Science, Technical University Munich, Germany*

5.3.1 Introduction

Recognition has identified increasing growth of the most important forest trees (European beech, Norway spruce, and Scots pine) at many sites in Europe (Chapter 4.2) and over the last decades (Chapter 4.1). The major cause of this increased growth seems attributable to the high nitrogen deposition over large areas of Europe, although increasing growing season temperatures and an elevated atmospheric CO_2 concentration may be contributing factors (Chapter 4.1). Despite the clear temporal and spatial trends in these environmental factors (Chapters 4.1, 4.2, and 4.4), such trends are directly reflected neither in the observations (Chapter 4.1, 4.2) nor in the modelling exercises (Chapter 5.1) within the project. The considerable noise in the observational data is probably masking existing trends, and also the observation period may be too short to identify multi-decadal trends. Anthropogenically increased nitrogen deposition was already prevalent in Europe in the 1960s, and the effects of the increasing trend in deposition from 1960 may be masked by the shift from low deposition in the period from the early 1900s to 1960.

Against this background of observed changes and difficulties in resolving causes and effects it is reasonable to ponder over a sustainable future of European forests. According to Merriam-Webster's Collegiate Dictionary the term "sustainable" relates to methods of harvesting or use of a resource so that the resource is not depleted or permanently damaged. Farrell et al. (2000) go further and distinguish between 'sustainability of forest yield' and 'sustainable forest management', where 'sustainability of forest yield' only includes traditional harvest of mainly wooden biomass, and 'sustainable forest management' refers to a much broader perspective of products and services from forest ecosystems. Recognition has focussed on tree growth, and sustainability of forest yield will therefore be the principal focus, although other aspects such as nutrient (e.g. nitrogen) losses to downstream ecosystems and hydrological budgets can be addressed from the modelling component.

An increased forest growth driven by increased access to nitrogen and carbon will also imply an increasing demand for other resources as well as a structural change in the forest ecosystem. It is clear that such an additional demand on other resources, as well as the effect of the changing forest on the environment, may jeopardise the sustainability of the forest. In particular, the potential problems may include insufficient supplies of water and elements other than carbon and nitrogen (potassium, phosphorus, magnesium, calcium, etc.). There could also be a question of the extent to which increasing growth of the extant genotypes could lead to a mismatch relative to the prevailing, but also changing, climate.

5.3.2 Spatial analysis of consequences for sustainability of observed growth changes

As has been pointed out, the major driving factor for increased growth has been the increased nitrogen deposition. The most critical areas will therefore be those with the largest discrepancy between the input of nitrogen through deposition and the specific absorption, uptake, and buffering capacity of the site. The simulations suggest, as we would expect, that changes are determined by a combination of rate of nitrogen deposition and initial soil fertility. Nitrogen deposition was high at the two Austrian sites, but initial soil fertility was already high at these sites, and so growth rate was not stimulated. High nitrogen deposition on poor ex-litter raking sites in Germany showed a larger effect on growth and smaller effect on leaching. The results from Recognition also show trends towards increased ratios of foliar nitrogen to other elements (Chapter 4.1). However, for the sites investigated within the project, growth at almost all sites has been clearly nitrogen limited, and the supply of other elements has been, in general adequate. Since the supply of other elements mainly depends on site-given properties, the potential for creating nitrogen excess will be largest in the areas with the highest nitrogen deposition – i.e. the northwestern part of continental Europe (see Figure 16 in Chapter 4.2). The problems associated with excess nitrogen deposition have been discussed extensively (e.g. Fenn et al., 1998) and will not be repeated here. It should, however, be noted that the first losses in sustainability due to nitrogen saturated forests will not be the forest ecosystems themselves, but rather surrounding ecosystems exposed to elevated nitrogen leaching from the forest.

Another effect of the increased growth is the increased evapotranspiration resulting from the increased leaf area. Where this increase is combined with an increase in evaporative demand from higher temperatures and/or decreased supply from a decrease in precipitation, the situation will be most noticeable. Such areas can be identified in the maps in Figure 1 in Chapter 4.2. These areas include southern Spain, a band along the border between Germany and Poland and extending through large areas of Austria, and a southwest-northeast band across southern Sweden. To what extent water deficits will become real problems has not been analysed, but as Mellert et al. indicate in Chapter 4.1, short-term variability in water supply is clearly evident in tree growth rates. However, when analysing the growth data from widely scattered sites across Europe, Kahle et al. (Chapter 4.3) did not find a significant relationship between the sensitivity of radial increment and the baseline level of climatic water balance for any of the species. The geographical coverage of stands within Recognition is, however, insufficient to identify with certainty which areas are likely to first encounter severe water shortage.

5.3.3 Temporal analysis of consequences for sustainability of observed growth changes

Sustainability is always a matter of time. Even drastic changes in environment over a short enough a time span (e.g. forest fertilisation with large single doses) may not change ecosystem behaviour radically, although in other cases episodic events (e.g. a forest fire) can have a devastating effect. From this perspective, sustainability must refer to the stability and resilience of the forest ecosystem and the maintenance of the forest over several rotation periods. However, the focus of Recognition has been on growth changes and hence covers only events during a single rotation period, and questions concerning regeneration are, therefore, largely outside the scope of the project – see however Chapter 5.4.

Within the rotation period, we can identify two types of influences from the changing environment: (1) changes that operate more or less instantaneously; and (2) changes that accumulate. Climate variables belong to the first category because at the instant the climate reverts to normal, the forest also experiences the normal climate. This does not exclude long-term effects of climate because a change in climate over a prolonged period will certainly lead to a different forest structure and development. However, changes in climate have to be compared to changes in nitrogen deposition, where, even if the deposition is completely stopped, a given cumulative deposition remains and exerts its influence. For this reason, resources basically depending on climate (e.g. water supply) can be brought outside the requirement of the existing forest when some critical limit is passed, and this is to a large extent independent of the time course of the change. On the other hand, the consequences of a changing nitrogen deposition do not so much depend on the instantaneous rates, but much more on the cumulative deposition. In addition, there are large pools of soil organic matter with slow turnover rates, which can buffer short-term changes in nitrogen deposition. It is, therefore, difficult to predict what types of environmental changes are likely to first cause changes in growth.

The temporal trends in nutrition that have been observed (Chapter 4.1) indicate that the increased nitrogen availability is not accompanied by a similar increase in the availability of other essential nutrients. In the long term this should eventually mean that nitrogen as a limiting element for growth will be replaced by some other element. The time when this will occur is difficult to predict because we do not currently understand to what extent an increased nitrogen availability also induces an increased release of the other elements, e.g through increased weathering. Some early predictions have indicated a fairly rapid exhaustion of other mineral nutrients (e.g. Sverdrup et al., 1992). The results from Recognition must, however, be applied cautiously on a European scale. Many of the sites used in the analysis are experimental sites and were originally chosen as experimental sites because they have been nutrient-poor, either as a result of former devastative management or on naturally poor sites. Such sites could, therefore, have a much larger capacity to absorb incoming deposition than more fertile sites.

5.3.4 Conclusions, outlook and summary

Recognition has identified ongoing changes in growth in European forests. The causes also seem clear: increased nitrogen deposition, and to a lesser extent, changes in climate can be identified as major causes. There are no indications from the project, at least in the short term, that the growth changes represent any threat to the sustainability of the forest ecosystems, and their provision of services. In a longer time perspective, the situation might be different and results from the project indicate possibilities of increased water stress and deficiencies of other nutrients than nitrogen. Nitrogen deposition depends also on precipitation; changing precipitation patterns will change the distribution of nitrogen deposition. The total amount of nitrogen deposition to European forests will on the other hand change less as it is determined by the amount of sources. As discussed by Kellomäki et al. in Chapter 5.4, the changing climate might also cause migration of species, which could move the forest tree species into regions to which they are not adapted. These latter aspects have, however, been outside the scope of Recognition.

With wise and foresighted management, it should be possible to cope with the problems arising from the factors investigated within Recognition, although there are other aspects of the changing environment that may turn out to be more difficult to manage (Chapter 5.4). The Pan-European Programme for Intensive and Continuous Monitoring of Forest Ecosystems is one of the tools, which can provide early warning signals of large-scale changes in European forests. Maybe we will also need additional monitoring systems.

References

Farrell, E.P., Führer, E., Ryan, D., Andersson, F., Hüttl, R. and Piussi, P. 2000. European forest ecosystems: building the future on the legacy of the past. Forest Ecology and Management 132: 5–20.

Fenn, M.E., Poth, M.A., Aber, J.D., Baron, J.S., Bormann, B.T., Johnson, D.W., Lemly, A.D., McNulty, S.G., Ryan, D.E. and Stottlemyer, R. 1998. Nitrogen excess in North American ecosystems: Predisposing factors, ecosystem responses, and management strategies. Ecological Applications 8: 706–733.

Sverdrup, H., Warfvinge, P., Frogner, T., Håøya, A.O., Johansson, M. and Andersen, B. 1992. Critical loads for forest soils in the Nordic countries. Ambio 21: 348–355.

5.4 Implications and consequences

Seppo Kellomäki, Heli Peltola and Aija Ryyppö

University of Joensuu, Faculty of Forest Sciences, Joensuu, Finland

5.4.1 Background

In Europe, the overall range of forests is limited primarily by climate as assumed in Recognition. On a south-north gradient temperatures decrease and water availability increases. On a west-east gradient, the maritime climate turns to the continental one with decreasing humidity and the increasing water limitation for forests (Figure 1). The occurrence and magnitude of changes in growth may represent temporal and spatial variability, which is related to the prevailing climate limitation (temperature, precipitation, atmospheric carbon) of growth. Concurrently, the availability of nitrogen and other nutrients may control the growth. In Recognition, the changes in growth were analyzed with regard to changes in nitrogen, moisture, temperature and atmospheric carbon. The effects of damaging abiotic factors (such as storm, frost, snow or noxious chemical agents) and/or biotic factors (such as insect, rodent or fungal attacks) influencing growth were excluded from the analysis. However, the impacts of these factors are discussed, when assessing the future risks.

Throughout Europe, forests have been utilised over centuries and management regimes have varied from country to country. The productivity of the forest ecosystem is related to the prevailing environmental conditions and species composition of forests. The environment ($E(j)$) and genotype ($G(i)$) may have single effects on the phenotype ($P(i,j)$), but they may also interact; i.e.

$$P(i,j) = G(i) + E(j) + G(i) \times E(j) \quad (1)$$

where i refers to the tree species (or ecotype of a species) and j to the site representing specific environmental conditions.

In Recognition, the changes in growth of Scots pine (*Pinus sylvestris*), Norway spruce (*Picea abies*) and beech (*Fagus sylvatica*) were studied. Each species has a wide geographical distribution, with ecotypic adaptation to varying site conditions. This may affect how sensitive each species is to any changes in the environmental conditions found in Recognition, with a consequent spatial variability in the changes in growth. Moreover, it is well known that many species grow better outside their natural boundaries (Zobel and Talbert, 1984, pp. 75–116). Unfortunately, it is not known if this is true for Europe, where Norway spruce, for example, is growing outside its natural boundaries in many places of the European lowland. The ecotypic variability in the response of trees to the changes in the environmental conditions was, however, excluded from the analysis, with an

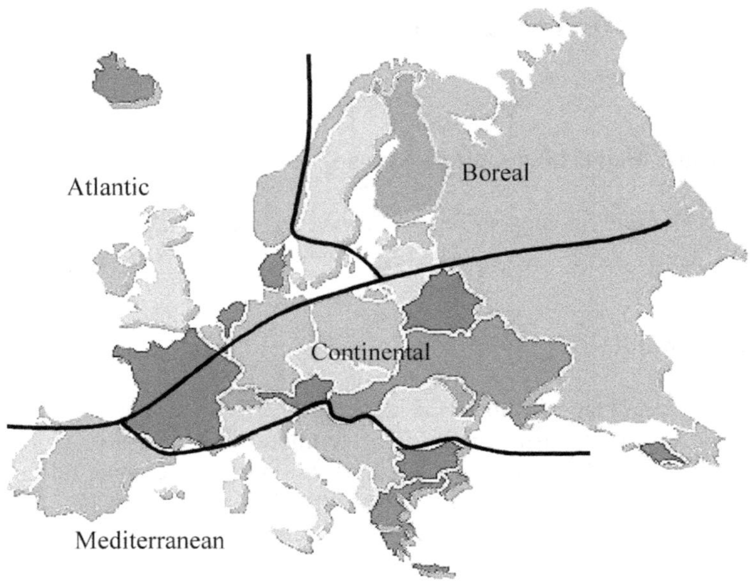

Figure 1. Forest regions of Europe used in analysing the implications and consequences of the changes in growth through Europe. Boreal forests are the forests in the European part of the boreal zone. The temperate forest in Central European are divided into Atlantic and Continental forests with emphasis on the differences in availability of soil water. Mediterranean forests are forests in the Mediterranean Basis.

assumption that the populations of Scots pine, Norway spruce and beech at each site were fully adapted to the prevailing climatic conditions. This implies that the spatial variability in growth is not expected to reflect the spatial variability in the genotypic variability within species populations occupying different sites.

In the above context, the aim of this Chapter is to address the long-term consequences and risks of observed changes for sustainability and to assess the implications of future management strategies, with the consequent effect on the success of forest throughout Europe. The concept of management refers to a set of measures, which are used to manipulate the environmental conditions in order to control the productivity of the forest ecosystem directly through the environment. Management is assumed to be species-specific with the specific changes in the environmental conditions; i.e. any changes in the environment may have single effects on the productivity. Recognition has focussed on tree growth, and therefore sustainability of timber yield will be the principal focus of this Chapter.

5.4.2 Findings for the current situation, with management implications

5.4.2.1 Changes in environmental conditions occurred in the past, with likely changes in growth

Temperature and precipitation

The analysis of the long-term weather records (1950–1999) showed that the annual air temperature has increased almost everywhere during the last 50 years. The seasonal allocation is variable through Europe, but more increase may be expected in winter time than in summertime. In Central Europe and the UK, the mean air temperature during the growing season (May through September) and the early growing season (April through June) was increased, but in Northern Europe the mean air temperature of the growing season remained unchanged, with a slight increase in the early growing season (Chapter 4.1). At the same time, the annual precipitation remained unchanged in Central Europe, with a slight decrease during the growing season. In Northern Europe, the precipitation tended to decrease, with no clear seasonal pattern.

Nutrients

In Europe the total annual nitrogen deposition varied from less than 5 kg N ha^{-1} in the most northern parts of Northern Europe to more than 30 kg N ha^{-1} in the central and central-west parts of Europe (Chapter 4.2). The macro-scale pattern of the nitrogen deposition closely resembles the spatial distribution of the mean foliar nitrogen concentration in Scots pine and in Norway spruce. However, the nitrogen status of Scots pine stands was poor even in Central Europe, whereas the nitrogen status of Norway spruce stands was optimal even in Northern Europe (Chapter 4.1).

Changes in growth

The general growth acceleration due to an increase in temperature is most likely to occur in Central Europe provided that the water supply is sufficient. An increase in air temperature connected with a reduction of water availability would rather lead to a growth decline than to improved growth. In Northern Europe the warming seems to have occurred mainly in the dormant season, with limited direct impact on the growth. In the growing season, temperature even slightly decreased at some sites. Therefore, the growth acceleration directly due to temperature increase is less likely in Northern Europe than in Central Europe.

However, the analysis by Kahle et al. (Chapter 4.2) showed that even in Central Europe the increase in the growth in large areas of Europe over the last decades is most probably related to the high nitrogen deposition rather than to the changes in climate. This conclusion is supported by the findings of Mellert et al. (Chapter 4.1); i.e. the growth at almost all sites was nitrogen limited, and the supply of other elements was adequate even under the highest nitrogen deposition. Other nutrients (e.g. K, P, Mg and Ca) may also be limiting growth. The enhancement of growth due to the increase in the atmospheric CO_2 is unclear, with no support from the results of the process-based models (Chapter 4.4) or from the measurements (Chapter 4.1). Nevertheless, the ample supply of nitrogen along with the increasing

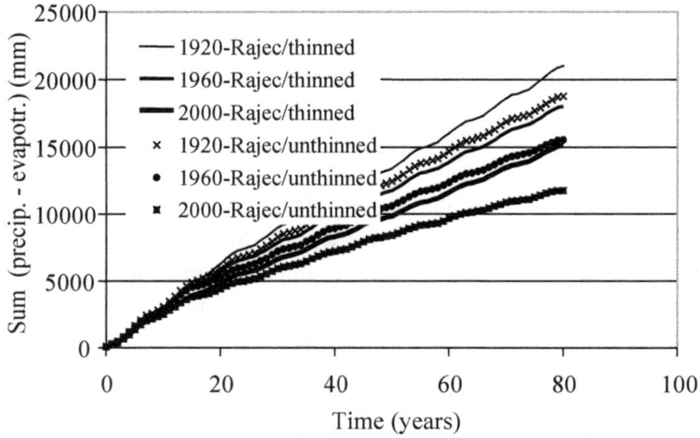

Figure 2. Water balance in unthinned and thinned stands of Scots pine in Rajec under the current and changing climate in the periods 1920–2000, 1960–2040 and 2000–2080 applying the climate scenarios used in Recognition.

CO_2 should increase growth, as has been demonstrated in numerous laboratory studies (Jarvis, 1998).

5.4.2.2 Management implications

Regardless of the underlying reasons, the enhancement of growth implies an enhanced turnover of carbon in the forest ecosystem. The life cycle of trees is more rapid; i.e. trees will mature more rapidly and stocking may increase if the management (e.g. rotation length) remains the same as applied in the past. This will feed back to the development of tree populations by affecting the availability of resources such as water and nutrients, with a risk of drought and imbalance between different nutrients. As tree populations mature, the crowding may lead to reduced growth, and finally death of some trees. Concerning the existing forests, this implies a need to modify the thinning practices and the length of rotation in order to avoid the adverse effects of the enhanced growth and to utilize the available benefits for forestry.

In Recognition the risk of drought was identified in many places in Europe. Mellert et al. (Chapter 4.1) found, that short-term variability in water supply may affect the growth, especially in Central Europe, where risk of drought with the reduction of growth was found to be quite clear. Figure 2 shows that wider spacing and regular thinning may increase the water yield 15–25% over the rotation (80 years), with the higher increase in the future than in the past. Wider spacing is especially useful in reducing drought risks, wherever an increased evaporative demand and decreasing precipitation is expected. Such areas are likely to be found in southern Spain, on the border between Germany and Poland, in large areas of Austria, and southern Sweden (Chapter 4.1).

The adaptation of thinning regimes to the changes in the climatic conditions and in the ecosystem dynamics may also help to increase timber yield. In Central Europe, the profitability of forestry may further be increased by the choice of shorter rotations, which the enhanced growth may facilitate. If the same criteria for defining maturity are applied as currently, the increase of 0.5–1°C in the annual temperature found for Central Europe implies an increase of 5–10% in the temperature sum, and a reduction of 10–15% in the length of rotations. The adaptation of the thinning regime and rotation length into the enhanced growth and turnover of tree populations is also of primary importance in order to avoid any increase in endogenous mortality, as will be discussed later.

5.4.3 Findings for the future, with management implications

5.4.3.1 Tree species composition and forest limits

The future development of the forest in different parts of Europe was investigated with the help of process-based models. The effects of the atmospheric CO_2, climate and nitrogen deposition were studied. In agreement with the empirical studies, the calculations showed the sensitivity of the past and current growth to nitrogen deposition. The nitrogen limitation may be of importance also in the future due to the expected reduction in nitrogen deposition throughout Europe. However, the changes in CO_2 and climate appeared to become increasingly important in the future, whereas nitrogen deposition will become less significant. This was especially clear for the northerly latitudes, where the simulations showed a clear increase of growth under elevated CO_2 and increased temperature.

The future increase of the climate impacts on the growth may be expected, since the overall range of forests in Europe is limited primarily by climate, either through limited moisture availability in the south and east, or through low temperature in the north. Climate projections suggest a displacement of climatic zones suitable for boreal forests by 150–550 km over the next century (IPCC 1996). The biggest change may occur at the boreal timber line, where the increase in temperature may enhance substantially seed crop (Figure 3). Similarly, the alpine timber line may be expected to move higher throughout Europe.

In Western and Central Europe, the changing climate may increase the competitive capacity of deciduous trees to such an extent, that natural conifers may be replaced by deciduous species in large areas. Associated with an increase in mean annual temperature of 0.3°C per decade by the end of the 21st Century, model computations by Lindner et al. (1997) indicate that Norway spruce may be replaced by beech and other broad-leaved species in northern Germany. Changes in the forests of Southern Europe are most likely to be driven primarily by changes in water availability; i.e. changes in precipitation will determine the relative importance of sclerophyllous and deciduous species. Water availability in the period April–June would be particularly important (Gavilán and Fernández-González, 1997).

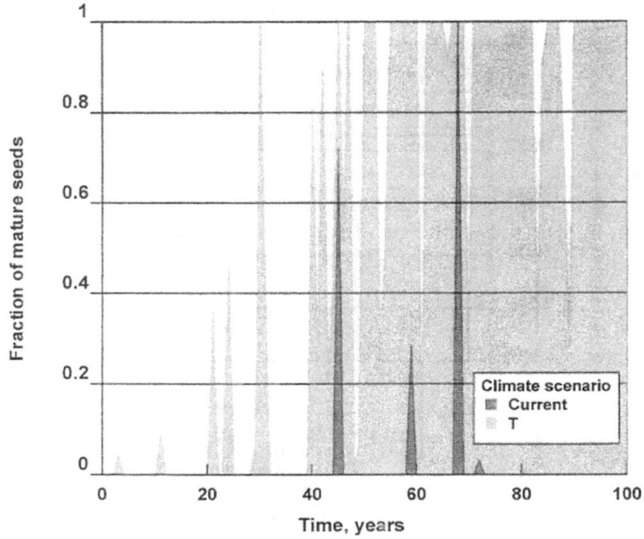

Figure 3. Effect of temperature increase on the fraction of mature seeds of Scots pine at the northern timber line in Finland (70°N) based on the results of FinnFor model simulations (Kellomäki and Väisänen, 1997) assuming a transient increase of temperature by 4°C over the next 100 years.

5.4.3.2 Growth

Previously, Spiecker et al. (1996) have recorded an increase in growth in Austria, Denmark, France, Germany, Slovenia, Sweden, and Switzerland, with possible increases also in Portugal and Spain. These findings are supported by the results of Recognition. The increased nitrogen deposition was mainly behind the increasing growth, but in the future the climate change with the increasing atmospheric CO_2 may further enhance the growth as the model computations in Recognition show (Chapters 3.2 and 4.4) (Figure 4). In spite of the fact that the future projections for growth are still uncertain, the physiological response of trees to the increases of temperature and atmospheric CO_2 suggest further increase in the growth of the main tree species of Europe.

The findings of Recognition are in line with the findings of several previous model-based studies. For the boreal forests, Kellomäki and Väisänen (1997) applied the FinnFor model by assuming a transient change of climate (elevating CO_2 by 3.5 μmol $mol^{-1}yr^{-1}$, and temperature by 0.04°C yr^{-1}) over 100 years. The temperature elevation alone increased the productivity of Scots pine by about 5–7% in southern Finland and by about 10–15% in northern Finland. The increasing CO_2 alone resulted in an increase in productivity up to 12–15% both in southern and northern Finland. The combined elevation of CO_2 and temperature further increased productivity, e.g. in northern Finland by 25–30%, and in southern Finland by 20–

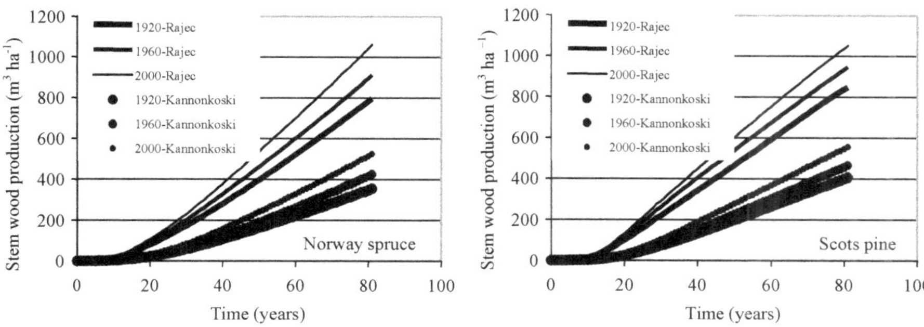

Figure 4. Simulations in *Recognition* of the total growth of Norway spruce and Scots pine in Rajec (49°N) and Kannonkoski (62°N) in the periods 1920–2000, 1960–2040 and 2000–2080 based on the FinnFor model.

25%. The increase in the total number of days in the growing season was responsible for much of the increase in productivity.

For the Atlantic forests, the site studies for conifer plantations in the northern UK have shown an increase in General Yield Class (GYC) of at least 1 m^3 ha^{-1} yr^{-1} per decade since the 1930s (Cannell et al., 1998). The analysis based on two process-based models showed that:

1. increases in nitrogen deposition (N), CO_2 concentration and temperature together might account for up to half of the observed increase in GYC in this century;
2. increased N deposition and CO_2, considered separately, probably increased forest productivity by a modest amount (7–14%), but their combined effect has been approximately additive;
3. increased temperature, even when combined with increasing CO_2 concentrations, promoted growth less than expected from site studies relating GYC to temperature; and
4. substantial further increases in productivity, GYC, leaf area index and standing biomass are forecast during the next century as a result of increasing CO_2 concentrations and continued N deposition, with or without climatic warming. The predicted increases in GYC could be large enough to have profound effects on the forest industry.

Similarly, Proe et al. (1996) have suggested, on the basis of an ecophysiological process model, that the GYC of Sitka spruce (*Picea sitchensis)* in Scotland would increase by 2.8 m^3 ha^{-1} yr^{-1} for each 1°C rise in temperature. This estimate was based on a uniform increase in temperature throughout the year. If there was greater warming during winter, the GYC increase was 2.4 m^3 ha^{-1} yr^{-1}. The reduction is attributable to the negative correlation between winter temperature and growth rate, which is believed to be caused by the increased activities of pests in warmer winters and the winter chilling requirements of the species. Negative relationships between

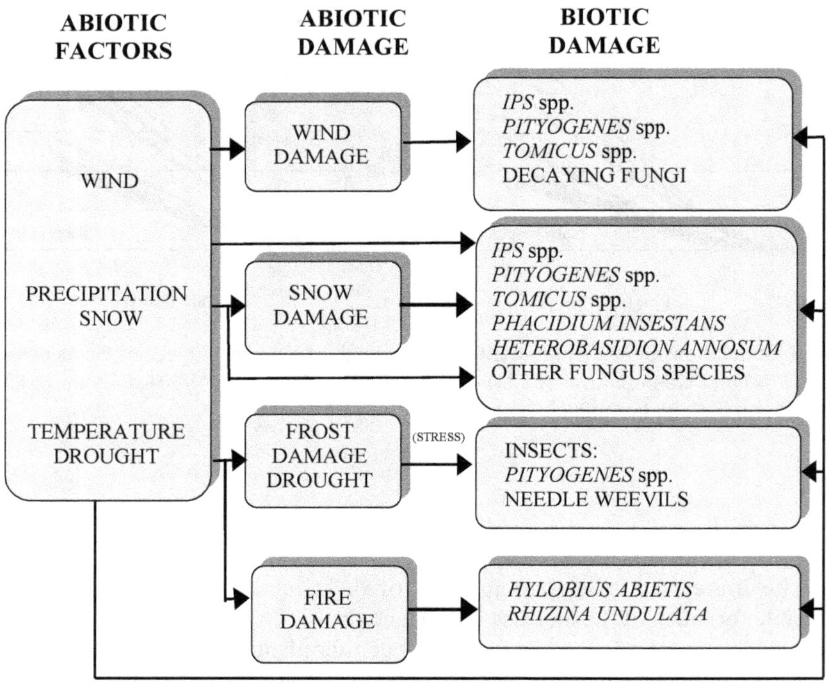

Figure 5. The effect of climate and specific weather events on the outbreaks of populations of damaging insects and fungi.

GYC and winter temperature have also been identified for Douglas fir (*Pseudotsuga menziesii)* and Scots pine (Tyler et al., 1996). The winter chilling requirements may be among the major factors in determining the distribution of many tree species. However, no major risk of spring and autumn frost is expected. In the Continental and Mediterranean forests, the productivity will probably decrease, mainly due to the more limiting soil moisture.

5.4.3.3 Risks of forest damage

Endogenous mortality

Tree mortality may be classified into two types: (i) endogenous mortality that is related to the growth and life span of trees; and (ii) exogenous mortality that is related to the abiotic (frost, wind, snow, fire) and biotic (insect and fungal pests) factors or damage.

Whenever the growth of trees is increased, an increase in endogenous mortality may be expected due to a shorter life span of trees. Thus, the increase in the mortality may be directly related to the increase in growth. Damage caused by abiotic and biotic factors also interact; e.g. dead, broken or uprooted trees may provide insect pests with favourable breeding conditions, which in some cases may enable the insect pest populations to reach such levels where the pest may also cause damage to healthy trees (Ravn, 1985; Schroeder and Eidmann, 1993) (Figure 5).

In Recognition, the risk of the forest damage was not studied, but the available models allow the changes in the endogenous mortality to be calculated. In central Finland, growth increased by 2.8% and mortality by 4.0% in 1960–2040 in comparison to 1920–2000, if no thinnings were applied. During 2000–2080, growth increased by 7.0% and mortality by 10.8%. In Central Europe, the increase in growth and mortality was smaller with the values 1.5% for growth and 1.8% for mortality in the period 1960–2040, and 3.0% for growth and 3.9% for mortality in the period 2000–2080. This implies that whenever there is a substantial increase in growth, the life span of tree populations will be shorter and there will be a need for more regular thinning and shorter rotations.

Risk of wind and snow damage

Throughout Europe, strong winds blow down and break trees with large economic losses in timber production and productivity of the forest ecosystems. These problems are exemplified by a storm episode in 1999, which blew down 175 million m^3 of timber throughout Western and Central Europe, and created problems with respect to regular timber supply. There is some evidence that the number of days with gale-like winds has increased in Central Europe (Dlugolecki, 2000). This is related to the higher frequency of milder winters. In the past, snow cover and stable cold high pressure systems used to develop as a barrier against the cyclones approaching from the Atlantic. The barrier may be weaker under the increasing temperatures, and may have shifted to the east. A possible consequence may have been the devastating storms experience in Central Europe in recent years. The mean return period of major wind damage in Europe is currently 2–3 years.

In the future, the risk of wind damage may be increased further by the large-scale maturing of forests beyond the economically feasible rotation length as occurring throughout Europe. Moreover, the preference of clear-cuts in timber harvest modifies the structure of forests susceptible to wind-induced damage. The risks of major wind damage may be further enhanced by the decrease in the duration of frozen soil, with the consequent reduction in the anchorage of trees (Peltola et al., 1999). At the same time, the share of snowfall from the total winter precipitation may be reduced with an obvious reduction in the risks of snow damage. However, the increasing winter precipitation may imply the increased accumulation of snow, with the increased risk of snow damage at higher altitudes and in the most northern areas. At lower altitudes and in areas outside the northern regions, the risks of snow damage may be decreased.

Risk of fire damage

In Europe in 1980–1993, fire damage affected nearly 8 million hectares of forests in 660,000 separate fires, the average size of the burnt area being slightly more than 10 hectares. Depending on the geographical location, there is a significant difference in the scale and frequency of fire. In Spain, in a single year (1990) fires damaged over 2 million hectares of land. In Italy, Portugal, Greece and the Mediterranean parts of France, forest fires are also a serious problem. There is no clear climatic trend in the occurrence of fire, but the fire risk seems to be mainly related to human activities. However, in the future high temperatures associated with drought will increase the intensity of fires and the area affected by fires.

In Northern Europe, forest fires are rare, the percentage of forest land burnt annually is less than 0.05%. The mean size of fires is less than 1 hectare (Zackrisson, 1977). In Northern and Western Europe, the return period of fires is much longer (mean return time longer than 50 years) than in the Mediterranean region (mean return time 5–15 years). Fire risk in maritime Europe and in mountain areas is small (mean return time more than 100 years). In continental Europe, the fire risk is probably larger than in Northern Europe, but still substantially less than in the Mediterranean region. In the Boreal region, even under the current precipitation (250–700 mm yr^{-1}), one to two weeks without rain is needed to significantly increase the fire risk.

The main uncertainties associated with the fire risk are related to the seasonal distribution of precipitation. Warmer temperatures in spring and early summer may lead to earlier melting of snow and drying of soil in summer, if the precipitation does not increase at the same time. The temperature increase of 3–5°C in June–August may increase the fire area by 15–50 times (Suffling, 1992), if no increase in precipitation occurs. This implies an increasing fire risk also for southern parts of the Atlantic forests and for the Continental forests. In the Mediterranean region the predicted increases in summer temperature and the reductions in precipitation will further increase the fire risk.

Risks of biotic damage

In Northern Europe, cold winters have protected forests from many of the insects and fungi that are common further south. Under a changing climate some pests will have opportunities to expand their distribution, e.g. *Ips typographus*. The occurrence of this bark beetle is related to the prevailing temperature conditions; the flight period is initiated when the daily mean temperature exceeds 18°C in spring, and the number of generations per year increases with increasing spring and summer temperatures. Similarly, the populations of *Neodiprion sertifer*, *Diprion pini*, and *Panolis flammea* may expand as a result of increasing summer temperatures. The temperature increase may also expand the occurrence of *Lymantria monacha* far above 60°N in Scandinavia. Moreover, new organisms (such as *Bursaphalenchus xylophilus*) with considerable potential to cause damage to trees may be expected to become established in Europe. Until now, low summer temperatures and short growing seasons have effectively limited the success of this pine nematode in the main part of Europe (except for Portugal). Climate change may remove these limitations, and this species may plague forests especially in Southern Europe (Evans et al., 2002).

High summer temperatures along with drought may decrease the epidemics of damaging fungi, but they may flourish in cool rainy summers. On the other hand, higher winter temperature may enhance epidemics of damaging fungi such as *Gremmenniella abietina* and *Lophodermella sulcigena*. The frequency of root rot induced by *Heterobasidion annosum* may also be larger, if the autumn, winter and spring temperatures are higher and the duration of soil frost is shorter. In Northern Europe, especially, the climate change may increase the occurrence of root rot with a consequent decrease in forest productivity.

5.4.3.4 Management implications

The results of Recognition demonstrate that forest growth has been enhanced in many parts of Europe. The climate change associated with the increasing CO_2, increased nitrogen deposition and changes in management are assumed to be behind the increased growth as supported by the model computations in Recognition (Chapter 4.4). The enhanced growth implies a need to modify the management practices in order to adapt the dynamics of forests to the changing environmental conditions, e.g. in terms of tree species choice, rotation length and thinning practises. The adaptive strategies are summarised in Table 1.

In Western, Central and Southern Europe, the introduction of new tree species or more southern provenances of the existing species have been among the main tools to increase forest growth. Unfortunately, it was not possible to study the role of these introductions in enhancing forest growth in different parts of Europe in *Recognition*, since the origin of seeds for the measured stands was not known. Until now, there has been no solid evidence that the introduced species would be less adaptable to the current climate, if the tree species choice is based on extensive provenance tests to identify phenologically suitable ecotypes (Lines, 1987).

In the boreal areas, where drought risk is less in comparison to other regions, several existing tree species (both native and exotic) are probably growing faster than previously because of the increase in temperature. Therefore these species will have a more rapid life cycle. This will require shorter rotations and more frequent thinnings in order to avoid biotic damage associated with diminishing tree growth. The reduction of rotation will be largest in Northern Europe, where the length of rotation may be reduced by up to 20–25 years (15–25% from that under the current climate), and more frequent thinnings will be needed in order to avoid self-thinning if the current thinning intensities are applied. In Central Europe, the reduction in rotation is smaller (up to 5–10 years, 10–15%) and the number of thinnings is increased to smaller extent than in Northern Europe. The need to modify the management practices is related to the magnitude of change in climate in order to find the balance between dynamics of the forest ecosystem and timber production.

Natural deciduous regrowth in coniferous plantations may require more precommercial cuttings to favour of coniferous species as demonstrated by Kellomäki and Kolström (1992). This is especially true for the boreal forests, but throughout Europe the competition balance between different tree species may be altered due to the changing environment. Consequently there may be an increased need for tending of stands to favour regeneration of the preferred tree species. More frequent and intensive thinnings may delay decline in tree performance and reduce risks of biotic damage. The regular thinnings will also increase the mechanical strength of trees due to the enhanced growth and thus reduce the risk of abiotic damage.

Under climate scenarios applied in Recognition, the dynamics of Scots pine populations in the boreal forests altered to a lesser extent. However, the timber yield will increase in the period 1960–2040 by 3%, and in the period 2000–2080 by 6% compared to the yield in the period 1920–2000. This increase implies a need to have more frequent thinnings or higher thinning intensities in order to avoid self-thinning. On the other hand, the enhanced growth may allow shorter rotations because trees will reach maturity sooner. This pattern holds throughout Europe, but it is related to the prevailing climatic conditions as indicated by the relationship

Table 1. Main adaptive measures for forest regions.

Region	Future conditions	Impact on forests	Management
Boreal forests and northern parts of Atlantic forests	Higher temperature and precipitation, short duration of snow cover and soil frost, enhanced decomposition of soil organic matter.	Enhanced growth and regeneration, increasing dominance of deciduous species, reducing fire risk, increasing risks of wind and snow damage.	Shorter rotation and regular thinnings, preference of natural regeneration and in plantation preference of more southern origin of existing tree species, more preference for deciduous species, in coniferous stands regular tending and precommercial thinning of deciduous species in favour of conifers, suppression of enhanced growth of herbs and grasses in seedling stands and regeneration areas, limiting the import of fresh timber from areas with pests potentially damaging forests under the climate change.
Southern parts of Atlantic forests and Continental forests	Higher temperature, reducing precipitation, increasing drought risk, reducing risk of spring and autumn frost, enhancing decomposition of soil organic matter, increasing fire risk.	Drought risk, reducing success of natural regeneration and growth of existing deciduous species, increasing frequency of endogenic and exogenic damage.	Shorter rotation and regular thinnings with wider spacing, preference of drought tolerant conifers in plantations with wider spacing, tree improvement programmes to increase the drought tolerance of tree plantations, limiting the import of fresh timber from areas with pests potentially damaging forests under the climate change.
Mediterranean forests	Higher temperature, strong seasonality of precipitation with high winter precipitation and extreme summer drought, severe fire risk.	Severe drought risk, reducing success of natural regeneration and growth of existing deciduous species, eliminating of tree species, increasing frequency of damaging fire.	Shorter rotation, soil management and plating technique need to be developed to meet the special conditions resulted in by the increasing drought, introduction of new species now successful in dry subtropics, management of forest plantations may need technique now proper in dry subtropics, large-scale tree improvement programmes in order to increase the genetic potentials of tree plantations to adapt higher temperatures with drought, preferring wider spacing in plantations and later in thinnings, the fire fighting needs special measures, limiting the import of fresh timber from areas with pests potentially damaging forests under the climate change.

between the temperature sum and the length of rotation applied in different parts of Europe.

5.4.4 Discussion and conclusions

Model computations carried out in Recognition support the previous findings that the growth of forests has increased in many sites in Europe, and it may increase further in the future. In the northern areas the increase of growth of Scots pine and Norway spruce may be related to supply of nitrogen but the elevation of CO_2 and temperature may also contribute. In Central Europe, the ample supply of nitrogen is a much more dominant factor but the influence of elevating temperature and CO_2 cannot be excluded. Regular thinnings will also increase the growth, and there is an interaction between management and the changes in climate; i.e. the growth of Scots pine and Norway spruce responds more vigorously to the elevation of CO_2 and temperature when thinned than when not thinned. The model simulations also indicate that the rotation of timber production may be reduced, especially in Northern Europe, where tree growth and consequent profitability of forestry may increase most.

Currently, there is a tendency in Europe to prefer native tree species with a more ecological approach to management. In the long term this may shift the tree species composition closer to that determined by the climatic and edaphic conditions prevailing in Europe. This is expected to increase the capacity of forests to adapt to the changing climate, but there is no solid evidence to support this assumption (Fries et al., 1997). On the other hand, the climate change may change the tree species composition, especially in northern areas, where deciduous species may become more abundant. The European forests will also be subject to a variety of other anthropogenic influences such as nitrogen deposition, sulphur emission, ozone injury, and changes caused by unbalanced game populations. These factors are likely to interact with climatic change to bring about a complex series of responses that will differ from place to place. Future forest communities are likely to be composed of tree species assemblages not necessarily occurring today (Lindner et al., 1997).

Planting in many European countries is based on improved stock – i.e. the genetic base has been narrowed. Where there is uncertainty about the conditions, it would be prudent to manage the forest to conserve the broad genetic base, rather than to narrow the genetic base further. In Northern Europe there has been a preference for natural regeneration in management. This provides more potential for the forests to adapt to unpredictable changes in climate. In Southern Europe, it is predicted that there will be an increase in the occurrence of droughts. Natural regeneration under such conditions could be unreliable. Under such conditions the preference of tree provenances of more southern origin, and also wider spacing in plantations may reduce the impacts of the predicted climate change. Throughout Europe, regular management with shorter rotations enhances the turnover of the current tree populations with faster introduction of more resistant tree species and provenances into forests. Deciduous tree species may become more abundant, even in Northern Europe.

More frequent storms, especially in Northern and Western Europe may result in large fluctuations of timber supply. The predicted shift of the tree species composition towards deciduous species requires a careful analysis how the European timber resources and future needs of wood-based products are matched in the future. The shift in the tree species composition may also alter the balance between the host tree species and the populations of forest pests. In some cases, the populations of these organisms are expected to increase but the extent to which they may be held in check by natural enemies or requires management control is not known. Conditions favouring the spread and epidemic status of pathogens are also poorly known.

Throughout Europe, the forested areas are the main source of the ground water, and they absorb precipitation and reduce the risk of excess surface flow, floods and erosion. In Central and Southern Europe a reduction of forest cover and the enhanced seasonality with increased role of the winter precipitation may increase surface flow and floods with a consequent need to reconsider water management. In Northern Europe, the increasing precipitation may also increase the risk of floods even though the forest cover buffers watersheds.

In many European countries, forests and forest reserves provide the main opportunities to maintain biodiversity and conserve rare and endangered species. If the tree species selection was expanded to include exotic species, or only a few resistant local species were preferred in order to maintain the productivity of forests, then this would reduce the opportunities to implement the national and pan-European conservation programmes. The most problems would probably occur at the forest margins in boreal areas, mountainous areas and Mediterranean regions. In mountainous areas, the current management of forests for reducing the risk of landslide or avalanches would be reconsidered due to the upward movement of timberline and changes occurring in the tree species composition.

References

Cannell, M.G.J., Thornley, J. H.M., Mobbs, D.C. and Friend, A.D. 1998. UK conifer forests may be growing faster in response to increased N deposition, atmospheric CO_2 and temperature. Forestry 71 (4): 277–296.

Dlugolecki, A. 2000. Insurance. In: Assessment of potential effects and adaptations for climate change in Europe. 2000. Final report of ACACIA Project. European Commission. Pp. 187–201.

Evans, H., Straw, N. and Watt, A. 2002. Climate change: implications for insect pests. In: Climate change: implications on UK forests. Broadmeadow, M. (ed.). Forestry Commission. Edinburgh. Pp. 98–118.

Fries C., Johansson, O., Pettersson, P. and Simonsson, B. 1997. Silvicultural models to maintain and restore natural stand structures in Swedish boreal forests. Forest Ecology and Management, 94(1–3): 89–103.

Gavilán, R. and Fernandez-González, E. 1997. Climatic discrimination of Mediterranean broad-leaved sclerophyllous and deciduous forests in central Spain. Journal of Vegetation Science 8: 377–386.

IPCC, 1996. Climate Change 1995: Impacts, Adaptations, and Mitigation of Climate Change: Scientific- Technical Analyses. Contribution of Working Group II to the Second

Assessment Report of the Intergovernmental Panel on Climate Change. Watson, R.T., M.C. Zinyowera, and R.H. Moss (eds.). Cambridge University Press, Cambridge, United Kingdom and New York, NY, USA. 880 p.

Jarvis, P.G. 1998. European forests and global change. The likely impacts of rising CO_2 and temperature. Cambridge University Press, Cambridge. 380 p.

Kellomäki, S. and Kolström, M. 1992. Computations on the management of seedling stands of Scots pine under the influence of changing climate in southern Finland. Silva Fennica 26: 97–110.

Kellomäki, S. and Väisänen, H. 1997. Modelling the dynamics of the boreal forest ecosystems for climate change studies in the boreal conditions. Ecological Modelling 97(1,2): 121–140.

Lindner, M., Bugmann, H., Lasch, P., Flechsig, M. and Cramer, W. 1997. Regional impacts of climatic change on forests in the state of Brandenburg, Germany. Agricultural and Forest Meteorology 84(1–2): 123–135.

Lines, R. 1987. Choice of seed origin for the main forest species in Britain. Forestry Commission Bulletin 66. HMSO, London.

Peltola, H., Kellomäki, S. and Väisänen, H. 1999. Model computations on the impacts of climatic change on soil frost with implications for windthrow risk of trees. Climate Change 41: 17–36.

Proe, M.F., Allison, S.M. and Matthews, K.B.1996. Assessment of the impact of climate change on the growth of Sitka spruce in Scotland. Canadian Journal of Forest Research 26: 1914–1921.

Ravn, H.P. 1985. Expansion of the populations of *Ips typographus* (L.) (*Coleoptera, Scolytidae*) and their local dispersal following gale disaster in Denmark. Zeitschrift fur angewandte Entomologie 99(1): 27–33.

Schroeder, L.M. and Eidmann, H.H. 1993. Attacks of bark- and wood-boring *Coleoptera* on snow-broken conifers over a two-year period. Scandinavian Journal of Forest Research 8: 257–265.

Spiecker, H., Mielikäinen, K., Köhl, M. and Skovsgaard, J.P. (Eds.). 1996. Growth trends in European forests. Studies from 12 countries. Springer Verlag, Berlin, Germany. 372 p.

Suffling, R. 1992. Climate change and boreal forest fires in Fennoscandia and Central Canada. In: Boer, N. and Koster, E. (eds.). 1992. Greenhouse impact on cold-climate ecosystems and landscapes. Catena supplement 22:111–132.

Tyler, A.L., Macmillan, D.C. and Dutch, J. 1996. Models to predict the General Yield Class of Douglas fir, Japanese larch and Scots pine on better quality land in Scotland. Forestry 69: 13–24.

Zaricksson, O. 1977. Influence of forest fires on the north Swedish boreal forest. Oikos 20(1): 22–32.

Zobel, B. and Talbert, J. 1984. Applied tree improvement. John Wiley & Sons. New York. 505 p.

6. Conclusions

6. Conclusions

Hans-Peter Kahle[1], Timo Karjalainen[2], Göran I. Ågren[3], Seppo Kellomäki[4], Marcel van Oijen[5], Jörg Prietzel[6] and Heinrich Spiecker[1]

[1]*Institute for Forest Growth, University Freiburg, Germany*
[2]*European Forest Institute and Finnish Forest Research Institute*
[3]*Swedish University of Agricultural Sciences, Uppsala, Sweden*
[4]*University of Joensuu, Faculty of Forest Sciences, Joensuu, Finland*
[5]*Centre for Ecology and Hydrology, Edinburgh Research Station, UK*
[6]*Chair of Soil Science, Technical University Munich, Germany*

Since the late 1980s, researchers in Europe have reported that trees seem to be growing faster. The results of the pan-European project 'Growth Trends in European Forests' (Spiecker et al., 1996) indeed showed that the rate of forest growth has increased on many sites in Europe. The causes for this increase have now been investigated in the Recognition project ('Relationships between Recent Changes of Growth and Nutrition of Norway Spruce, Scots Pine and European Beech Forests in Europe). The results indicate that increased nitrogen nutrition, thought to be mainly driven by the interaction of elevated atmospheric nitrogen deposition and natural recovery of the sites from former devastative land-use practices, contributes most to the observed increases in forest growth. Climatic changes and CO_2 fertilization have not been the key factors for forest site productivity changes during the 20th century, but will, according to model predictions, be of increasing importance in the future.

For the future sustainable management of European forests, it is important to know what is causing the growth increase. The main objective of the Recognition project was to assess the relationship between recent changes in forest growth, stand nutrition, and changes in atmospheric CO_2 concentration, air temperature and precipitation regimes in Europe. Altogether 24 European research teams contributed to the identification of the potential causes of observed increased forest growth, focusing on the relative importance of nutrients, CO_2 and climate. The teams conducted growth analyses at selected sites, ran process-based model simulations, and discussed potential long-term consequences and risks of observed changes for sustainability and for future forest management.

The project was unique in combining experts from several disciplines and using different methods in order to obtain reliable results. The best available data, including data from the EC-UN/ECE forest intensive monitoring programme, the so-called Level II programme, were used as a basis of the analyses.

Besides analysis of past growth changes, the project also aimed to determine whether the observed trend of increased forest growth is likely to persist. Researchers also studied whether different tree species and sites would sustain projected increased forest growth in the future.

When testing the hypotheses about the causes for increased growth, the project analysed data from long-term research plots as well as from current and recent growth investigations. Computer models developed in the UK, Finland and Sweden were used to analyse the interrelations between site productivity and potential causes of increased growth, and also to make projections of the potential consequences.

Analysis of temporal trends from the sampled material shows that during the period 1960 to 2000 height growth rates of the investigated tree species were significantly and consistently larger than expected. Average annual height growth of Scots pine and European beech has accelerated by 25% and that of Norway spruce by 23% as compared to the height increment of nowadays old trees growing on the same sites when they had the same age. According to yield level models – that is the functional relationship between total wood volume production and stand height – of conventionally used yield tables the observed increases in site index indicate substantially larger increases in total wood volume production potential between 1960 and 2000 of up to +56% for spruce, +45% for pine, and +74% for beech. The few relevant literature references do not give evidence for a change in the yield level models in recent time. The observed accelerations in height growth and the deduced increases in wood production potential are driven by changes in site productivity, hence are not attributable to changes in forest area, changes in mean age structure, species composition or density of the forests. However, we found evidence for isometric change of growth, i.e. age related ratios of tree height to diameter in breast height did not differ between the age cohorts for Norway spruce and Scots pine, indicating that trees nowadays not only are taller but also have larger diameter at a given age. The mean annual radial increment of the cohort of younger trees was indeed significantly larger than that of the older trees for all species. Compared to the average growth rate of the reference trees acceleration in radial growth was +15.5% for spruce, +12.8% for pine, and +26.8% for beech (mean over the age period 1 to 50 years). Since diameter growth of trees strongly depends on the crown size and therefore on competition, we hypothesize that tree diameter growth was favoured by earlier thinnings in the young than in the old stand.

In addition to long-term changes in radial growth, short- and medium-term fluctuations were investigated. Despite of the large geographical diversity of sites in the study material, significantly synchronized high-frequency as well as medium-term variation was detected in the radial increment time series. The large extent of synchronous variation suggests a strong climatic control over radial growth of the sample trees. And indeed, cross-spectral analysis revealed significant relations between variation in radial growth and in climatic water supply for all species. The fact that no significant relation was found across the sampling sites between sensitivity in annual radial increment and seasonal climatic water balance is interpreted in terms of adaptation of the trees to the mean level of water supply at any given site.

How can we explain the observed increase in site productivity? The clearest signal is that of foliar nitrogen. Analysis of time trends in foliar nutrition reveal that the N-nutrition status of Scots pine stands in Central Europe in most cases has improved considerably and often significantly. However, the investigated Scots pine stands in Northern Europe and Great Britain show a heterogeneous pattern:

improved N-nutrition may have occurred at some sites, but N seems to be less important as potential driving factor for long-term growth changes of Scots pine stands in Northern Europe and Great Britain than in Central Europe. In contrast to Scots pine, most investigated Norway spruce stands in Central Europe were optimally supplied with N. Even though foliar N-concentration of Norway spruce stands in Central Europe slightly decreased during the period of investigation, foliar N-content increased parallel to increased needle masses. For Northern Europe trends in N-nutrition were heterogeneous. The N-nutritional status of all except one studied European beech stand improved considerably during the observation period. The observed increases in N/P-, N/Mg-, and N/Ca-ratios for most of the investigated Norway spruce, Scots pine and European beech stands indicate that increased nitrogen supply is not accompanied by an equivalent increase in the supply of other essential nutrients. The more abundant supply of nitrogen can in certain circumstances induce deficiencies of other nutrients with the risk of nutrient imbalances.

With regard to the role of former site index as a proxy for the "native" site productivity it is shown for all species that the higher the site fertility level, the lower the level of change in height growth. The fact that trees on sites with stronger limitations in the supply of natural resources showed larger response in terms of changes in height growth is in line with the hypothesis that in many areas N nutrition is the main driving force. A mixed spatially explicit geostatistical model for the relative height growth rate of Norway spruce and Scots pine using foliar N concentration as auxiliary variable was developed. The model indicates largest increase in height growth rate in an area extending from Central Norway in the North-West over Eastern Germany, Western Poland, and Czech Republic to Slovakia in the South-East.

We note that theoretically each of the investigated putative causal factors is expected to leave a unique 'fingerprint' in the forest growth changes, as drivers directly enforce different processes, have different distributions in space and time and may affect some forest sites more than others, depending for example on soil fertility. However, attribution of particular changes in forest growth to particular causes across the multiple observational datasets was seriously restricted by the quantity (e.g. number of observations, length of time series), quality (e.g. temporal and spatial coverage, comparability of data in time and among different sites) and integrity of the available data (e.g. largely missing data on needle mass).

The results obtained with different process-based models were in line with the findings from field observations – that is, increased nitrogen nutrition, mainly driven by the interaction of elevated nitrogen deposition and recovery of the sites from former devastative land-use practices contributed most to the increased forest growth to date. The growth enhancing effect of nitrogen deposition operated through an increase in nitrogen supply and in gross nitrogen uptake rate. Compared to increased nitrogen nutrition, changes in climate and increased atmospheric CO_2 concentration seemed to be of secondary importance in causing the observed growth acceleration. The model results suggest, however, that in the future the nitrogen effects will diminish – also because of an expected decrease in nitrogen emissions – and that further changes in forest growth are more likely to be caused by increasing atmospheric CO_2 concentration and, especially in northern latitudes, by increasing temperature. There are uncertainties regarding the response of nitrogen uptake and nitrogen use efficiency to environmental change evident in diverging model results that need to be resolved in further experiments and modelling exercises.

Empirical modelling is a method in which observational data are analysed directly, and it uses fewer assumptions about the functioning of forest ecosystems than process-based modelling. If ample field data are available it thus is the best tool for identifying the variables that should be part of an explanation of changes in forest growth. Process-based models, on the other hand, embody more assumptions, and require calibration of more parameters – all adding to output uncertainty – but this allows them to disentangle the various correlations and interactions in the forest ecosystem, and thereby provide mechanistic explanation of growth change. We concluded that using both modelling approaches together may be the best way to provide reliable explanations well supported by the data and mechanistically understood. If the combined approach shows consistent results, as it did in the Recognition project, the process-based models can be used to explore potential changes in future forest growth.

Project results did not indicate that the growth changes represent any threat to the sustainability of the forest ecosystems and their provision of services, at least in the short term. In a longer time perspective, the situation might be different and results from the project suggest possibilities of increased water stress and deficiencies of nutrients other than nitrogen. Concerning anticipated nutrient imbalances, it is concluded that most critical areas will be those with the largest discrepancy between the input of nitrogen through deposition and the specific absorption, uptake, and buffering capacity of the site. The changing climate might in the future also cause migration of species, which could push forest tree species out of those regions to which they are not anymore adapted. Such aspects were, however, outside the scope of this project.

The potential problems arising from the factors investigated within the Recognition project should be possible to handle with wise and foreseeing management, although there are other aspects of the changing environment that can turn out to be more difficult to manage. Increased growth and accumulating wood volume in the forests indicate that if the structure of the forests should be maintained, changes in the intensity of management practices, e.g., more frequent or intensive thinnings, are needed. This also refers to ensuring the stability of the stands, e.g. with respect to storm. Since the risk of storm damage increases with stand height, the risk at a given stand age is higher when height growth is accelerated. The model simulations also indicate that the rotation of timber production can be reduced; a finding which especially refers to Central Europe where increases in forest site productivity have been largest during recent decades, whereas in the future tree growth and consequent profitability of forestry may increase most in Northern Europe. Sustainable forest management in the future may require more attention to the nutrient balance of forest ecosystems. Optimized management options have to be developed to mitigate impacts of anticipated environmental changes. Sustainable development of forests requires integrated measures to reduce risks due to changed frequencies and intensities of biotic and abiotic hazards and to increase adaptive potential, resilience and stability of

Europe's forest ecosystems. The Pan-European EC–UN/ECE Programme for Intensive and Continuous Monitoring of Forest Ecosystems is one of the tools, which can provide us with early warning signals if large scale changes in European forests are occurring.

Reference

Spiecker, H., Mielikäinen, K., Köhl, M. and Skovsgaard, J.P. (eds.). 1996. Growth Trends in European Forests – Studies From 12 Countries. European Forest Institute Research Report 5. Springer-Verlag, Berlin. 372 p.

www.ingramcontent.com/pod-product-compliance
Lightning Source LLC
LaVergne TN
LVHW081515300926
849161LV00006B/1184